Projectile Dynamics in Sport

- How can we predict the trajectory of a baseball from bat to outfield?
- How do the dimples in a golf ball influence its flight from tee to pin?
- What forces determine the path of a soccer ball steered over a defensive wall by an elite player?

An understanding of the physical processes involved in throwing, hitting, firing and releasing sporting projectiles is essential for a full understanding of the science that underpins sport. This is the first book to comprehensively examine those processes and to explain the factors governing the trajectories of sporting projectiles once they are set in motion.

From a serve in tennis to the flight of a 'human projectile' over a high jump bar, this book explains the universal physical and mathematical principles governing movement in sport, and then shows how those principles are applied in specific sporting contexts. Divided into two main sections, addressing theory and application respectively, the book explores key concepts such as:

- friction, spin, drag, impact and bounce;
- computer and mathematical modelling;
- variable sensitivity;
- the design of sports equipment;
- materials science.

Richly illustrated throughout, and containing a wealth of research data as well as worked equations and examples, this book is essential reading for all serious students of sports biomechanics, sports engineering, sports technology, sports equipment design and sports performance analysis.

Colin White began his career as a software engineer, working on various military modelling projects. In 1986, he joined the Department of Physics at the University of Portsmouth as Senior Lecturer before transferring, in 2000, to the Department of Sport and Exercise Science where he taught sports technology and modelling.

Projectile Dynamics in Sport

Principles and applications

Colin White

Routledge
Taylor & Francis Group

LONDON AND NEW YORK

First published 2011
by Routledge
2 Park Square, Milton Park, Abingdon, Oxon, OX14 4RN

Simultaneously published in the USA and Canada
by Routledge
270 Madison Avenue, New York, NY 10016

Routledge is an imprint of the Taylor & Francis Group, an informa business

© 2011 Colin White

Typeset in Times New Roman by Glyph International Ltd.
Printed and bound in Great Britain by TJI Digital Ltd, Padstow, Cornwall

British Library Cataloguing in Publication Data
A catalogue record for this book is available from the British Library

Library of Congress Cataloging in Publication Data
White, Colin.
Projectile dynamics in sport : principles and applications/Colin White.
    p. cm.
Includes bibliographical references and index.
1. Sports sciences. 2. Human mechanics. 3. Biomechanics. 4. Athletics–Equipment
and supplies. 5. Sporting goods. I. Title.

GV557.5W45 2010
612.7′6–dc22   2009051137

ISBN 13: 978-0-415-47331-6 (hbk)
ISBN 13: 978-0-203-88557-4 (ebk)

Contents

Appendices

Figures

Tables

Excel models

These models are available at the book's website: www.routledge.com/9780415473316

Acknowledgements

So many have contributed in terms of time and intellectual effort to this project that I thought it best to approach my gratitude of all their undertaking chronologically!

Having taught simple Newtonian trajectory analysis and modelling to second-year undergraduate sports scientists for some years, I happened upon the inspiring work of Dr Jane Blackford of Edinburgh University in the field of curling. It was she who enlightened me to this whole fascinating field of projectile dynamics and how similar mathematical analyses and techniques operate across widely differing sports disciplines. Early on in the development of this book, Professor Lesley-Jane Reynolds provided me with the scholarly and physical space to allow me to progress the project from my initial enthusiasms through to a completed piece of work.

Thanks also to Dr Mike McCabe for allowing me to bounce (oh dear!) my mathematical ideas off him and for the loan of his superball/table tennis ball trick apparatus, which formed the basis of Question 1 in Chapter 4. I will, of course, return it one day. Andy Barrow, Subject Librarian in Portsmouth University Library was my saviour on more than one occasion, pulling the most obscure of papers from the archives in less time than it takes to fill out an interlibrary loan slip. I must also thank Alex Mellor of the ExPERT Center, University of Portsmouth. When it comes to scanning diagrams, he's second to none. I am also particularly indebted to Bob Healey for stepping in at the last minute to draw (and redraw) the many images which could not be scanned.

I realized that I would not be able to write meaningfully about all the individual sports that make up Part 2 of this book without a range of experts to check over my scribblings. First into the breach was James Lockrose, PGA professional coach, who enthusiastically proofread the golf chapter. Next, I really do have to record my immense gratitude to Dr Alan Neuff of Neuff Athletic Ltd, who went through all three 'throwing' chapters with the finest of tooth combs, and reported back many essential edits; this despite his personal health issues at the time.

The tennis and squash chapters were reviewed by Kevin Baker, Coach of Portsmouth Indoor Tennis Club and Professor Paul Hayes, respectively. I considered cricket to be such an important element of the book, and both Dr Richard Thelwell and my son, Alan White, reviewed those pages with great care. I taught Alan everything I know about the game – it happened one afternoon on Paignton Beach! I'm grateful to Dr Clare Hencken for being my 'American' consultant, covering baseball, softball and American football. Chris Hughes tackled rugby (pun intended!) and Phil Ashwell soccer. Thanks to you both.

Dr Nick Ball took one look at an early draft of basketball, and then promptly emigrated with his family to Australia! I did take on board his advice though. Dr David Franklin scrutinized archery for which I am most indebted.

I also have to thank all those working in the Department of Sport and Exercise Science and the ExPERT Center at the University of Portsmouth, as well as all my Twitter followers, for having to suffer my trials and tribulations, on an almost daily basis, of the 'tortured artist at work'!

And finally … at home, there's 'The Management'. You've had the worst of it, love. Sorry; and … yep, the decorating starts here!

Symbols

A	Cross-sectional area
a	Acceleration
C	Ballistic coefficient
C_D	Drag coefficient
C_L	Lift coefficient
e	Coefficient of restitution
F	Force
F_D	Drag force
F_L	Lift force
G	Universal gravitational constant
g	Acceleration due to gravity
H	Altitude (above Earth)
h	Height
I	Moment of inertia
K	Energy, kinetic energy
k	Radius of gyration, resistance coefficient per unit mass, thermal conductivity
L	Angular momentum, characteristic length
l	Length
m	Mass
N	Force acting normal to surface
$\hat{\mathbf{n}}$	Unit vector normal to projectile travel
P	Power, potential energy
p	Momentum
q	Distance from axis of rotation to center of percussion (striking implement)
R	Range of flight
R_e, R_{ecrt}	Reynold's number, critical Reynold's number
r	Distance between two bodies
$\mathbf{r}$	Position vector
s	Displacement
S_m	Maximum distance
T	Torque, total time of flight, temperature
t	Time
u	Old velocity
v	Velocity
v_s	Velocity of fluid flow
v_0	Launch velocity
W	Work

$\mathbf{w}$	Wind vector
x	Displacement
α	Angular acceleration, angle of attack (or yaw angle) of aerodynamic projectile
ε	Drag-to-weight ratio
θ	Angle
θ_0	Launch angle
θ_m	Optimum launch angle
κ_1, κ_2	Tait coefficients
μ	Absolute dynamic fluid viscosity
μ_k	Coefficient of kinetic friction
μ_s	Coefficient of static friction
$\hat{\boldsymbol{\tau}}$	Unit vector in direction of projectile travel
ρ	Fluid density
φ	Earth's latitude
ψ	Angle of slope, angle of attack (discus)
ω	Angular velocity
υ	Kinematic fluid viscosity given by μ/ρ
ξ	Wind speed
ς	Wind coefficient (force)

Part I

The theory

Chapter 1

Sports projectile modelling
Why, how and ... so what!

1.1 Introduction

Science and the study of the motion of bodies have been inextricably linked since the earliest foundations of academic studies. When the cave dwellers threw stones in battle, or to obtain food, they must have had some gut feeling for the trajectory path the stone would take once it left their hand. At the very least, they must have known that the harder it was thrown, the further it might go. One suspects that, by means of pure practice alone, they may have achieved superbly accurate and powerful shots, in much the same way as South American rainforest tribes wield a blowgun with such devastating precision. It is most unlikely, in either of these cases, that a profound understanding of the mathematical equations that underlie the motion of the projectiles would improve their skills.

As time progressed and weaponry evolved, ingenious lever-type devices came into play, utilising a range of mechanical catapult or slingshot actions. Projectile paths were now replicable and could be varied by means of mechanical adjustment. A mathematical study of the projectile motion, once it left the appliance, became an essential component in the overall weapon design strategy, allowing predictions of performance and target accuracy prior to the machine's operation in anger (so to speak!). The earliest catapults could fire a 20 kg stone maybe up to 40 m. Later, the Trebuchet was developed; a highly advanced slingshot machine designed to destroy fortifications. It could fire stones of 120 kg a distance of up to 200 m. Most significantly though, a good degree of accuracy was attainable by means of a counterbalancing torque action, which ensured the weapon's stability during the enormous forces created through the firing action.

As a point of interest, stones were not the only missile to be flung from these magnificent contraptions; sometimes, disease-infected corpses were flung into cities in an attempt to infect the people under siege. This may be the earliest record of biological warfare!

Today, modern superguns can fire unpowered shells over a hundred miles with some degree of accuracy. Actual up-to-date specifications are hard to come by for obvious reasons, but it is thought that work is afoot to produce weapons that can fire unpowered ammunition into space where it will orbit and return where and when required.

The earliest recorded analysis of ballistic motion, known as 'Impetus Theory', was derived from the field of Aristotelian dynamics. Its main premise stated that an object shot from a cannon, for example, follows a straight line until it 'lost its impetus'; at which point, it fell abruptly to earth (coincidentally, rather similar to the flight path of a back-spun, driven golf ball). The argument stated that: all motion against resistance requires some 'mover' to supply a continuous motive force. Of course, in the case of a launched projectile, such a mover is not present, and so Aristotle postulated an alternative auxiliary theory, stating that the mover was provided by the launcher itself. It was then transferred into the projectile at the point of launch, ready to operate once it was in flight. Key to this is the erroneous assumption that the projectile movement is not mediated by the medium through which it moves.

Various modifications to the Aristotelian Impulse Theory ensued over many centuries in an attempt to bring the mathematical analysis in line with observations, but it was not until Galileo created his seminal work, published in 1638, the *Dialogues of the Two New Sciences*. Here, in the second half of the book, he took up the question of projectile motion. Aristotle had approached the problem from the point of view of the origins, causes and effects of projectile motion. By contrast, Galileo ignored these characteristics, in favour of focusing on the measurable quantities, such as weight, time, velocity and acceleration. When he dropped a stone from his hand, for example, he did not concern himself with *why* it fell, more that it fell such that the square of the time of flight was proportional to the height of fall. He knew that, for this to happen, a vertical force must be acting on the stone. He ignored why that should be and calculated its value, by experiment, to be $9.8\,\mathrm{m\,s^2}$. Perhaps this was his first major insight into the field of trajectory modelling; the ability to model nature's movement by pure mathematical symbols. In *The Assayer* he wrote:

> Philosophy is written in this grand book, the universe ... It is written in the language of mathematics, and its characters are triangles, circles, and other geometric figures.

His second essential insight in the field came from his prior work on connected bodies. He had stated that, if one body carries another, the carried body shares the motion of the carrier body. Extrapolating this concept, he concluded that a projectile shot from a cannon, is not influenced by one force, but by two independent motions: the first motion acts vertically under the force of gravity, and this pulls the projectile down by the same times-squared law stated above. The second motion moves the projectile forward, horizontally, with uniform velocity. He set out to prove this and he considered a ball moving down an incline. He showed that the ball was indeed subject to two forces from which he could calculate the acceleration for differing angles of incline.

His next stage was to consider what happens if the incline comes to an abrupt end and the ball is allowed to free-fall with some horizontal velocity and subject to gravity. He, thereby, proved the ball moved in a curved arc, and when he did the calculations, he successfully (and essentially correctly) derived the equation of a parabola; an equation he was familiar with, both from its mathematical function ($y^2 = 4fx$) and from conic section analysis. The fact that he also derived the equation for maximum range (see Equation 3.11) indicates the appreciation he possessed of the equations' importance.

The mysteries of gunpowder and the tremendous force it can generate also attracted such scientists as Leonardo da Vinci and, significantly, Isaac Newton to the field. Gunpowder was becoming commonplace in battle scenarios, and firearms now operated faster than the human eye could see, and so determination of the flight of projectiles was considered essential.

Running parallel, though not apparently directly related to this study, was the work of Kepler on planetary motion. Just as Galileo had derived the equation of fired projectiles, Kepler had derived an equation that could calculate the path of planets and showed the movement to be elliptical around the Sun. Different equations, certainly, yet they both modelled natural movements by mathematical means, and both can still be usefully applied today as predictive tools.

Newton realized that, just as planet movement was caused by external forces acting on the bodies, projectiles must be acted on by an external force, which is directed downward. The genius step was to reason that they were both *one and the same force*. Thus the Universal Theory of Gravitation was born (see Equation 2.6). This, together with his three famous laws of motion were stated formally in his *Philosophiae Naturalis Principia Mathematica* (commonly referred to as *Newton's Principia*) published in 1687, although the first two laws were known to Galileo. This publication also contained his equations of motion (Equations 2.2–2.5), together with his statement of the conservation of momentum and conservation of angular momentum.

It is fair to say that the triad that was da Vinci, Galileo and Newton, were virtually completely responsible for creating the correct mathematical analysis of simple (i.e. not including drag or spin) trajectory motion which has proved over the ensuing years to be sufficiently accurate for the analysis of many sporting projectile trajectories. What later followed, therefore, were merely refinements to this solid foundation, to be applied to only some special projectile trajectory cases.

It is thought that the first book that used physics to describe ball motion (or maybe it was; used ball motion to help explain the physics) was written by Gaspard-Gustave Coriolis in 1835. It was entitled *Mathematical Theory of Spin, Friction and Collision in the Game of Billiards*. This is the same Coriolis who studied movements in circular reference frames that led to his meteorology theories concerning the circular flow of wind around low pressure areas and cyclones. He also analysed the trajectory of a cannon ball fired from a cannon mounted on a rotating turntable. This model described clearly, powerfully and in a visual manner, the complex effects of both rotating and inertial frames of reference.

Spinning balls were first studied by Newton, but Lord Rayleigh published his paper, 'On the irregular flight of a tennis ball' in 1877. This was the first rigorous treatment of spin forces and was the first paper to cite 'The Magnus Effect' as the cause of the force. As well as the more famous studies of this Nobel prize winning scientist in such areas as, the discovery of argon, Rayleigh waves (surface waves associated with sea waves and earthquakes), Rayleigh scattering (leading to an explanation of the colours of the sky) and the Rayleigh criterion (which describes the resolving power of telescopes), he also found time to consider some more esoteric fields. Papers with such titles as, 'Insects and the colour of flowers' (1874), 'The soaring of birds' (1883), 'The sailing flight of the albatross' (1889) and 'The problem of the Whispering Gallery' (1910) give an indication of the wide-ranging nature of his interests.

Another Nobel Laureate, Sir J. J. Thomson, famous for his work with cathode ray tubes and credited with the invention of the electron as a particle, wrote 'The Dynamics of the Golf Ball' in 1910 in *Nature*. His article was read by Professor P. G. Tait, who, with the support of C. G. Knott advanced the field, publishing a number of papers in the field. Tait's seminal paper 'The Path of a Rotating Spherical Projectile' rather conceals his interest in the game of golf, opting instead for perhaps a more academically appealing title.

However, although J. J. Thomson's son, P.G. Thomson, continued his father's research with electrons (albeit from a wave, rather than particle, perspective), one of Tait's six sons, F. G. Tait, took up golf with a vengeance, winning the Amateur Golfing Championship twice. He was a strong man who could drive a ball with both distance and precision. At the Royal and Ancient Golf Club of St Andrews, he hit a ball some 250 yards. The ball then rolled on frozen ground and finally came to rest 341 yards from the tee, thereby refuting his father's calculation that 190 yards was the maximum possible flight for any golf ball. Sadly, F. G. Tait, a military man, was killed in action in the Second Boer War in 1900.

Sir Barnes Neville Wallace, the inventor of the bouncing bomb and swing-wing aircraft, gained a solid understanding of projectile airflow that he applied, again, to golf ball trajectories. In fact, in order to persuade the British Government to invest into his 'dam-buster' concept, it is reported that he fired spun golf balls into model dams at a test tank belonging the Ministry of Shipping at Teddington, London.

In the world of cricket, the great astronomer, Professor R. A. Lyttleton succeeded in combining his astronomical research interests with his profound love of cricket. As well as boasting a low golfing handicap, he played cricket for Cambridgeshire C.C.C., as well as Warwickshire 2nd XI. In the late 1950s, he gave a radio broadcast on the swerve and swing of a cricket ball, which caught the attention of Sir Donald Bradman. It is worth noting that, of the 126 papers Professor Lyttleton published during his life, many were actually concerned with the spin of galaxies, stars, planets and gases. Perhaps it was not such a big jump to transfer his attention to the analysis of the spin of cricket balls? Professor Lyttleton collaborated with 'The Don' on his classic 1959 book *The Art of Cricket* and the two kept up a regular correspondence following this popular volume's publication.

While in the field of cricket, most recently, it is worth bringing to attention Bob Woolmer's great masterpiece, *The Art and Science of Cricket* (2008) which successfully brings together all of this trajectory science, and places it into an essential practical setting of the cricket square.

1.2 Computer models, simulations and animations

Building and playing with models has been a fun activity since prehistoric times; archaeological digs have discovered children's toys, miniature representations of humans (i.e. dolls), tools and weapons, in the vicinity of many ancient sites. Today, of course, our toys are more sophisticated with their use of radio and/or computer control and advanced materials to heighten the illusion of reality. Over those ensuing epochs, however, the aim of the model has been the same; to give humans apparent control over something we would like to command, but, for whatever reason; be it cost, size, danger or rarity, we cannot.

A computer model copies the behaviour of a dynamic system (i.e. one that has parameters, such as space and time, which vary in some non-trivial manner) that may be difficult or impossible to operate in the real world. Yet, if we consider, for example, the javelin model outlined in Chapter 9, we could not possibly argue that throwing a real javelin is, in any way, so restrictively expensive, large or rare, as to prevent 'real world' experimentation. So what, you may ask, do we gain from developing a computer model of something so commonplace? This question will be considered in some detail in the next section.

One common source of confusion and misinterpretation will be addressed here; the difference between computer modelling, computer simulation and computer animation. In fact, the terms 'simulation' and 'model' are often used interchangeably within modelling contexts, and it would appear that there are as many 'definitions' of these terms in the literature as there are numbers of models (or is it, simulations?) in existence.

The simplest definition of a computer model might be a computer program which attempts, from a set of parameters or initial conditions, to find analytical solutions to problems, and/or to permit predictive 'what-if' analyses to be performed. A computer simulation, on the other hand, attempts to copy or mirror a dynamic system as accurately as possible, by taking various input variables – the initial conditions – and then operate on them, mathematically, to create a system of variable values that change in consonance with the reality of the model it is trying to simulate. To do this, of course, it is necessary to obtain mathematical solutions and predictive responses in line with our definition of the model. It is, therefore, a truism that the term simulation is, necessarily, broader than that of the term model.

Before moving on, it is worth briefly mentioning an alternative, commonly stated and succinct pair of definitions for these two phrases: modelling is an *activity*, where as simulation is a *process*. Perhaps the relationship between the former and the latter definitions can be recognised.

Finally, a computer animation is the result of the creation of moving images on the computer screen. Now, in terms of sport trajectory modelling, the 'output' from our simulation is going to be represented by a set of answers that correspond to values in space and time. It makes sense, therefore, to model these two variables as objects on a computer screen which also vary in space and time, i.e. produce an animation as a modelling response. So, for our purposes, we might like to consider that a computer animation is the matrix of space–time answers output by the sports trajectory model, and represented on the screen as an object moving in equivalent space–time. The object of the animation, the projectile, knows nothing of the underlying process that created the parameter values that form its movement, in much the same way as Mickey Mouse has no knowledge of *his* creation, or indeed that he is a rodent. Our animation is merely a convenient means of outputting values from a simulation in an easy to interpret, recognisable and visual manner.

An animation does not have to be created by a mathematical model. After all, Mickey is nothing more than an aesthetic creation produced by a group of artists. However, there is an overlap of expertise here, since modern day cartoon animations are often supported by mathematical models. As an example, Computational Fluid Dynamics is a complex mathematical modelling technique that can accurately predict airflow patterns around sporting projectiles. And yet, the same technique can also generate the series of pictures that represent the time-varying movement of magic carpets, clothing fabric, or hair on cartoon characters, without the need for human-created rendering of each individual image in the series that makes up the animation.

While on this subject of cross-disciplinary activities, it is clear that modelling and simulation are far from singularly associated with sports science. In fact, although ball prediction models such as Hawk-Eye have become very popular in recent years, sports modelling of any form still constitutes but a small fraction of models in existence compared to, say, models which predict weather, land erosion, disease proliferation or traffic flow.

However, this can be used to our advantage. When we enter the field of sports trajectory modelling we can utilise accepted wisdom from the more established modelling fields; we are eligible to join their club, so to speak. At the most fundamental level we can use their language to describe our models, so a climatologist or volcanologist modeller understands what we are attempting to accomplish when we mathematically model, say the spin of a ball. This ability to transfer knowledge across disciplines must be to all modellers' advantage. A good example is the modelling of a javelin when thrown into a strong headwind. It is actually quite easy to construct a model in which the forward energy of the javelin is partially cancelled by the opposing force from the headwind. However, what is the profile of that headwind? Is it, for example, constant with height, so the javelin is subject to a constant opposing force throughout its flight? Perhaps the headwind increases with height (the so-called laminar flow model). If so, does it increase in a linear manner, or perhaps in some logarithmic, or other non-linear relationship? Answer? Ask a climatologist modeller. They will possess a whole series of models of wind flow to cover every eventuality and they will know precisely which model to apply for which application. It is all about not reinventing the wheel.

1.3 Why model?

Models represent some simplification of reality and a means of understanding systems whose spatial and temporal scales or complexity might otherwise render the system too complex to comprehend. In spite of this inherent simplification process, it is important that the model retains the significant features or relationships of the real activity. So, all models will necessarily be subjective, as the modeller chooses those features of the real world that they think should be included in the particular model, as well as the model's mode of representation of the output.

Rather like sculpturing – itself a form of modelling – the modeller is forced to look at the reality they are attempting to model with a detective's eye for detail. They look closely at the visible structure, possibly discovering aspects that would elude the casual observer. But, as in sculpturing, the end result only reflects the modeller's interpretation of their perception of reality. Furthermore, modelling can provide a stimulus for thought. Models are used to describe, explore and analyse how complex systems work. From here, then, we may enter the world of predictive activities; the 'what-if' scenarios.

The modelling process is often one of stepwise refinement: comparing the model with reality (where possible) and making adjustments or tweaks to the model to work towards the expected result. Taking the case of the javelin model, we might begin with a basic application of Newton's equations of motion and see how that approximates with a real javelin trajectory. As it turns out, in this particular case, if we input realistic values for launch velocity and angle of launch, the most basic model is actually surprising accurate with errors of typically ±10 m in range. Underlying this, however, are a number of approximations which produce errors, some which create undervalued results, and some overvalued results: these errors partially cancel giving a more optimistic notion of accuracy than is actually justified.

One aspect that we have, at this point, ignored, for instance, is that the javelin is not launched from ground level, but from about 2 m above the ground-level launch point. Add this into the equations, and for most typical javelin launch angles, it adds about 1 m to the range.

We may next include into our model an air drag factor which, of course, will reduce the range. We then have to consider how we are going to mathematically treat this drag. Do we just think of it as a constant value over the duration of the trajectory, or do we say that drag varies with velocity, and therefore since the javelin's velocity varies over the duration of the flight, the drag must also vary correspondingly? In reality, the drag does, indeed, vary quite considerably over the flight duration, because, unlike the spherically symmetric ball, its relative orientation compared to its trajectory direction changes over the duration of the flight.

Next to consider, is the aerodynamic effect. The aim is to launch the javelin so that it points about 10° above the launch angle. This allows the air to get 'under' the projectile, providing a lift effect in a similar manner to an aircraft wing. This causes the javelin to 'hang' longer in the air and. Now, if we add this correction into our model, this time the range increases.

We are now getting quite close to an accurate model of reality – so long as the javelin is thrown in windless conditions. So now, we might want to add in a wind factor, which is dependent on strength and direction. As stated earlier, there are several climatologically derived models that, for instance, represent wind variations with height. Now, the issue complicates because there will also be short-term spatial and temporal variations which are unpredictable random fluctuations in nature. Will we ever be able to model this?

Up until now, we have described what are known as deterministic models: we input all our parameters (launch angle and velocity, launch height, aerodynamic and drag factors) and we get one series of answers; the unique and, hopefully, accurate solution. But when we consider the last refinement we cannot, with any accuracy, specify an exact and constant value that represents the wind elements. However, we *can* specify a range of realistic values and how the wind may fluctuate between those values; a statistical probabilistic approach to the problem. Inputting variables as a set of values that statistically vary in this manner will lead to answers which represent a probable range of values. This is now what we refer to as a stochastic model. It works like this. We could state that we are throwing our javelin in a precisely northerly direction into an approximate tailwind but whose direction varies by say, ±10° in a statistically normally distributed manner, and a Beaufort wind force scale of value 2 (corresponding to a light breeze). This represents a normally distributed wind variation of 6–11 km h^{-1}. Our stochastic model will then calculate the most probable range, the possible upper and lower extremes of that range and the statistical distribution of the complete set of range of values.

We are now reaching the end of the iterations towards the hypothetically perfect javelin model. Each step resulted in a more accurate model, and yet the law of diminishing returns has definitely been in evidence: the nearer we get to that illusive perfect model, the greater the complexity and effort is required in each step. However, one thing we can state with confidence about the process is that, if the model is tried and tested, and is proved to be accurate under a range of realistic conditions, then the mathematical theory underlying the process must be correct.

As stated, it is important that, no matter how much we simplify the system to allow it to be modelled, it must retain those significant features or relationships of reality. Hence, all models are subjective, often there is no, one 'correct' model, but only the modeller's interpretation of what they feel are the important factors to be retained. The question then arises, on what basis are the features/relationships selected: intuition, experimentation or theory? In modelling a system, we are attempting to characterize very complex structures. If some aspect of the overall system can be modelled and the results relate to reality, then the assumption can be made that at least that section of the process is 'understood' (to some extent). The relationship between model and theory has, indeed attracted a lot of attention.

With regard to mathematical complexity in relation to the specific case of modelling trajectory dynamics, the mathematical derivations often generate pairs of linked equations of a differential nature. These are frequently quite difficult to solve, one reason being that they may not, in fact, possess exact solutions. However, these equations usually contain the two key, linked variables of distance and time, in two, or sometimes three, dimensions. Now, in the case of trajectory modelling we can usually divide these two parameters into a discrete series of values; a technique known as discrete-time modelling. For example, a javelin is usually in flight for about 4 seconds. It might be adequate to divide the time parameter into, say, 0.2 sec time intervals, and then calculate the javelin positions at only these individual 20 time-slots or windows (often known as 'bins'). Alternatively, if that granularity is too great, then 40×0.1 sec bins may be chosen. Either way, the results of the equations are easier to obtain, since we have exchanged the differential equations derived by the continuous time model, for the simpler *difference* equations of the discrete time model. These values, although not providing a complete and continuous range of solutions, can be chosen to supply as many results as is required for a given level of resolution and accuracy.

Let us now assume that we have a model that appears to be sufficiently accurate for purpose within the range of parameters over which it can be tested. We might say, for instance, that in our model, a javelin thrown at 30° to the horizontal and with a launch velocity of 40 m s^{-1} under some wind condition has a range of 95 m, which, when compared to a real throw, is accurate to, say, 2 per cent. Furthermore, as we vary the parameters of launch angle and velocity, accuracy is maintained. We are now in a position to use this model in its predictive mode. We can examine how the javelin's all-important range varies over a selection of launch angles, or indeed, velocities. For want of a better term, and referring to our initial 'toy' analogy, we might like to refer to this process as 'playing' with the model. Without doubt, it is the fun part of the process. Notwithstanding, some very important aspects can be uncovered by this activity.

First, we can gain an appreciation of the *sensitivity* of the variables. In fact, this can be illustrated even with our simple 'Newton's Equations of Motion' model, without any of the refinements, as delineated above. If we investigate the javelin's range over a series of angles, say, from 35° to 55°, for a fixed throw velocity, we discover that the maximum range occurs at 45° (only true for this simplest of models), but it actually varies little between the two stated extremes. If, by contrast, we keep the angle fixed at 45°, and increase the launch velocity by but a few metres per second, we will find the range increases quite dramatically with launch velocity. This procedure is termed sensitivity analysis and, at least in the case of a javelin throw, the analysis tells us that the launch velocity is a much more sensitive variable than the launch angle.

Now, in the real world of javelin throwing, it is easier, for biomechanical reasons, to launch the javelin at angles somewhat lower than the, theoretically, optimum angle of 45° for maximum range. Lowering the angle allows the athlete to increase the launch velocity, which, as we have stated, is the more sensitive variable. So even with this, simplest of model, we have learnt that it is definitely worth sacrificing optimum angle (which admittedly will reduce the range slightly) for increased velocity (which will increase the range quite considerable).

In carrying out this analysis, the need to alternately fix the launch angle and then the launch velocity was stated. Only on a mathematical model is this possible. In the real world, we could ask our athlete to *try* to throw a successive series of javelins at exactly 42°, with differing launch velocities, so that we can measure the ranges in each case, but the degree of repeatability simply is not that good; and on a windy day, impossible.

Another issue is the possible financial advantages acquired from modelling activities. Although there will be initial design, test and validating costs associated with creating a model in terms of hardware and man-power in the software design, once created, the 'playing' phase is comparatively cheap when compared with the real-world dynamic system. On a computer, we can throw 20, 100, 1,000 javelins very quickly and cheaply compared with the hiring of facilities, and employing an athlete's skill and time. Indeed, if the modelled throws do not have to take place in 'real-time', such as, if only the range or maximum height value is sought, there is no reason why 10,000 throws with varying parameters (launch angle, launch velocity, drag, wind etc.) could not take place in a second or two.

Finally, let us extend the discussion regarding the predictive capacity of the model, or those 'what-if' scenarios. We now understand that the model can emulate the dynamic responses as a consequence of the equivalent of varying the real-life inputs. Now consider the possibilities of extrapolating the values of those input parameters into regions that may be impractical to reach. For example, will the model still hold true if we set the launch angle equal to 90°, and if so, how high will the javelin go? In this scenario, the model will probably remain accurate and a maximum altitude may be found. But one thing is certain: this is not an option for a real-life test if we value the life of our athlete! Another 'what-if' may involve the effects of throwing a javelin into very strong winds. The real-life option involves waiting for a suitable day when there is a constant wind of the prerequisite strength. A question may be asked: what is the strength of headwind required for the javelin to be blown off course to the extent where its direction fully reverses, and the athlete's health is again in jeopardy. In reality, not a practical option, but for our computer model, it is merely a matter of choosing the right values for the variables (according to one of the models presented with this book, a headwind velocity of $85 \, \text{m s}^{-1}$ will do it!).

However, it must be noted that, even though we may have tested and validated our model for accuracy, repeatability and robustness using easily obtainable real-life values, often applying 'what-ifs' involve plugging in parameter values close to the extremities of their range, where the model, in all likelihood, has not, and cannot be tested. Reliability and accuracy of these results may be questionable in these outermost regions.

1.4 How to model

Broadly speaking, models may be classified chronologically in relation to their theory of development as either 'a priori' or 'a posteriori' models. For a priori models, and with

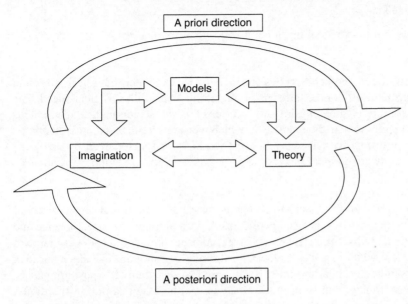

Figure 1.1 A priori and a posteriori modelling theory.

reference to Figure 1.1, an idea or construct results in the development of a model, which may, in turn, lead to the formulation of a theory through the process of hypothesis testing (in Figure 1.1, clockwise starting at 'Imagination'). In such circumstances, the model can be constructed in the absence of a complete theory although one may be subsequently developed out of the results of the model. The alternative situation is one in which observation has already resulted in the development of a theory that may be explored using an a posteriori model (anti-clockwise in Figure 1.1, starting at 'Imagination'). However, as knowledge progresses, a model may change from an a priori to an a posteriori type.

The stages in modelling are best indicated by enumerating them:

1   Decide whether a model is actually necessary, and which type is most appropriate. This decision is based on such matters as: the aims of the exercise, the nature of the system to be modelled, the level of understanding of the processes and the availability of data to be modelled.

2   Identify the appropriate model structure. This is a 'high-level' process and must be correct. Any misjudgment here can lead to a major re-engineering operation later.

3   Estimate the parameters that will characterize the model. The more complex the process, the more difficult it can be to select suitable parameters. A process of iteration, working inward from 'realistic' extremes may work. It is certainly a fact that the ability to choose accurate parameters shows the level of understanding of the process to be modelled. Parameters may be obtained by laboratory or field measurement (empirically derived), or obtained mathematically (theoretically derived).

4   Run and validate the model. This involves comparing the model with the real world, ideally using a different set of data to that used to generate the model. Of course, once the model is 'running', it may be adjusted or tuned to match observations more closely, and make the model a more accurate representation of the real world.

The modeller's toolkit

To carry out sports trajectory modelling, a number of items are required.

1   *A computer*
For some years now, the computer power of standard personal computers have been sufficient to handle the mathematical rigors of this type of modelling – and with sufficient speed. The area that has been more challenging in hardware terms is the screen-update speed. This requirement only applies to real-time models in which we want to study smoothly animated images moving in a manner which scales the real-world velocity. However, by using the correct, optimised software on modern PC or MAC computers, this is seldom a problem.

2   *Data collection*
This is a necessary process if we want to compare our model with a real throw or a ball hit. There are a number of ways data can be collected, depending on the size, distance and speed of the projectile. Balls, for instance, can be traced using high-speed (or even standard speed, in some cases) video cameras. Camera positioning is important to reduce parallax distortion, and a suitable background must be arranged so that the ball's position can be clearly seen. Precise timing, and from this, velocities, can be obtained during the digital editing phase, using the time-frame number marked on each frame. This increments on a standard-speed camera at a rate of 24 frames per second. Range is seldom a problem to measure.

An alternative approach is to use wireless, miniature, on-board telemetry equipment. Such equipment can be attached to javelins without unduly impairing performance. They are used inside the larger balls; footballs, basketballs and rugby balls, and make use of accelerometers and gyroscopic direction finding devices to deliver orientation and spin information. Position in time can be derived by calculating and resolving the combined output from these instruments.

3   *The software language*
In many respects, this is the most important modelling aspect to consider. Software can be categorised by its position on what is known as the low level/high level scale. Low-level languages tend to look more symbolic, and unreadable as standard text with meaning. They are closer to the characters that the computer operates on to carry out its commands. The advantages of low-level languages are that the programs are quite compact, requiring little memory, they are efficient to run and they run very quickly. They are also the most general of languages, allowing just about any aspect of the hardware to be controlled in all its respects. The disadvantages of low-level languages are that the programs are slow and difficult to construct, and also difficult to debug. For our trajectory modelling purpose, in all probability, this family of languages will not be appropriate.

A higher-level language (say, mid-way up the scale) is easier to write, understand and debug, as it uses a more grammatically understandable syntax. On the negative side, the equivalent model will require a larger program, more memory and it will be slower to run. However, these disadvantages are largely negated by the sheer power and memory capacity of the modern computer. These languages are quite general in purpose, allowing most aspects of the hardware to be controlled, though sometimes, it does not allow the fine control that the lower level languages can accomplish. However, some of these languages, such as; C++, Visual Basic, Borland's Delphi (my personal favourite!), C#, Java and Javascript (if you want to run the models in a web browser), make very suitable and powerful modelling languages for sports trajectory analysis.

Finally, towards the top end of the high level language options are the languages (or now, possibly, applications) that are designed for a particular narrow purpose, and, maybe, even usable by a total non-programmer. There are a number of such languages designed specifically for computer modelling, and some even only applicable for certain types of model. For example, LabSim is an application from which one can construct complete laboratory test setups to allow virtual experimentation. VRML is a language which can create three-dimensional space models inside a web browser. Fluid flow dynamics is such an important aspect of projectile modelling and there are a number of high-level languages designed specifically for this purpose, including FLOW-3D and FIELDVIEW.

Indeed, MS Excel can be considered a high-level language. It has a very specific purpose, and its uses are restricted to operation within that defined realm of grid/cell calculation and graphical plotting. It just happens that trajectory modellers can utilise this functionality to their advantage. The programs, or cell functions and calculations, are intuitive and easy to understand. The mathematical functionality is, for the most part, adequate in terms of power and speed. The input and output responses are correspondingly easy to handle and interpret. They are always visible in an array format and changes to any of the inputs can instantly recalculate and adjust the output array as well as the corresponding graphical display. It is such a popular program that many will understand the functioning of the models and be able to edit and 'play' with them. For this reason, MS Excel is the application of choice for the models provided with this book.

That is not to say that it is not without its limitations. As an example, take the ubiquitous javelin throw. We can attach the model's control equations to the required cells; we can specify the input cells (say, just launch velocity and angle), and specify the array of output cells (say, the height and horizontal distance over suitable incremental discrete time-periods). Once we have that resulting output array, we can then plot it on a scatter chart in order to study the shape of the trajectory and assess the range. But, what happens if we want to actually watch a real-time animation of the javelin as it flies through the air? Rather than just plotting the static graph, we want the position of the javelin marked in x–y space *at the appropriate moments in the flight*. We would also like to see this animation run smoothly. Maybe we want to depict the actual orientation of the javelin at each point in the flight; approximately horizontal at the peak and with appropriate angles at launch and landing. We might even want a realistic sounding computer generated 'thud' to mark the point where it hits the ground, although maybe that is going a little too far! For these sorts of computer activity we would need a lower level, more general language than MSExcel. Such functionality can be provided by the modern general-purpose visual languages such as Delphi or the MS Visual Studio Suite. Employing languages like these, we could even choose to superimpose a movie clip of a real javelin throw over the top of our computer-generated model, so that any discrepancies between the model and the real-world throw would be readily visibly apparent.

1.5 Trajectory diversions: adding a touch of chaos to your model

We have seen how a variety of modelling techniques from a range of disciplines can be utilised in trajectory model construction. Examples mentioned include: computational fluid dynamics to study airflow past the airborne projectiles, and wind flow models taken from the field of climatology to incorporate into our drag analyses.

An analysis known as Fractal Geometry has found applications in many models where complex and apparently random shapes appear in nature. Coastline, river and mountain range patterns can be modelled with the help of this theory; as can lightning bolt shapes and even the patterns found on cauliflower and sunflower heads. However, interesting though the subject of fractal geometry is from a broader modelling perspective, the subject has yet to find widespread use in our field of sports trajectory modelling, although the methods are certainly utilised as part of the investigation into the vibrations set up in rackets, bats and clubs.

By contrast, a field that is closely related to fractal modelling, but which has certainly proved most useful in stochastic trajectory modelling activities, is that of Chaos Theory. Like fractal theory, this method also supports the modelling of complex, naturally occurring phenomena, by applying comparatively simple functions.

But first, a philosophical dilemma. The computer is often considered the most perfect of deterministic machines. Assuming it is fully working, it always does exactly what you tell it to do: it is impossible for it to do otherwise. You tell it to do something stupid and it will do it 'in spades'. The world is littered with evidence of computers reminding us how stupid we are: gas bills for £1 m, goldfish being put on an electoral register, tax demands issued to babies; the list is endless. So the dilemma is this: how do we tell our perfect deterministic machine to generate a random number; a number we cannot predict; a number that will surprise us, if you like? Further, how could a computer generate a whole sequence of random numbers that may represent, say wind variations over a period of time? Remember, we need our trajectory model to handle such chaotic random inputs so that we can assess the *probabilistic* effect of such parameters as height and range. In order to emulate the variations in wind velocity, we will have to consider that the required random sequence conforms to some statistical requirements involving range, mean value and probability distribution. For example, wind on the Beaufort Scale 2, corresponding to a light breeze, generates velocities in the range $1.6 - 3.3 \, \mathrm{m \, s^{-1}}$. However, at any instant, the velocity is much more likely to be close to the middle of that range, $\sim 2.4 \, \mathrm{m \, s^{-1}}$ than at either of the extremes. We would like our random sequence to generate numbers that mirror this profile, with a much more likely occurrence of numbers in that mid region.

This last aspect of generating probability distributions is beyond the scope of this 'diversion'. Now, let us take a little time out and ask ourselves this question: how can our computer – that perfect little machine – generate this representation of randomness? The facetious answer might be: badly! Certainly, compared to resolving other mathematical functions, computers are pretty poor at producing numbers out of the blue.

Poor? Surely the pesky deterministic machine can either do it – or it can't! How can it generate random numbers *badly*?

Three considerations:

1   How 'random' do you want your random to be? Philosophers will tell you that the word 'random' *itself* cannot exist. It is a single-word oxymoron; a contradiction in terms; a paradox. To be random you have to define limits on your randomness, even if the stated limit is 'it has to be some number'. Therefore, by definition, it is not random; it could not, for example, be a daffodil.

    Even if we specify a normal distribution for our wind model (ensuring that the wind spends more time in the middle of its allowable range), if we have created that rare, slight puff of wind at the lower end of the scale (at the $1.6 \, \mathrm{m \, s^{-1}}$ velocity), it is unlikely that the next discrete time step will result in a sudden jump up to the middle of the range, even though this is where the highest probability lies. Yes, we want our random number to be attracted to that region, but slowly. We must place, what is known as, a low-pass filter on the numbers to restrict their unnatural jumping about.

2   How quickly do you need your sequence? This can be important in real-time modelling – you do not want to be waiting for the wind velocity to change in mid-flight.

3 How much do you want to pay? Some of the best random number generators are add-on boxes specifically feeding the numbers to the computer. The numbers are not created mathematically, but by noise in electronic circuits or, for even more money, in some quantum mechanical way. These can produce very random numbers with any filter or probability distribution you might like. But they do not come cheap.

For many years, random numbers were generated by the so-called middle-square method, invented by the great computing mathematician, John Von Neumann in 1946. It worked like this. He provided the computer with a 10 digit sequence to get it started; the seed. Say we input:

5 7 7 2 1 5 6 6 4 9

That is the first 10 numbers in the sequence. The computer now squares this number:

3 3 3 1 7 7 9 2 3 8 0 5 9 4 9 0 9 2 3 0 1

It then removes the five most significant figures and the five least significant figures, leaving the 10 digits from the center of the result:

7 9 2 3 8 0 5 9 4 9

This is not only your next 10 random numbers; it also forms the seed for the next squaring, whereupon the process is repeated. Of course, the sequence is not random; it is fully deterministic, but it *looks* random; it is known as a pseudo-random sequence. Even Von Neumann, himself, in 1951 stated:

Anyone who considers arithmetical methods of producing random digits is, of course, in a state of sin.

There are two problems with the scheme, as described. First, squaring 10 digit numbers is a comparatively slow process even for modern day computers, so the sequence will be created at quite a sluggish rate. One solution for time critical applications is for the sequence to be pre-created and saved in a look-up table, ready to be picked up as required by the model. Second, the random numbers are not very 'natural' in character. The sequence generated just does not look quite like the typical 'randomness' that appears in nature.

However, the chaos theory method of generating the random number sequence overcomes both these problems; and it creates some pretty intriguing graphical patterns, to boot!

The chaotic example described below is largely attributed to an economist, Thomas Malthus (1766–1834) who devoted much of his time to the study of animal population growth. However, the theory is based on the research of many before him, including such luminaries as Jacob Bernoulli and Henri Poincaré.

Malthus' equation is stated simply as:

$$x_t = ax_{t-1}\left(1 - x_{t-1}\right)$$

where x_t is the current number being generated, x_{t-1} is the previous number that has been generated and a is a value known as the Malthusian factor.

Obviously this equation is fully deterministic in the sense that, for any fixed value of a, given one value of x_t, it is possible to calculate all future values of x_t. But, when we iterate

their corresponding values of x_t, even though a is incrementing in a linear and equally spaced manner. By choosing suitable values of the quantity and range of Malthusian factors, a; the starting point, x_0, and the number of iterations, it is possible to efficiently create packages of random numbers which emulate many naturally occurring random events, such as wind variations.

Perhaps the weirdest aspect of this whole chaos theory business is the phenomenon known as self-similarity. To illustrate this, enter values of 3.35 and 3.6 as the lower and upper values of a, respectively in Model 1.1, and then alter the y-axis so that it only displays x_t values of between 0.7 and 0.9. You will now obtain a pattern which is disconcertingly similar to that of Figure 1.2b.

Then you can repeat this dual operation of zoom in/shift axis … again … and again … and again; all the way to infinity!

It's a bit like applying a magnifying glass to the plots of Figure 1.1, only, instead of just seeing a thicker series of straight lines, you reveal previously hidden significant deviations in the plots which you really should have noticed before! So, you try and increase the zoom on your magnifying lens again, only to find even more deviations within the previous deviations. It is this self-similarity/deviation characteristic that is so often found in naturally occurring random patterns and events, and why chaos theory is such a useful modelling technique.

Summary

In this chapter, we looked at the history of computer modelling of projectiles and showed how it lies at the very foundations of science and deterministic theory; the ability to determine the outcome of an event from knowledge of the initial conditions. We considered computer modelling activities, simulations and animations in their general context and showed how sports trajectory modellers can learn from the 'old hands'. In particular, we highlighted the influence of the computer model in research.

We learnt of the essential hardware and software requirements to perform meaningful and useful models, and delineated the pros and cons of the different types of languages available. Finally, the topic of chaos theory was introduced; a technique which has found much application in many computer modelling areas, and which may be usefully applied to certain sporting scenarios.

Chapter 2

Launching projectiles into motion

2.1 Introduction

Imagine if you will, a golf ball sitting on a tee awaiting your drive towards the green. For a moment cast your eyes away from the lure of the fairway and look down at that ball you are about to send on its way. Sure, it's white and spherical, and covered in dimples, but what other physical parameters can we ascribe to the projectile?

It possesses a number of important intrinsic physical properties, even while it sits there balanced, motionless on its tee. The ball's mass and weight are two obvious examples. Located at some point within the body of the ball are the centers of mass and gravity. Furthermore, the projectile must possess a moment of inertia, and its surface has a certain frictional value.

Then you line yourself up and impel the ball with a perfect (imaginary) driving swing. The ball has now suddenly acquired a new range of dynamic (i.e. changing with time) parameters. Its direction and speed (or its velocity) are perhaps the most obvious. Undoubtedly, you applied just the optimum amount of backspin, thereby imparting an angular velocity to the ball, a value that, in turn, leads to yet another of its properties: its angular momentum.

Over the duration of its flight, the ball is subject to a number of external forces: the main one, of course, being gravity. However, there will also be a whole range of lesser, or second-order, forces acting on the ball; perhaps caused by crosswind, sidespin (a slice shot) or air resistance. It is the interplay of all these forces on the intrinsic properties mentioned above that defines the in-flight trajectory which is the theme of this book.

Then finally, the ball lands strategically on the green; it bounces three times, rolls along the verdant turf and then disappears into the hole at the foot of the flag. The height and the number of times the ball bounces are determined by yet another property of the ball: its coefficient of restitution. When the ball runs out of bounces, it will roll smoothly along the green to the hole. The fact that the ball reaches the hole, rather than stopping short or overshooting, is down to the combination of its linear and angular momentums. These values are, in turn, determined by the parameter values possessed by the ball at the instant just before it first made contact with the ground.

This chapter begins by defining and explaining many of these fundamental, inherent properties of sporting projectiles. Once the projectile is hit, however, a number of principles and laws come into play, most notably those laid down by Newton. The remainder of this chapter describes these tenets, and discusses how the equations can be used to calculate the dynamic, time-varying nature of some of the projectile's parameters throughout its flight. These theoretical foundations will then form the basis for the fundamental projectile trajectory path models described in the next chapter.

2.2 Vectors and scalars

Before we commence a study of the parameters associated with sporting projectile movement, it is important to recognise that all the relevant parameters fall into one of two distinct classes: they are either vector or scalar quantities. Vector quantities have both magnitude and direction, while scalar quantities possess only the magnitude component. It will become clear as the book, and indeed this chapter, progresses why it is important to understand this distinction. For now though, Table 2.1 shows examples of each of the types.

Table 2.1 Examples of vector and scalar quantities

Scalars	Vectors
Distance	Displacement
Speed	Velocity
Mass	Force (weight)
Energy (work)	Acceleration
Volume	Momentum
Temperature	Torque

So, for example, the volume of a football is simply the size of it. We would not say it is $8,580 \, cm^3$ in the direction of the corner flag. In contrast, we might state that the velocity is $4.5 \, m \, s^{-1}$ in the direction of the goal. Its speed, however, would be completely defined as simply $4.5 \, m \, s^{-1}$.

One difference between vectors and scalars is worth investigating at this point. It is fundamentally easier to cancel out a vector quantity than a scalar one. If we take the displacement of a tennis ball as an example, it may be served in a north to south direction with a velocity of $50 \, m \, s^{-1}$. A successful return may lead the ball, in a south to north direction, right back to the point of service. The speed in both directions may be equal, but the velocity is of the opposite sign, caused solely by the direction reversal on return. This leads to a total velocity cancellation (i.e. $+50 - 50 = 0 \, m \, s^{-1}$), and the ball is back at its starting point. This cannot be done with say, volume. The football will not actually cease to exist by placing it into a volume of $-8,580 \, cm^3$.

One interesting result of the vector definition concerns projectiles travelling through some curved trajectory. In the case of a swing ball serve, for instance, the ball may spin around the center pole with a constant speed of say $10 \, m \, s^{-1}$. However, its direction is ever-changing as it encircles the pole. Consequently, its velocity is also changing, so the swing ball is in a constant state of acceleration as long as it is moving around the pole.

2.3 Mass, weight, force, velocity and acceleration

2.3.1 Mass

The mass of a body is a measure of its resistance to acceleration and is measured in kilograms (kg) in SI units.[1] It is a measure of the inertia of the body. Its value does not depend on where on Earth, or indeed 'off-Earth', we choose to measure it. So, in baseball, for example, if a ball approaching the striker at, say, $80 \, m \, s^{-1}$ is hit between the second and third bases to

---

1 The international system of units known as SI units (Système International d'Unités).

the outfield, then, for a given, fixed strike power, the impact force apparent to the batter will be the same whether the baseball diamond is in New York, the Swiss Alps or, indeed, the Moon.

2.3.2 Weight

The weight of a body is the force acting on its mass due to the gravitational attraction of the Earth (or planet). Its SI unit is the newton (N) and a body's weight is simply calculated by multiplying its mass by the appropriate acceleration due to gravity. The acceleration due to gravity is usually taken to be $9.81\,\mathrm{m\,s^{-2}}$, although it can vary between about $9.789\,\mathrm{m\,s^{-2}}$ and $9.823\,\mathrm{m\,s^{-2}}$ depending on the angle of latitude and altitude at which the measurement is taken.

The comparative invariance of the acceleration due to gravity across the Earth's surface was demonstrated by Galileo's famous experiment at the Leaning Tower of Pisa in which he *didn't* drop a tennis ball and a cricket ball at exactly the same time. If he had though (and it's doubtful that he ever actually carried out the famed experiment anyway), he would have found that, ignoring drag effects, both balls would have hit the ground at exactly the same moment in time, implying they were subject to the same acceleration.

2.3.3 Force

Qualitatively speaking force is any push or pull exerted on a body. We will see later that such a force will always cause a change in velocity to a moving body. Weight is but one type of force, so we know that the standard SI unit of force is, again, the newton.

Force is a vector, so combinations of forces acting on a body can be added or subtracted, depending on each of their individual magnitudes and directions.

Consider the flight of a Frisbee at a point when it is in the region of the middle of its trajectory. It is subject to at least three significant forces; the drag force, caused by air resistance which tends to act in opposition to its direction of motion, the gravitational force acting downwards (i.e. its weight), and the lift force, caused by its aerodynamic shape and air getting 'trapped' in its underbelly. Now, we know that around the middle of its flight path, the Frisbee is flying approximately horizontally. This means that the Frisbee's lift force is approximately cancelling out its downward weight: similar magnitudes, but opposite directions.

2.3.4 Velocity

Velocity is defined as the rate of change of position. As stated, it is a vector physical quantity, so both speed *and* direction are required to fully define it. In the SI system, it is measured in metres per second: (m/s) or $\mathrm{m\,s^{-1}}$. The scalar absolute value (magnitude) of velocity is called speed. For example, '8 metres per second' is a scalar value and not a vector, whereas '8 metres per second due east' is a vector value. The *average velocity* $\bar{v}$ of an object moving through a displacement (Δx) during a time interval (Δt) is described by the formula

$$\bar{v} = \frac{\Delta x}{\Delta t}$$

where Δx is the total change in distance, or displacement over the averaging range, and Δt is the total time to cover that distance.

It is difficult to ascertain what might be the slowest contender for a sporting projectile, but the puck in underwater hockey is certainly a sporting projectile which is subject to one of the greatest drag resistance forces. Above water, curling stones are a contender, which travel at anything from 6 down to $0\,\mathrm{m\,s^{-1}}$. At the other end of the scale, Olympic shooting rifles with pellet calibers of 5.6 mm can achieve muzzle velocities in excess of $400\,\mathrm{m\,s^{-1}}$.

Although the average velocity can be an important variable in our trajectory modelling calculations, seldom do sporting projectiles move at a constant velocity for long; over the duration of their flight, their speed will most certainly change, and their direction might well also. For this reason, sometimes it is important to consider the *instantaneous* velocity. For example, in Chapter 15, consideration is given to the modelling of a badminton shuttlecock in terms of the skirt turnover process. In this case, knowledge of the shuttlecock's instantaneous velocity at the time precisely following the stroke from the racket is an essential requirement to derive an accurate model.

In differential calculus terms, the instantaneous velocity may be written in terms of the limiting case as $t \to 0$, as:

$$v = \lim_{\Delta t \to 0} \frac{\Delta x}{\Delta t}$$

which, in differential calculus notation, is usually written as:

$$v = \frac{\mathrm{d}x}{\mathrm{d}t}$$

Table 2.2 shows some typical speeds from a range of projectiles.

Table 2.2 Some interesting speeds

Description	Value ($m\,s^{-1}$)
Speed of light	3.0×10^8
Muzzle velocity of 10 m air rifle (Olympic standard)	200
Typical arrow fired from longbow	66
A good serve in squash	40
Typical release velocity of a javelin	30
Rate of growth of human hair	3.0×10^{-9}

2.3.5 Acceleration

Any motion involving a change of velocity is termed acceleration. So, in a manner, analogous to velocity, we can define the average acceleration as:

$$\bar{a} = \frac{\Delta v}{\Delta t}$$

and the instantaneous acceleration as:

$$a = \frac{\mathrm{d}v}{\mathrm{d}t}$$

Table 2.3 shows some typical accelerations of interest.

Table 2.3 Some interesting accelerations

Description	Value ($m\,s^{-2}$)
High velocity rifle bullet on exit from the barrel	2×10^5
Baseball struck by a bat	3×10^4
Soccer ball struck by a foot	3×10^3
Loss of consciousness in man	70
Gravity on the surface of the Earth	9.81
Gravity on the surface of the Moon	1.7

2.4 Friction, coefficient of friction

Frictional forces are always present where there is movement of objects in any form. However, since this text is primarily concerned with sporting projectiles moving through the medium of air, our focus of analysis will be on how the frictional forces between a propelled sporting projectile and the surrounding air molecules lead to the retarding force we refer to as drag. The details of the effect of the drag so produced will be addressed in Chapter 5.

However, drag is not the only important consequence of friction within the study of trajectory dynamics. The ways in which balls bounce depend on the frictional properties between the balls and surfaces. The friction between bat, or racket, and ball dictate the amount of spin that can be induced. Obviously, in curling, the game dynamics are totally dependent on the frictional values that exist between the curling stone (granite) and the ice. Towards the end of this book, ski-jumping is discussed. In this sport, the projectile (the competitor) is capable of changing their shape to minimize frictional drag as well as to provide as much aerodynamic lift as possible. Waxing the skis will reduce a very different kind of friction, and lead to the highest possible speed on the in-run to take-off.

At a fundamental level the concept of friction is easy to grasp. When the surface of a body moves, or tends to move, over that of another, each of the bodies experience a frictional force. This force acts tangentially along the common surface and in such a direction as to *oppose* the relative motion of the bodies. There are two types of friction to consider: kinetic friction (also known as sliding or dynamic friction), and static friction. The former is the friction experienced once the body is in motion; the latter is the friction experienced in order to commence the movement. If you had to slide a table-tennis table into its playing position, you would have to overcome the static friction in order to first set the table into motion. Once it is moving, however, the reduced reaction force you would feel against your hands corresponds to the lower dynamic friction.

Leonardo da Vinci first enumerated the simple empirical law which quantifies friction, which for the most part, provides an adequate level of accuracy:

> The magnitude of the force of friction between unlubricated, dry surfaces, sliding one over the other is proportional to the normal force pressing the surfaces together and is independent of the area of contact and of the relative speed.

So, to a good approximation, a ski jumper will not suffer any variation in friction as they accelerate down the *constant* slope section of a ski-jump, nor will the frictional force alter much if a snowboard was used instead of jump-skis, even though the respective contact areas of the two types of ski are quite different. Note: Use of snowboards for such activities are not recommended!

These invariances can be explained by considering the causes of friction at a microscopic level. Frictional forces arise from adhesion between the two adjoining surfaces. The contacts between the atoms of the two materials, in effect, form bonds. In fact, in some cases the bonding is so strong that atoms are actually exchanged between the two materials, and a form of spot welding occurs. The natural surface roughness that exists at this microscopic level means that contact is only made where the peaks of each surface align together. To picture this, imagine two sheets of sandpaper in contact with each other, face-to-face. The sum of all the contacting points will be only a small fraction of the total contacting area, and is bound to be dependent on the contacting pressure. So, if the surface area were to increase, although the number of contacting points would increase proportionally (in fact, by a power of 2), the pressure would reduce by the same amount, as the force is now spread out over the wider area. This serves to offset the increase in the number of pressure points, and so maintains the 'sticking' effect at an approximately constant value. It must be emphasised that this is only an approximate invariance; otherwise Formula 1 cars may as well be shod with mountain bike tyres! Although to be fair, the reason for the enormous Formula 1 car tyres has little to do with friction, and more to do with stability of ride and the tyre's ability to shift water from the road under wet conditions.

The relative invariance of the frictional force with velocity is explained by the fact that the microscopic gluing between adjacent sets of atoms happens very quickly. The friction effect cannot be 'beaten' by surfaces sliding at the relative speeds encountered under normal sporting conditions. However, in the case of the friction between a solid body and air, the mechanisms that lead to drag are quite different, and, in this case, the drag force is most definitely dependent on the velocity in quite a complex manner. This topic is addressed in Chapter 5.

It is perhaps worth reiterating that the frictional force always acts in opposition to the direction of travel. Now, in the case of friction between two solid bodies, such as ski and snow, or tennis racket and ball, the motion direction is fairly obvious, and the direction of the frictional force vector equally so. By contrast, in our main area of interest, the drag friction between the projectile and air will be dynamically changing over the duration of the flight in terms of both magnitude and direction, in the same way that the projectile itself is continually changing speed and direction as the flight progresses.

Kinetic friction can be written mathematically as:

$$F_k = \mu_k N$$

where μ_k is the coefficient of kinetic friction and N is the force acting normal to the surface. Note that, when discussing earlier, the friction applied to a ski jumper on the slope, it was emphasised that the friction was essentially invariant down the *constant* slope section of the ski-jump. It should be clear now that friction will increase where the slope flattens out as the skier approaches the jump point, because the normal force will increase to the maximum value of the skier's weight (= mass of skier × the acceleration due to gravity) where the movement is horizontal. Further back up the slope, however, the normal force is reduced by a factor of $\cos\theta$, where θ is the angle of the slope to the horizontal.

The equivalent equation for static friction is:

$$F_s \leq \mu_s N$$

where μ_s is the coefficient of static friction. Note that the inequality symbol is used to indicate that this equation applies even when the body's static friction is greater than the applied force and the body has not started moving. The two sides of the equation equalize at the point where movement commences and just before the dynamic friction takes over. Static friction is never less than dynamic friction; it can be significantly more, but in the case of steel on lead, they both have similar values at 0.9.

Table 2.4 shows values of kinetic and static friction coefficients for a range of materials used in sport.

Table 2.4 Some interesting kinetic and static friction coefficients

Materials	μ_k	μ_s
Rubber on asphalt dry (wet) (motor racing)	0.8(0.75)	0.5(0.25)
Tennis ball on synthetic carpet	0.61	
Tennis ball on hard court	0.49–0.53	
Leather on wood (cricket)	0.3	0.4
Waxed ski on snow		
At 10°C	0.2	0.2
At 0°C	0.05	0.1

2.5 Newton's laws of motion

In 1687 Sir Isaac Newton published his *Philosophiae Naturalis Principia Mathematica*. In it he stated the three laws on which the science of all dynamic motion, including trajectory motion, is based.

2.5.1 Newton's first law of motion

> Every body continues in its state of rest or of uniform (unaccelerated) motion in a straight line unless acted on by some external force.

This law expresses the concept of inertia which can be described as being its reluctance to start moving, or to stop moving once it has started.

In terms of our sporting projectiles we can rephrase the law as follows. If no forces are acting on a projectile which is at rest, or moving with constant velocity, then it will always remain at rest or at that same constant velocity. Conversely, a projectile which is at rest, or constant velocity, experiences no net resultant force. That is, there is either no forces acting on the projectile at all, or all the forces acting on it work to oppose each other and, as a consequence, cancel out in all directions.

If a projectile's speed or direction changes we know that a net force is acting on it. We know that a ball competitively thrown from the hand follows an approximately 'n' shaped path (later proved to be approximately parabolic). Its direction and its speed are, therefore, continually changing over the duration of the flight. As a consequence, we can conclude that it is being acted on by a force. Of course we know that force to be gravity.

2.5.2 Newton's second law of motion

> The rate of change of momentum of a body is directly proportional to the external force acting on the body and takes place in the direction of the force.

In mathematical terms the second law may be written as:

$$F \propto \frac{\mathrm{d}}{\mathrm{d}t}(mv)$$

where F is the applied force, and $d(mv)/dt$ is the rate of change of momentum where m is mass and v is velocity.

The proportionality sign can be removed by introducing a constant of proportionality, k, thus:

$$F = k\frac{d}{dt}(mv)$$

The SI unit of force (the newton) is defined such that $k = 1$ provided the mass is given in kg and velocity in m s^{-1}. So now:

$$F = \frac{d}{dt}(mv)$$

and if we assume m is constant, we derive the most common representation of Newton's second law:

$$F = ma \tag{2.1}$$

where a is the acceleration which results from the application of a force.

So in terms of projectile motion, if a force acts on a projectile, then the projectile accelerates in the direction of the force. The magnitude of this acceleration is equal to that of the applied force divided by the projectile's mass.

2.5.3 Newton's third law of motion

Whenever a body exerts a force on another body, the latter exerts a force of equal magnitude and opposite direction on the former.

In tennis, the action of successfully striking a ball results in equal and opposite forces acting on the ball and on the strings of the racket. It is one of the jobs of the racket to attempt to absorb these forces before they travel up the handle and into the player's body with the potential for injury (e.g. tennis elbow). In performing a shot put, the force on a shot as it is launched is exactly balanced by an equal and opposite force felt on the hand of the athlete as they push the shot put on its way.

2.5.4 Newton's equations of motion

Newton's second law will prove to be vital in many analyses performed throughout this book. However, it will, often as not, be complemented by one or more of the so-called 'equations of motion'. These are simply derived from our definitions of distance, velocity and acceleration, and the relationships between them:

$$v = u + at \tag{2.2}$$

$$v^2 = u^2 + 2as \tag{2.3}$$

$$s = ut + \tfrac{1}{2}at^2 \tag{2.4}$$

$$s = \tfrac{1}{2}(u + v)t \tag{2.5}$$

where u is the velocity when $t = 0$, v is velocity at time t, a is constant acceleration, and s is the displacement from the starting point at time t.

It is worth noting that, with the exception of time, t, all the variables in Equation 2.2 through to Equation 2.5 are vectors: their sign is important and denotes direction. If, say, the positive direction is taken as up, then a projectile heading down to earth has a negative velocity. Points below the start point of the projectile will have a negative value of displacement, s. Negative accelerations act downwards to the earth. Acceleration acting in the opposite direction to the velocity will result in retardation.

Worked Example 2.1

The Aeron B99 air pistol has an 8 inch (20 cm) barrel and a muzzle velocity of 44.3 ft/s ($13.5 \, \mathrm{m \, s^{-1}}$). Calculate the average acceleration of the pellet in the barrel. If the mass of the 0.22 calibre pellets are 15.6 g, ignoring the frictional and drag resistances in the barrel, calculate the force delivered from the pump-up mechanism of the air pistol to achieve the stated muzzle velocity.

From Equation 2.3:

$$v^2 = u^2 + 2as$$

where $u = 0 \, \mathrm{m \, s^{-1}}$, $v = 13.5 \, \mathrm{m \, s^{-1}}$ and $s = 0.2 \, \mathrm{m}$

$$a = \frac{(v^2 - u^2)}{2s} = \frac{13.5^2}{2 \times 0.2} = 455.6 \, \mathrm{m \, s^{-2}}$$

Note that this is the average acceleration of the pellet over the length of the barrel. In reality, the pellet is subject to a much greater acceleration than this immediately following the explosive firing action, and it is already slowing down by the time it reaches the end of the barrel.

From Newton's second law, Equation 2.1, the force applied to the pellet is given by

$$F = ma = 0.015 \times 455.6 = 6.83 \, \mathrm{N}$$

Furthermore, Newton's third law tells us that the recoil force on the firing hand will also be 6.83 N.

2.6 Newton's law of universal gravitation

Every particle in the Universe attracts every other particle with a force which is proportional to the product of their masses and inversely proportional to the square of their separation.

In mathematical terms, this may be stated as:

$$F = G \frac{m_1 m_2}{r^2}$$

where F is the force of attraction between two bodies whose masses are m_1 and m_2, and whose centers are separated by a distance r, and G is a constant of proportionality known as the universal gravitational constant ($= 6.67 \times 10^{-11} \, \mathrm{N \, m^2 \, kg^{-2}}$).

The force, F, is a vector that points between the centers of masses of the two bodies concerned. In standard vector notation, we may denote this as:

$$F = G\frac{m_1 m_2}{r^2}\hat{\mathbf{r}}$$

(2.6)

where, now, $\mathbf{r}\ (= r\,\hat{\mathbf{r}})$ is the vector from the center of the mass of the larger body to the center of mass of the smaller body and $\hat{\mathbf{r}}$ is the corresponding unit vector.

It is clear that it is this force of attraction, F, which is responsible for the motion of a sporting projectile which is fired in an upward direction heading back to Earth. In such cases, the larger mass, m_1, will be the mass of the Earth $(= m_e = 5.98 \times 10^{24}$ kg), the smaller mass, m_2, will be the mass of the projectile, and r can simply be taken as the radius of the earth $(= r_e = 6.38 \times 10^6$ m), since any additional height achieved by the sporting projectile will be negligible compared to the earth's radius.

Equation 2.6 may now be written as:

$$F = -G\frac{m_e m_2}{r_e^2}\hat{\mathbf{j}}$$

(2.7)

where $\hat{\mathbf{j}}$ is the unit vector in the upward vertical direction, which may be considered constant in both magnitude and direction. This assumption that F has a constant direction over the flight of the projectile is known as the 'flat earth' assumption and is accurate enough for all but high power rifle shells shot over very, very long distances. As an indication of the invariance of F, it can easily be shown that its value reduces by only 3 per cent for a projectile which is shot 100 km into the air!

Equation 2.7 now simplifies to:

$$F = mg$$

where $m = m_2$, $g = Gm_e/r_e^2$, and with our values of G, m_e and r_e as stated above, $g = 9.81$ m s^{-2}, where g is known as the acceleration due to gravity. In such cases, F is known as the weight of the projectile and, as stated in Section 2.3, is equal to the mass of the projectile multiplied by g, the acceleration due to gravity.

Worked Example 2.2

Two curling stones of mass 20 kg lie on the ice at a distance 40 cm apart. Calculate the force of attraction between them. If the static friction between the stones and ice is $\mu_s = 0.1$, prove that the stones will not move towards each other and collide under the force of attraction.

$$F = G\frac{m_1 m_2}{r^2}$$

where $G = 6.67 \times 10^{-11}$, $m_1 = m_2 = 20$ kg, $r = 0.4$ m

$$F = 6.67 \times 10^{-11} \times \frac{20^2}{0.4^2} = 1.67 \times 10^{-7}\,\text{N}$$

The force required to overcome friction on the ice is given by:

$$F_s = \mu_s N$$

where N on the horizontal ice surface is simply given by mg, the weight of each of the stones.

$$F_s = \mu_s mg = 0.1 \times 20 \times 9.81 = 19.62\,\text{N}$$

So, as expected, the force of attraction between these two stones is orders of magnitude less than that required to overcome the frictional resistance of the stones on the ice. However, it is worth pointing out that even this tiny force of attraction is easily accurately measurable using present-day equipment.

2.7 Conservation of linear momentum

We have, thus far, discussed an assortment of parameters, properties and variables that may have a bearing on, or in some way alter, the way a projectile launches into motion. One such quantity is momentum. We have stated that momentum is a vector, rather than a scalar quantity. Newton's second law concerns the rate of change of momentum, and we stated in Section 2.5 that the momentum of an object is equal to the product of its mass and its velocity. We say:

$$\mathbf{p} = m\mathbf{v}$$

where $\mathbf{p}$ is the momentum of the body, a vector whose direction is the same as that of the velocity $\mathbf{v}$.

Where momentum becomes fundamental in our analyses is when bodies collide. This may be the collision of two projectiles such as billiard balls or curling stones, or the collision between a striking implement and a projectile such as a golf club and ball.

When two bodies collide, it is the combined momentum of the bodies, immediately before and after collision, which is conserved.

Newton's second law may be written as:

$$\mathbf{F} = \frac{d\mathbf{p}}{dt}$$

and for completeness, the kinetic energy of the body, which is the amount of work required to accelerate the body from being stationary up to its velocity, is given by:

$$K = \tfrac{1}{2}mv^2 = \frac{p^2}{2m}$$

Consider the collision between, say, a bowling ball and a jack ball on a crown green. Let F_{12} denote the force exerted on the jack by the bowling ball, and F_{21} be the force exerted on the bowling ball by the jack ball. Newton's third law tells us that in the case of a perfect collision, with no energy lost through heat, sound or absorption of energy into either of the bodies:

$$F_{12} = -F_{21}$$

So:

$$\frac{d\mathbf{p}_1}{dt} + \frac{d\mathbf{p}_2}{dt} = \mathbf{F}_{12} + \mathbf{F}_{21}$$

$$\frac{d(\mathbf{p}_1 + \mathbf{p}_2)}{dt} = \mathbf{F}_{12} - \mathbf{F}_{12} = 0$$

Therefore $\mathbf{p}_1 + \mathbf{p}_2 = $ a constant.

Thus, the conservation of linear momentum states that

> The total linear momentum of a system of interacting (e.g. colliding) bodies, on which no external forces are acting, remains a constant.

Worked Example 2.3

The Aeron B99 air pistol of Worked Example 2.1 has a mass of 2.2 lb (1.0 kg). As stated, the muzzle velocity is 44.3 ft/s (13.5 m s^{-1}) and the mass of the 0.22 calibre pellets are 15.6 g. Calculate the recoil velocity of the pistol.

The analysis of the shooting of a gun may be considered a collision problem in the sense that the two bodies (the pistol and the pellet) are static prior to firing ($u_1 = u_2 = 0$), while, following the shot, the two velocities may be taken, respectively, as the pellet velocity immediately after firing (assumed in this case to be approximately equal to the muzzle velocity), v_1, and the recoil velocity of the pistol, v_2. So:

$$m_1 u_1 + m_2 u_2 = 0 = m_1 v_1 + m_2 v_2$$

where m_1 is the mass of the pellet, v_1 is the muzzle velocity of the pellet, m_2 is the mass of the gun, and v_2 is the recoil velocity of the gun. Therefore,

$$v_2 = -\frac{m_1 v_1}{m_2} = -\frac{0.156 \times 13.5}{1.0} = -2.106 \text{ m s}^{-1}.$$

As a final thought on the subject of momentum, it is clear that, the larger an object's momentum is, the more difficult it is to change it by means of altering its velocity (and, of course, altering its mass in mid-flow is usually impossible!). There is an inherent inertia associated with the magnitude of the momentum.

A practical application of this phenomenon can be seen when comparing a golfing drive from the tee with a putt on the green. In the former case, once the full driving action is under way, the mechanical momentum will keep the swing going, and there is little a player can do to interrupt the inherent rhythm in that swing. There are only two points in the swing where the momentum is zero; at the commencement of the up-swing and at the top of the swing, just before the driving swing commences. It is, therefore, imperative that the head of the club is positioned correctly at these points with the target in mind, since error correcting is almost impossible when the club head is actually in motion.

However, in the case of less powerful shots such as putts or chip shots, the momentum in the swing is less and the player may exert more corrective influences at all stages in the stroke. However, there is a danger that less able players interrupt the essential smooth action of the stroke in an effort to correct errors, which may result in a weak, nervous jab at the ball.

By contrast, high-speed camera images show that elite players maintain a slow deliberate action throughout the stroke, with the club's momentum gradually increasing through the striking phase of the stroke, to reach a maximum at the point of impact with the ball, and then slowly reducing on the follow-through.

2.8 Conservation of angular momentum

There is a considerable equivalence in mechanics between bodies moving in straight lines (translational motion) and those moving in curved trajectories. Many of the translational laws and equations previously stated have a correspondence in circular motion. Quite simply, all the equations correlate precisely if the appropriate substitutions for the variables are carried out in accordance with Table 2.5.

Table 2.5 Rotational analogues of translational motion

Translational quantity		Rotational analogue	
Mass	m	Inertia	I
Velocity	v	Angular velocity	ω
Force	F	Torque	T
Acceleration	a	Angular acceleration	α
Distance	s	Angle	θ
Momentum	p	Angular momentum	L

So, for instance, the relation defined by Newton's law of motion stated in Equation 2.4:

$$s = ut + \tfrac{1}{2}at^2$$

transcribes directly to:

$$\theta = \omega_0 t + \tfrac{1}{2}\alpha t^2$$

where ω_0 is the angular velocity at time $t = 0$ and the other symbols are as stated in Table 2.5.

In general biomechanical sporting scenarios, these rotational and circular motion equations are important, especially in relation to the striking of a ball with some instrument. However, the focus of this book is really only about what happens to the projectile *once it has been launched* following the strike or throw.

Nonetheless, one area of rotational analysis which remains relevant is the revolving of the projectile through the air; whether it is the spin of a ball which will change the trajectory path, the gyroscopic action of a spinning discus or the 'rifling' effect important in a javelin throw. To aid in these types of analyses the principle of conservation of angular momentum must be understood. From Table 2.5, we can see that the angular momentum of a body is given by

$$L = I\omega$$

(cf. $P = mv$). The principle of conservation of angular momentum states that:

The total angular momentum of a system is constant unless an external torque acts on it.

The usual demonstration of this is an ice skater who, while spinning, brings her arms in closer to her body. In so doing, she concentrates her mass closer to her axis of rotation and, thereby, reduces her moment of inertia. This results in an increase in angular velocity as, according to the principle, the angular momentum must remain constant. Incidentally, since her angular velocity increases, so does her rotational kinetic energy. This extra energy comes from the work done in bringing her arms in towards her body against the centripetal force tending to pull her arms outwards.

So, the moment of inertia is the reluctance of a body to rotate around some given axis in much the same way as the mass of a body is the reluctance of a body to move in a linear motion. Also like mass, it is a scalar quantity but with units of kg m^2. A more formal definition of the moment of inertia is a measure of the object's resistance to accelerated angular motion about an axis. It is calculated by summing the products of all the elemental masses of the body with the squares of their distances from the appropriate axis of rotation.

Table 2.6 gives calculated moments of inertia for common simple shapes although for our purposes it is especially worth noting that for a solid sphere (i.e. many of our sports balls):

$$I = \tfrac{2}{5}mr^2$$

It is clear that I, for a given body, is a function of both the body's overall weight, its shape and the way in which the weight is distributed around the axis of rotation. The total-weight parameter can be removed by defining the radius of gyration, k, by means of the relation: $I = mk^2$. The radius of gyration is a parameter which is purely based on the shape of the body and the way in which its mass is distributed around the axis of rotation. The radius of gyration is a particularly useful property in terms of sports striking implements; they tend to have an approximately constant value for each *type* of sporting equipment (i.e. all hockey sticks have roughly one value, while all cricket bats have a different one, etc.).

These concepts of rotational analysis become important in sports trajectory modelling, not only in the study of spinning balls colliding with surfaces or other balls (see Chapter 4), but also in the study of the modification of the standard trajectory path by spin effects (see Chapter 6).

Table 2.6 Some common moments of inertias

Shape	Formula	Comment
Solid sphere (rotation around any axis through center)	$I = \tfrac{2}{5}mr^2$	Most common solid balls such as snooker balls, bowling balls or cricket balls.
Hollow sphere (rotation around any axis through center)	$I = \tfrac{2}{3}mr^2$	Most inflated balls such as squash balls, tennis balls footballs.
Solid cylinder (rotation through long axis)	$I = \tfrac{1}{2}mr^2$	Spinning discus or rifling bullets.
Hollow cylinder with inner radius $= r_1$ and outer radius $= r_2$ (rotation through long axis)	$I = \tfrac{1}{2}m\left(r_1^2 + r_2^2\right)$	A closer approximation to a spinning javelin than the solid example above.
Thin rod rotated around one end	$I = \tfrac{1}{3}mL^2$	Approximation of the shaft of a striking implement

$m =$ mass of the shape
$r =$ distance between center of rotation and spinning surface
$L =$ length of rod

Note: Solid objects are assumed to have a constant density throughout the body.

2.9 Work, impulse, power and energy

These parameters are discussed more for the sake of completeness within a chapter which includes an overview of most of the other basic physical properties associated with projectile launch and consequential motion. Nevertheless, as will be shown, the principle of conservation of energy turns out to be a most useful tool, particularly in calculating such quantities as trajectory altitudes.

2.9.1 Work

If a body moves as a result of a force being applied to it, the force is said to be doing work on the body. The work done is given by:

$$W = Fs$$

where W is the work done in joules (J), F is the applied force in newtons (N) and s is the distance moved in meters (m).

In many sporting situations, the projectile is launched by means of a sharp impact with an appropriate implement, so the calculation of work done is not practical. However, see Worked Example 2.4.

2.9.2 Impulse

We know that Newton's second law may be written as:

$$\mathbf{F} = \frac{d\mathbf{p}}{dt} = \frac{d(m\mathbf{v})}{dt}$$

For an object of constant mass and summing over time, we obtain:

$$\int \mathbf{F}dt = \int d(m\mathbf{v}) = m \int d(\mathbf{v}) \tag{2.8}$$

The term $\int \mathbf{F}dt$ is known as the impulse (units Ns) and is equal to the change in momentum, or if the mass is constant, simply the change in velocity. Equation 2.8 is known as the impulse-momentum equation and it may be written, for constant mass as $F\Delta t = m\Delta v$, where F is the mean value of the force acting over a time interval Δt, during which the speed of the object changes by Δv.

Although in many sporting science texts much importance is given to properties such as velocity and force, in fact, it is often the impulses created by athletes that lead to the large forces and velocities. For example, in analysing the bowling of a cricket ball, one may think in terms of the ball release velocity, but what actually accounts for that velocity is the series of impulses resulting from, in turn; the ground reaction forces in the run up, the delivery stride and placing of feet and finally the bowling arm action.

Each impulse will have a force and time associated with it, which is directly proportional to the changes in velocity at each stage. In catching the ball in the slips, the impulse required to stop the ball is determined by the mass ($m = 0.16$ kg) and the change in speed (Δv) of the ball. The catcher can reduce the mean force (F) acting on their hands by increasing the duration of the contact time (Δt) by 'giving' with the ball.

2.9.3 Power

Power is the rate at which work is done and is measured in watts:

$$P = \frac{dW}{dt}$$

where P is the instantaneous power (W) and dW/dt is the rate of working (J s^{-1}).

Worked Example 2.4

The force applied to a javelin on launch is a constant 1000 N. Calculate the power in the throw and the work carried out by the athlete if the athlete's launch action is 1.6 m long and lasts for 0.6 s.

The work done in launching the javelin is $1.6 \times 1000 = 1600$ J.

The power the athlete puts into the launch is $1600/0.6 = 2666.67$ W.

2.9.4 Energy

If a body is capable of carrying out work it is said to possess energy. The amount of energy a body possesses is equal to the work it is capable of doing. It is a scalar quantity. There are two types of energy to be considered; kinetic and potential energy.

Kinetic energy

This is the energy a body possesses solely because it is moving. If a body of mass m has a velocity v, then the kinetic energy, K is given by:

$$K = \tfrac{1}{2}mv^2$$

Potential energy

Potential energy is the energy a body possesses by virtue of its position or due to the arrangement of its component parts. In essence, potential energy is about the storage of energy. A discus resting on a shelf possesses potential energy because work had to be done in raising it against the force of gravity, and the discus will retain and store that energy until such time as it leaves the shelf and returns to earth. However, potential energy is also stored in any resilient ball when it is hit by a striking implement, or when it bounces. At the point of impact, the ball distorts from its spherical shape against the elastic forces of the material. The ball has a tendency to return to its spherical shape in the same sense that the discus on the shelf wants to return to the ground. While the ball is in its distorted state, potential energy is stored, which is then released as the ball recovers to its natural shape.

In the case of an object which has been placed in a position against the influence of gravity, if the mass of the body is m and it has been displaced by a height h, then:

Gravitational potential energy $= mgh$

Conservation of energy

The conservation of mechanical energy states that:

> In a system in which the only forces acting are associated with potential energy, (either of a gravitational or elastic nature), the sum of the kinetic and potential energies are a constant.

In the case of the flight of a sporting projectile, the sum of the kinetic and potential energies will remain constant over the whole duration of the flight. As an example, take the simple case of a ball being thrown vertically up, and then being caught on its way down at the same height. The ball leaves the hand with a certain kinetic energy based on its throw velocity, and no potential energy. As the ball rises, it loses its kinetic energy and gains potential energy until, at the top of the motion, the ball has zero velocity. At this instant, the ball has its maximum potential energy and zero kinetic energy (maximum *h* and zero *v*). As the ball falls back down to the hand, the reverse process occurs. The ball loses height and so loses potential energy, while it gains velocity and gains kinetic energy. It arrives back in the hand with the same kinetic energy as when it was thrown, and the same magnitude of velocity, but of course, with the opposite sign, reflecting the opposite direction of travel.

Worked Example 2.5

A juggler throws a ball vertically upwards at $3.5\,\mathrm{m\,s^{-1}}$. Use the law of conservation of energy to calculate how high the ball will go.

By the conservation of energy, the kinetic energy lost while the ball is going up is equal to the potential energy gained. At all times in the flight

$$\tfrac{1}{2}mv^2 = mgh$$

Therefore,

$$h = \frac{v^2}{2g} = \frac{3.5^2}{2 \times 9.81} = 0.625\,\mathrm{m} = 62.5\,\mathrm{cm}.$$

Summary

This chapter covers all the fundamental physics required to commence our trajectory calculations in the next chapter, including scalar and vector quantities, and the physical quantities such as mass, volume, force and work. We defined the term coefficient of friction, Newton's laws, Newton's equations of motion and his gravitation law. We then solved our first moving body calculations and progressed onwards to consider power, momentum and kinetic and potential energy, which we then used to analyse the motion of vertically thrown balls.

Problems and questions

1   From a vector perspective, if an athlete runs a 400 m race at a constant speed, why do they have to work harder to maintain that speed on the bends of the track?

2   The angle of the sloping section of a ski jump is 60° to the horizontal. If the coefficient of friction between the skis and the snow is 0.1, calculate the frictional drag on a skier of mass 80 kg when on the slope of the run, and also on the horizontal jump-off section.

3   Two snooker balls weighing 160 g each are spaced 60 cm apart on a snooker table. Calculate the force of attraction between them and consider, what is stopping the balls moving together. You may wish to contrast your deliberations on this with the reason given in Worked Example 2.2 for the curling stones not coming together.

4   An archer draws a bow to a point where its potential energy is 64 J. What will the release velocity of the arrow be if its mass is 300 grains (19.5 g)?

5   The bow in the previous question weighs 4 kg. Calculate the recoil velocity of the bow when the arrow is released.

6   Sketch a graph showing both the kinetic and potential energies against time, for a shotput from the time when it is released from the athlete's hand to when it lands on the ground. Remember that, in the case of putting the shot, the height of release and the height of the ground are not the same.

Chapter 3

Motion of projectiles under the influence of gravity

3.1 Introduction

There are many ways of representing a projectile's motion under the influence of gravity. One of the simplest methods involves deriving the motion of the projectile in two orthogonal directions, and then recombining these movements into a single resultant motion. For sporting projectiles, these directions are usually taken to be horizontal and vertical with respect to the Earth's surface, which is presumed, for these purposes, to be flat.

Other assumptions we will make, at least for the duration of this chapter, are:

- The projectile is small, heavy and round and, therefore, is subject to negligible drag and lift.
- Its surface is smooth and there is no spin on the projectile, so it is not subject to any form of swing.
- It is considered to be moving through still air; there is no cross, head or tailwind to distort the trajectory.
- Unless otherwise stated, the projectile is impelled from ground level in some upward direction, and arrives some time later at a point also at ground level: the ground is assumed to be horizontal as well as flat.
- Having established this 'flat-earth' assumption, we can further state that gravity acts normal to the Earth's surface, in parallel lines of flux, throughout the duration of the flight.

With such assumptions established, we can state that the only force influencing the trajectory, following the initial impact, is the force due to gravity. Furthermore, the only influencing factors with regard to the impact are the velocity and angle of the projectile immediately following that impact. Therefore, if values for gravity, initial trajectory velocity and initial trajectory angle are known, we can fully describe the motion of the projectile in terms of distance, time, maximum height, etc., following impact.

If the initial velocity of the projectile immediately following impact is v_0 and its initial angle of trajectory immediately following impact is θ_0 to the horizontal, the velocity can be resolved into two orthogonal components, v_x and v_y, as follows:

$$v_x = v_0 \cos \theta_0 \tag{3.1}$$

$$v_y = v_0 \sin \theta_0 \tag{3.2}$$

where v_x is the component of the velocity parallel with the Earth's surface, and v_y is the component orthogonal to it and parallel to the force of gravity.

We may consider some particular values of θ_0. These are as follows:

- $\theta_0 = 45°$. In this case both $\sin \theta_0$ and $\cos \theta_0 = 0.7071$. Therefore, both the horizontal and vertical components of the velocities are the same, and are 0.7071 of the actual launch velocity.
- $\theta_0 = 60°$. The horizontal velocity component is exactly half the launch velocity; the vertical velocity component is 0.866 of the launch velocity.
- $\theta_0 = 30°$. The inverse of the 60° case. This time the vertical velocity is half the launch velocity, while the horizontal velocity is 0.866 of it. We will see later that both these angles result in the same projectile range for a given value of v_0.
- $\theta_0 = 0°$. In this case, the horizontal velocity component is equal to the launch velocity and there is no vertical velocity component. The projectile is rolling along the ground, and, for the assumptions stated above, will do so forever.
- $\theta_0 = 90°$. The projectile is now being launched directly vertically at the launch velocity. Obviously, there is no horizontal component. The projectile will decelerate, reach a peak height and then fall to earth, landing precisely at its launch point.

3.2 Horizontal motion

To understand better the physical significance of the resolved component in the x direction, imagine a soccer ball being kicked high in the air and down the pitch. The component v_x would then represent the velocity that the kicker would have to run in order to stay directly under the ball over the whole duration of the flight. This may sound tricky, but consider the Percy Grainger feat discussed in Section 3.7.

On the basis of the assumptions delineated in the previous section, after the initial impelling impact, there is clearly no force acting in the x direction. And so, in accordance with Newton's second law, the velocity cannot change along this direction; v_x remains constant for the duration of the flight. Of course, once it lands, other factors come to bear, some of which are discussed in later chapters. Furthermore, as we start to consider other more complex aspects of flight, such as drag and dynamic lift, the horizontal velocity is indeed subject to change over the duration of the flight.

The position of the projectile along the x direction at any point in time t, after the initial launch, and while the projectile is in flight, is simply given by:

$$s_x = v_x t = v_0 t \cos \theta_0 \tag{3.3}$$

Worked Example 3.1

Jan Zelezny attained a 98.48 m world record javelin throw in 1996 at the World Track and Field Championships in Jena, Germany. If the time of javelin flight was 3.82 s and the throw angle was 36° what was his throw velocity? You may assume the conditions stated in Section 3.1 apply.

From Equation 3.3:

$$v_0 = \frac{s_x}{t \cos \theta_0} = \frac{98.48}{3.82 \times \cos 36°} = 31.872 \text{ m s}^{-1}.$$

3.3 Vertical motion

The vertical component of the velocity is subject to a constant gravitational force a, which, on the earth's surface, is usually taken to be of value 9.81 m s^{-2}, and usually takes on the symbol, g.[1] The resulting vertical motion is equivalent to an object being thrown directly upwards; it reaches its peak, at which point the velocity is zero, and it then drops back to earth. The governing equation of motion in this resolved direction is, from Newton's second equation of motion, Equation 2.4:

$$s_y = v_y t - \tfrac{1}{2}at^2 = v_0\, t \sin \theta_0 - \tfrac{1}{2}at^2 \tag{3.4}$$

This is of a quadratic form in t. It tells us that, for every allowable value of s_y, there are potentially *two* values of t. In practical terms, this means the projectile passes through the point s_y on the way up, at time, say, t_1, and then passes through the same point s_y again, on the way down, at some later time, say, t_2.

It can easily be shown that the vertical motion is symmetrical about the projectile's peak point. They say that, 'What goes up, must come down', but we can now further assert that, for this simple trajectory model with the assumptions as stated, the projectile takes exactly the same amount of time to go up as it does to come down. Furthermore, at each position between release and peak points, the speed of the projectile on the upward path is the same as the speed on the downward path through that point – only the sign changes on the velocity vector. So, if we were to throw a tennis ball in the air such that it leaves our hand at, say, 1 m s^{-1}, and catch it at the same height, it will land in our hand at a velocity of -1 m s^{-1}.

If we plot, for a typical sporting projectile (say a soccer ball being kicked half way down a football pitch), the time of duration of the flight against the vertical distance s_y we will get the familiar parabola of the quadratic function of Equation 3.4 as shown by Figure 3.1.

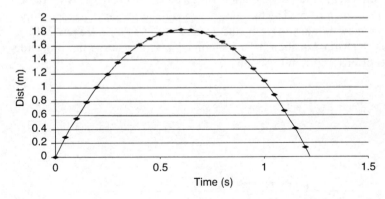

Figure 3.1 Distance–time graph of projectile thrown with a vertical component velocity of 6 m s^{-1}. It reaches a height of 1.83 m and is in flight for 1.25 s.

---

1 Two main factors affect the variation in the acceleration due to gravity, g, around the planet's surface.

  (i)   The centrifugal force due to the planet's rotation, is greater at the equator than at higher latitudes. This factor alone would result in a variation in value of g of between 9.789 m s^{-2} at the equator, and 9.823 m s^{-2} at the poles.

  (ii) There is a pronounced equatorial 'bulge' in the earth's sphere. This serves to accentuate the discrepancy, resulting in an overall variation in g of 0.052 m s^{-2}.

Worked Example 3.2

In order for a juggler to juggle four balls at the same time, the balls must be in the air for 0.8 s. Assuming the balls move in a vertical direction only (he's juggling with one hand!), calculate, first, the throw velocity, and then the height the balls must attain.

As the balls are thrown and leave the juggler's hand, the velocity reduces from the throw velocity, v_y, to $0 \, \text{m s}^{-1}$ in half the flight time $= 0.4 \, \text{s}$. It does this as it decelerates at a rate of $9.81 \, \text{m s}^{-2}$. It, therefore, must be thrown at $v_y = 9.81 \times 0.4 \, \text{m s}^{-1} = 3.924 \, \text{m s}^{-1}$.

Now, using Equation 3.4:

$$s_y = v_y t - \tfrac{1}{2}at^2$$

$$= 3.924 \times 0.4 - \frac{9.81}{2} \times (0.4)^2 = 0.785 \, \text{m}.$$

3.4 Combined motion: the trajectory path

So long as the assumptions listed in Section 3.1 apply, a projectile which is launched horizontally along the ground with initial velocity v_x, will have a motion described by that of Section 3.2, while a projectile launched vertically with initial velocity v_y will have a motion which is described by Section 3.3. A projectile launched at some general angle θ_0 to the horizontal, with a velocity v_0, will have velocity components, v_x and v_y, defined by Equations 3.1 and 3.2. Its resultant motion at all times in its flight will be a combination of the two orthogonal motions; a vertical motion governed by the launch velocity component v_y, and a simple linear motion with unchanging velocity given by the velocity component v_x (Figure 3.2).

Combining Equation 3.3 with Equation 3.4 by eliminating t, we derive the projectile equation in Cartesian coordinates. From Equation 3.3:

$$t = s_x/v_x$$

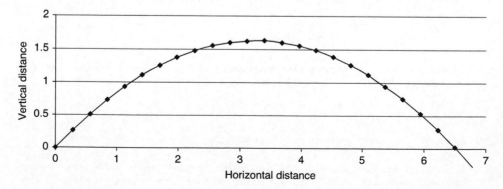

Figure 3.2 Trajectory path of a projectile launched at $8 \, \text{m s}^{-1}$, at an angle of 45° to the horizontal. Components v_x and v_y are both $5.66 \, \text{m s}^{-1}$ and the peak height is 1.63 m. Its range is 6.5 m.

Substitution into Equation 3.4:

$$s_y = v_y \frac{s_x}{v_x} + \frac{1}{2}a\left(\frac{s_x}{v_x}\right)^2 \tag{3.5}$$

By trigonometry on the velocity vectors

$$\tan \theta_0 = \frac{v_y}{v_x}$$

$$s_y = s_x \tan \theta_0 - \frac{as_x^2}{2v_0^2 \cos^2 \theta_0}$$

$$s_y = s_x \tan \theta_0 - \frac{as_x^2 \left(1 + \tan^2 \theta_0\right)}{2v_0^2} \tag{3.6}$$

(See Model 3.1 – Standard trajectory model – Standard trajectory model worksheet tab.)

Equation 3.6 represents a quadratic function. The trajectory of a projectile, influenced only by gravitational forces, launched at an angle θ_0 to the horizontal and with a velocity v_0 will trace out a parabolic shape which passes through the point of launch with a positive gradient of θ_0 and lands with a gradient of $-\theta_0$.

3.4.1 Some further results

The speed of the projectile at any point in its trajectory, v is given by:

$$v^2 = \left(\frac{ds_x}{dt}\right)^2 + \left(\frac{ds_y}{dt}\right)^2$$

$$= v_0^2 \cos^2 \theta_0 + \left(v_0 \sin \theta_0 - at\right)^2$$

$$= v_0^2 \cos^2 \theta_0 + v_0^2 \sin^2 \theta_0 - 2v_0 at \sin \theta_0 + a^2 t^2$$

$$v^2 = v_0^2 - 2as_y \tag{3.7}$$

The angle made by the tangent of the projectile's path to the horizontal at any time following launch, θ is given by:

$$\tan \theta = \frac{ds_y/dt}{ds_x/dt}$$

$$= \frac{v_0 \sin \theta_0 - at}{v_0 \cos \theta_0}$$

$$\tan \theta = \tan \theta_0 - \left[\frac{at}{v_0}\right] \sec \theta_0 \tag{3.8}$$

This result could equally well have been derived by differentiating Equation 3.6 with respect to s_x:

$$\tan \theta = \frac{ds_y}{ds_x} = \tan \theta_0 - \frac{as_x \sec^2 \theta_0}{v_0^2} \tag{3.9}$$

The point at which either Equations 3.8 or 3.9 equate to zero represent the apex of the trajectory (i.e. the gradient of the trajectory is zero):

$$0 = \tan\theta_0 - \left[\frac{at_{1/2}}{v_0}\right]\sec\theta_0$$

where $t_{1/2}$ is the time of flight from launch to apex (half the total time of flight):

$$t_{1/2} = \frac{v_0\sin\theta_0}{a}$$

By symmetry, therefore, the total time of flight t_m will be twice $t_{1/2}$:

$$t_m = 2\frac{v_0\sin\theta_0}{a} \tag{3.10}$$

In a similar manner, we can work on Equation 3.9 to obtain the range of the projectile. At the apex, again $ds_y/ds_x = 0$. And so:

$$0 = \tan\theta_0 - \frac{as_{x_{1/2}}\sec^2\theta_0}{v_0^2}$$

where $s_{x_{1/2}}$ is the horizontal distance from launch to apex:

$$s_{x_{1/2}} = \frac{v_0^2}{a}\sin\theta_0\cos\theta_0$$

$$= \frac{v_0^2}{2a}\sin 2\theta_0$$

By symmetry, therefore, the total range of the flight s_m will be twice $s_{x_{1/2}}$:

$$s_m = \frac{v_0^2}{a}\sin 2\theta_0 \tag{3.11}$$

In summary, we are now in a position to state that a projectile which is launched at a coordinate position (0, 0) with a launch velocity v_0 and a launch angle of θ_0, will land at coordinates:

$$\left(\frac{v_0^2\sin 2\theta_0}{a}, 0\right)$$

Furthermore, it will pass through a peak altitude with coordinates of:

$$\left(\frac{v_0^2\sin 2\theta_0}{2a}, \frac{v_0^2\sin^2\theta_0}{2a}\right)$$

An interesting question may be asked: what angle will ensure that the projectile's peak height is equal to its range? This can be calculated by equating the two relevant coordinates from above:

$$\frac{v_0^2 \sin 2\theta_0}{a} = \frac{v_0^2 \sin^2 \theta_0}{2a}$$

$$2 \sin 2\theta_0 = \sin^2 \theta_0$$

So, $\sin \theta_0 = 0$ or $\tan \theta_0 = 4$ are the two possible solutions. The first is trivial and may be disregarded (the projectile does not leave the ground!). But, from $\tan \theta_0 = 4$, we obtain that $\theta_0 = 76°$. Note that this is independent of v_0. So, the cricket commentator who is reported to have announced that 'It was such a big six that the ball went as high as it went forward' was not really stating anything special; the ball could still have done this without it even reaching first slip!

In sporting applications, whether it is the javelin or the long jump, often the most important parameter to be maximized is the distance travelled, or the range of flight. The question may be asked: what is the optimum release angle to ensure this maximum range? This can be obtained by inspection of Equation 3.11. The s_0 will reach its maximum when $\sin 2\theta$ achieves its maximum value of unity, i.e. when $\sin^{-1} 1 = 2\theta$. This will be when $2\theta = 90°$. Therefore, the best angle to launch a projectile to ensure maximum range, θ_m, is $45°$.

It must be stressed, however, that this result only applies under the conditions outlined in Section 3.1. In particular, the launch and landing points must be at the same height. In most sporting applications (e.g. javelin, tennis serve) this is patently not a condition that may be assumed. A more complex analysis must be performed in order to derive the optimum release angle under conditions of differing launch and land heights and this is the subject of the following section.

On further inspection of Equation 3.11, it can be noted that the projectile range is dependent on v_0^2, and on $2\theta_0$. This would imply that the all-important range is very sensitive to variations in the release velocity v_0, but less so to variations in the release angle θ_0. Table 3.1 indicates this clearly. By changing the launch angle by $14°$, from $30°$ to the near 'perfect' angle of $44°$, the range of a projectile with a launch velocity of 8 m s^{-1} only increases from 5.65 m to 6.52 m, this is an increase in range of merely 0.87 m. By contrast, if we keep the launch angle fixed at $30°$ but increase the launch velocity by just 2 m s^{-1} from 8 m s^{-1} to 10 m s^{-1}, the range increases to 8.83 m s^{-1}; an improvement in range of 3.18 m.

This sensitivity of the launch velocity parameter compared to that of the launch angle parameter goes some way towards explaining why the typical launch angle for a javelin is nearer to $35°$ than the supposedly ideal launch angle of $45°$. Put simply, biomechanical constraints limit the launch velocity achievable at the higher angle, and it is worth sacrificing this optimum angle to attain the increased throw velocity available at the lower angle. In reality, many other factors come into play when considering the dynamics of a javelin throw, some of which are discussed in Chapter 9.

It can further be noted from Table 3.1 that, for a fixed launch velocity, the range attainable is symmetrical around the strategic $45°$ launch angle, although Equation 3.10 indicates that the times of flights will be different. This fact was utilized in medieval battle bombardment strategies. By firing two projectiles from similar weapons (which ensured similar values of v_0) placed adjacent to each other, with appropriate timings but at differing angles spaced symmetrically either side of the $45°$ angle, the target could be attacked by two projectiles from two different elevations at the same time: obviously a difficult situation to defend.

Table 3.1 Range of a projectile (in metres) for different values of launch angle (θ_0 in degrees) and launch velocity (v_0 in $\mathrm{m\,s^{-1}}$)

Range (m)

Angle(deg)	Release velocity (m/s)				
	4	6	8	10	12
30	1.41	3.18	5.65	8.83	12.71
32	1.47	3.30	5.86	9.16	13.19
34	1.51	3.40	6.05	9.45	13.61
36	1.55	3.49	6.20	9.69	13.96
38	1.58	3.56	6.33	9.89	14.24
40	1.61	3.61	6.42	10.04	14.46
42	1.62	3.65	6.49	10.14	14.60
44	1.63	3.67	6.52	10.19	14.67
46	1.63	3.67	6.52	10.19	14.67
48	1.62	3.65	6.49	10.14	14.60
50	1.61	3.61	6.42	10.04	14.46
52	1.58	3.56	6.33	9.89	14.24
54	1.55	3.49	6.20	9.69	13.96
56	1.51	3.40	6.05	9.45	13.61
58	1.47	3.30	5.86	9.16	13.19
60	1.41	3.18	5.65	8.83	12.71

Worked Example 3.3

In the case of Jan Zelnezy's world record described in Worked Example 3.1, now calculate the maximum height of the javelin, and the velocity and angle with which the javelin hits the ground. How much further would the javelin have flown if it had been thrown at an angle of $45°$?

The coordinates at the peak are given by:

$$\left(\frac{v_0^2 \sin 2\theta_0}{2a}, \frac{v_0^2 \sin^2 \theta_0}{2a} \right)$$

So,

$$
\begin{aligned}
v_{y\text{max}} &= \frac{v_0^2 \sin^2 \theta_0}{2a} \\
&= \frac{(31.872)^2 \sin^2 (36)}{2 \times 9.81} = 17.89\,\mathrm{m}
\end{aligned}
$$

The angle that the javelin makes on impact with the ground will be given by Equation 3.8:

$$
\begin{aligned}
\tan \theta &= \tan \theta_0 - \left[\frac{at}{v_0} \right] \sec \theta_0 \\
&= \tan 36 - \left[\frac{9.81 \times 3.82}{31.872} \right] \bullet \frac{1}{\cos 36}
\end{aligned}
$$

$$= 0.7265 - \frac{1.176}{0.809} = -0.7269$$

$$\theta = -36.012° \quad \text{(as expected)}$$

Two practical points to note regarding this calculated angle.

1 The negative sign reflects the negative tangent of the parabolic curve. i.e. the javelin makes an angle of 36° with the ground on impact, having been launched from left to right.
2 This is the angle that the javelin, as a moving projectile, makes with the horizontal. This has no bearing on the angle that the actual line of the javelin makes as it hits the ground. If this angle were positive, the back of the javelin would hit the ground first and it would be a false throw (even if θ were equal to $-36°$).

Equation 3.11 allows us to calculate the maximum range if $\theta = 45°$:

$$s_m = \frac{v_0^2}{a} \sin 2\theta_0 = \frac{(31.872)^2}{9.81} \sin 90 = 103.55\,\text{m}$$

A range yet to be achieved in competition!

3.5 Projection to different levels and up/down an inclined plane

3.5.1 Projection to different levels

All the theory discussed so far has assumed that the height of the projectile launch is the same as the height of landing of the projectile. Patently, this is seldom a valid condition in sporting applications. It may apply in the case of a soccer kick which hits the ground before it reaches another player, or possibly a golfing pitch in an area where the course is level. However, in the vast number of cases, whether it is throwing a basketball from approximately head height into the basketball net (regulation height $= 10\,\text{ft}$), or a tennis serve, or, indeed, a discus/javelin/hammer throw, the launch and landing heights are going to be different. By way of another example, we may wish to calculate the optimum angle of kick in the case of a rugby conversion, from ground level to the goal bar level, at the regulation height of 3 m (see Worked Example 3.4).

We can extend the theory above to include differing levels of launch and landing for simple cases. After all, the equations simply define the trajectory of a parabola. If the landing point P is a vertical distance h above the launch point ($h = 0$) we can simply calculate the time of flight and the horizontal distance covered (Figure 3.3).

It is obvious that, if the projectile lands at a point P which is below the origin, then $h < 0$.

First, let us find the time that the projectile is in flight. From Equation 3.4 and putting in the value h for s_y, we obtain:

$$h = v_0 t \sin \theta_0 - \tfrac{1}{2}at^2$$

In quadratic form this yields:

$$\tfrac{1}{2}at^2 - v_0 \sin \theta_0 t + h = 0$$

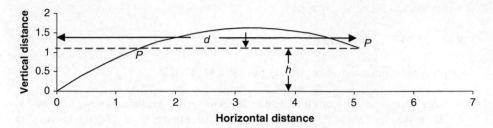

Figure 3.3 Projectile launched at origin ($h = 0$) and landing at point P, a distance h above the origin.

The solutions for t are therefore:

$$t = \frac{v_0 \sin \theta_0 \pm \sqrt{v_0^2 \sin^2 \theta_0 - 2ah}}{a}$$

Now, when $v_0 \sin \theta_0 < \sqrt{2ah}$, the solution is a complex number. This implies that the projectile never reaches the height h with the chosen values of v_0 and θ_0. However, when $v_0 \sin \theta_0 > \sqrt{2ah}$ there are two possible solutions for t; one for the projectile going up, and the other for the projectile coming down. When h is negative, the lesser value of t is also negative and can be neglected.

Now that the time of flight has been calculated, it is a simple matter to derive the range by inserting this value of t into Equation 3.3. This yields a value for the range, s_m of:

$$s_m = \frac{v_0 \cos \theta_0 \left[v_0 \sin \theta_0 \pm \sqrt{v_0^2 \sin^2 \theta_0 - 2ah} \right]}{a} \tag{3.12}$$

We are now interested in the optimum launch angle and resulting maximum range for a given v_0 when the launch and landing heights differ by h. We could derive this from Equation 3.12 by differentiating it with respect to θ_0 and then equating the result to zero. However, one can do a similar process with the simpler Equation 3.6 to obtain the same result. We begin by equating s_y to h:

$$h = s_x \tan \theta_0 - \frac{a s_x^2 \left(1 + \tan^2 \theta_0 \right)}{2 v_0^2} \tag{3.13}$$

Differentiating with respect to θ_0 yields:

$$0 = \frac{\mathrm{d} s_x}{\mathrm{d} \theta_0} \tan \theta_0 + s_x \sec^2 \theta_0 - \frac{a s_x \, \mathrm{d} s_x}{v_0^2 \, \mathrm{d} \theta_0} \sec^2 \theta_0 - \frac{a s_x^2}{v_0^2} \sec^2 \theta_0 \tan \theta_0$$

As before we set $\mathrm{d} s_x / \mathrm{d} \theta_0$ to zero to locate the point of maximum $\mathrm{d} s_x$, so:

$$0 = s_x \sec^2 \theta_0 - \frac{a s_x^2}{v_0^2} \sec^2 \theta_0 \tan \theta_0$$

Simplifying gives:

$$0 = s_x \left[1 - \frac{a s_x}{v_0^2} \tan \theta_0 \right]$$

which gives two solutions: $s_x = 0$, trivial and is neglected, and:

$$s_x = \frac{v_0^2}{a} \cot \theta_0 = s_m \tag{3.14}$$

Equation 3.14 can be substituted into Equation 3.13 to obtain the maximum distance the projectile travels in the horizontal direction when it lands at a different height to the launch level:

$$s_m = \frac{v_0 \sqrt{v_0^2 - 2ah}}{a} \tag{3.15}$$

From Equation 3.14, the optimum angle of projection is:

$$\theta_m = \arctan \left[\frac{v_0}{\sqrt{v_0^2 - 2ah}} \right] \tag{3.16}$$

Equations 3.15 and 3.16 show the allowable angles and subsequent maximum attained distances, as h varies.

First, let us consider positive values of h; that is, firing up onto a raised platform, for example. As h increases from 0 (launch and landing heights are at the same level) up to a 'platform' height of $v_0^2/2a$, s_m decreases from its 'level' value as described by Equation 3.11, down to zero, when $h = v_0^2/2a$. This is the maximum height the projectile can reach with a velocity of v_0. Meanwhile, the optimum angle, θ_m increases from $45°$ up to, theoretically, $90°$. That is, to achieve this maximum height of $v_0^2/2a$, the projectile must be launched vertically

As expected, launching a projectile onto a lower 'platform' shows no such limitation: the lower the level, the further the projectile goes ($s_m \rightarrow \infty$). Furthermore, as $|h| \rightarrow -\infty$, the optimum angle, $\theta_m \rightarrow 0°$.

3.5.2 Projection up/down an incline

Assume the projectile is launched with an initial velocity v_0 at an angle θ_0 up an incline of angle ψ. We are interested in deriving an expression for s_i, the distance the projectile will fly up the slope (Figure 3.4).

By simple trigonometric expression, $\cos \psi = s_x/s_i$ and $\sin \psi = s_y/s_i$.

Combining these with Equations 3.3 and 3.4 at the impact point P gives:

$$\frac{v_0 t \cos \theta_0}{\cos \psi} = \frac{v_0 t \sin \theta_0 - \frac{1}{2} a t^2}{\sin \psi}$$

$$- v_0 t \cos \theta_0 \sin \psi + \cos \psi \left(v_0 t \sin \theta_0 - \frac{1}{2} a t^2 \right) = 0$$

$$t \left(-\frac{1}{2} a t \cos \psi + v_0 \sin (\theta_0 - \psi) \right) = 0$$

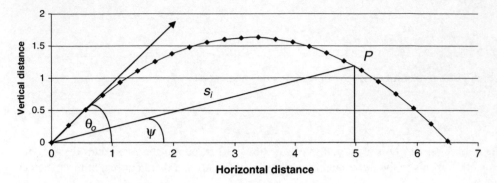

Figure 3.4 Coordinates for projection up an inclined plane.

From which it is clear that the time of flight up the incline, t_i, is either zero or

$$t_i = \frac{2v_0 \sin (\theta_0 - \psi)}{a \cos \psi} \qquad (3.17)$$

Now, since $\cos \psi = s_x/s_i$:

$$s_i = \frac{v_0 t_i \cos \theta_0}{\cos \psi} = \frac{2v_0^2 \sin (\theta_0 - \psi) \cos \theta_0}{a \cos^2 \psi}$$

Rewriting, using the 'sum-to-product' formula for $(\sin u - \sin v)$, the expression for s_i produces:

$$s_i = \frac{v_0^2 \left[\sin (2\theta_0 - \psi) - \sin \psi \right]}{a \cos^2 \psi}$$

So, for given values of v_0 and ψ, the range s_i will be a maximum when:

$$\sin (2\theta_0 - \psi) = 1$$

And so:

$$s_{i_m} = \frac{v_0^2}{a(1 + \sin \psi)} \qquad (3.18)$$

The angle of launch to ensure maximum range, θ_m, is given by $2\theta_m - \psi = \pi$, or when:

$$\theta_m = \psi + \frac{1}{2} \left(\frac{\pi}{2} - \psi \right) \qquad (3.19)$$

It can be seen from Equation 3.18 that the optimum angle for launch is the bisector of the angle between the inclined plane and the vertical (just as it is with the special case of the 'inclined' plane actually being horizontal!); that is, $\theta_m = 45°$.

Worked Example 3.4

In rugby, in order to convert a try, the ball must be kicked over the goal bar the regulation height of which is 3 m. If the kicker is 30 m from the goal, what will be the angle to ensure minimum kick velocity, and what will that velocity be?

From Figure 3.5 it can readily be seen that $\psi = \tan^{-1}(3/30) = 5.71°$.

From Equation 3.19 $\theta_m = 5.71 + 0.5(90 - 5.71) = 47.855°$.

From Equation 3.18 and with $s_{i_m} = \sqrt{30^2 + 3^2} = 30.15$ m (Pythagoras theorem):

$$v_0^2 = s_{i_m} a(1 + \sin \psi) = 30.15 \times 9.81 (1 + \sin 5.71)$$

$$= 295.77(1 + 0.1) = 325.35$$

$$v_0 = 18.04 \, \text{m s}^{-1}$$

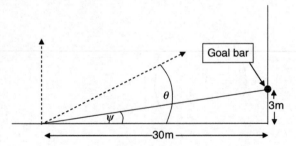

Figure 3.5 Dimensions for goal kick worked example.

3.6 The parabola of safety

(See Model 3.1 – Standard trajectory model – Envelope of safety worksheet tab)

If we investigate how a projectile flies under the conditions of a fixed launch velocity v_0, and varying launch angle θ_0, it is obvious that there are some points in space through which the projectile will never travel regardless of angle (known as the 'safe' zone). Likewise, there are other regions of space through which the projectile could fly although maybe only at certain specific angles.

The envelope that describes the boundary between these two conditions is of a parabolic shape and is variously known as: the bounding parabola, the enveloping parabola or the parabola of safety (Figure 3.6).

The proof that the boundary is indeed a parabolic follows.

If we rearrange Equation 3.6:

$$s_y = s_x \tan \theta_0 - \frac{a s_x^2 (1 + \tan^2 \theta_0)}{2 v_0^2}$$

$$\frac{a s_x^2}{2 v_0^2} \tan^2 \theta - s_x \tan \theta + s_y + \frac{a s_x^2}{2 v_0^2} = 0$$

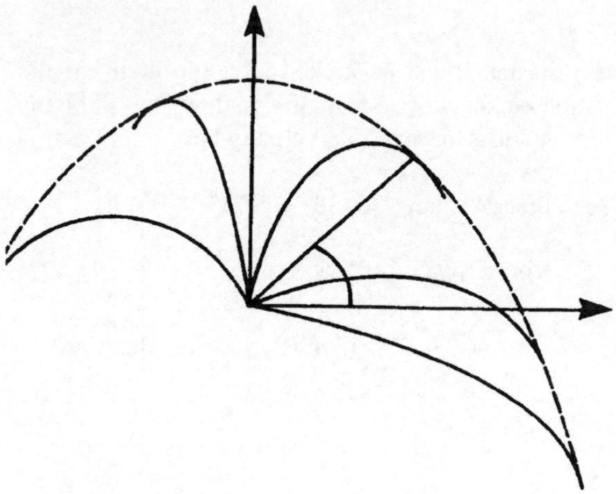

Figure 3.6 The parabola of safety (dashed curve).

Solving this quadratic equation for $\tan \theta$ we obtain:

$$\theta_{\pm} = \tan^{-1}\left[\frac{v_0^2}{as_x}\left(1 \pm \frac{2a}{v_0^2}\sqrt{\frac{v_0^2}{2a}\left[1-\left(\frac{s_x}{v_0^2/a}\right)^2\right]-s_y}\right)\right]$$

$\theta_{\pm}$ will be imaginary for:

$$s_y > \frac{v_0^2}{2a}\left[1-\left(\frac{s_x}{v_0^2/a}\right)^2\right]$$

$$s_y > \frac{v_0^2}{2a} - \frac{as_x^2}{2v_0^2}$$

that is, s_y values are unobtainable for cases beyond:

$$s_y = \frac{v_0^2}{2a} - \frac{as_x^2}{2v_0^2} \tag{3.20}$$

And, of course, Equation 3.20 is of a standard quadratic form, or of a parabolic shape. For any given value of v_0 (and the acceleration due to gravity, g), it is now possible to ascertain, if you were to stand at a particular coordinate position, (x, y), whether or not the projectile is capable of reaching you for any launch angle.

Worked Example 3.5

The futuristic Investec Media Center at the Lords Cricket Ground is approximately 120 m from the batting crease at the Pavilion End (opposite the Media Center). It is approximately 15 m above the pitch level. What velocity of batting drive is necessary to hit the Media Center?

Rearranging Equation 3.20 we get:

$$v_0^4 - 2as_y v_0^2 - a^2 s_x^2 = 0$$

$$v_0^2 = \frac{2as_y \pm \sqrt{4a^2 s_y^2 + 4a^2 s_x^2}}{2}$$

Substituting our values for a, s_x and s_y:

$$v_0^2 = \frac{294.3 \pm \sqrt{86612.5 + 5543200}}{2}$$

$$v_0^2 = \frac{588.6 \pm 2372.7}{2} = 1480.7 \quad \text{or} \quad -892.0$$

Ignoring the complex root:

$$v_0 = 38.47 \, \text{m s}^{-1}$$

This is considered an achievable launch velocity from a professional cricketer using the modern bat.

3.7 The Percy Grainger Feat

Percy Grainger was born in 1882 in Victoria, Australia. He was a renowned pianist and a close friend of Edvard Greig. His broad canon of work includes many notable examples of English, American and Danish folk music, and latterly early electronic 'synthesizer' music. Perhaps his most notable piece is the ballad 'Early One Morning'.

He was famous on at least three other counts. In no particular order; he was an enthusiastic sado-masochist, a believer in the racial superiority of the blond-haired, blue-eyed northern Europeans, and finally, putting his xenophobia to one side, he was a great athlete. In fact, he was referred to as the jogging pianist. It is this last facet that interests us.

The story goes that he was able to throw a cricket ball over the roof of his house, and then run right through his house and catch it on the way down. We are now going to ascertain if this feat is indeed possible, and, if so, under what practical conditions it may be achieved (Figure 3.7).

Let us take some typical values:

X_1, the distance from the throw to the front wall of the house = 10 m;

X_2, the width of the house;

H, the height of the house (or more accurately, the difference in height between the throw/catch height and the top of the front wall of the house) = 5 m;

L, the height of the roof ridge above the top of the house wall = 2 m;

v_0, the velocity of throw = 20 m s^{-1}.

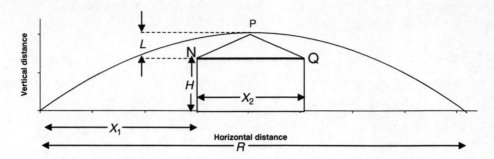

Figure 3.7 The Percy Grainger Feat.

We then ask ourselves the question, is there any value of throw angle, θ_0, which will result in a realistic, attainable value of v_x, the horizontal velocity, such that the ball can be caught on its down-flight?

Intuitively, we can say that there must be a window of allowable values of θ_0 which may be successful between: (i) a lower value, when the ball hits either, the side wall of the house below N, or the front face of the roof in front of P, and (ii) an upper value of θ_0, in which the ball fails to clear the back side of the roof at Q.

Now Percy's running speed would have to have been greater that $v_0 \cos \theta_0$ to enable him to arrive before the ball.

In addition, there are further obvious constraints:

1   when $s_x = X_1, y > H$
2   when $s_x = X_1 + X_2/2, y > L + H$
3   when $s_x = X_1 + X_2, y > H$

Applying condition 2 to Equation 3.13:

$$s_y = s_x \tan \theta_0 - \frac{a s_x^2 \left(1 + \tan^2 \theta_0\right)}{2 v_0^2}$$

we obtain the quadratic inequality:

$$2.8125 \tan^2 \theta_0 - 15 \tan \theta_0 + 9.8125 < 0$$

The solution for this inequality is:

$$0.763 < \tan \theta_0 < 4.57$$

which means that θ_0 must lie between $37.34°$ and $77.66°$.

Similarly, applying condition 3 to Equation 3.13, we obtain the quadratic inequality:

$$\tan^2 \theta_0 + 4 \tan \theta_0 + 2 < 0$$

which means that θ_0 must also lie between $30.34°$ and $73.68°$.

So, applying both these conditions would imply that the allowable throw angle must lie between $37.34°$ and $73.68°$.

Now, consider condition 1. Putting the lower extreme angle, $\theta_0 = 37.34°$ into Equation 3.13 produces a value of s_y of 5.688 m; the ball will therefore clear the front face of the house. Putting the higher extreme value, $\theta_0 = 73.68°$ into Equation 3.13 produces a value of s_y of 18.62 m. Therefore, condition 1 is automatically met within these angles.

The question remains: will Percy have been able to run through his house and catch the cricket ball in time? Now that the possible angles are known, we can simply apply $v_x = v_0 \cos \theta_0$ for the two extreme angles. His slowest required velocity will be for the largest angle, when $v_x = 20 \cos 73.68°$ or $v_x = 5.62\,\mathrm{m\,s^{-1}}$. This average speed is quite attainable from a competent athlete. However, at the lower extreme angle $v_x = 15.9\,\mathrm{m\,s^{-1}}$, this is well beyond the realms of practicality.

So, the answer to the question; 'is the Percy Grainger Feat possible?' is yes. The conditions being that the throw angle is as close as possible to $73.68°$, without going beyond it, and then Percy would have to have run at quite a speed – truly an athlete!

Summary

The trajectory of a simple projectile, neglecting drag and lift, is analysed in this chapter, by first considering only the horizontal motion, and then the vertical motion. The final trajectory equation is derived and plotted by combining these two orthogonal motions. Key features such as range, height and flight time are derived. An important optimization process is performed which results in statements of optimum range for given initial conditions. The next stage in the process was to derive similar trajectory equations for a projectile being fired up and down an incline; an important process when, for example, optimizing the trajectory over a rugby goal bar.

Next, the concept of the parabola of safety is explained, which describes, for a given launch velocity, the boundary beyond which a projectile cannot proceed regardless of its launch angle. This derivation is then used in a practical application known as the Percy Grainger Feat.

Problems and questions

1   A shuttlecock is hit at an angle of 60° to the horizontal with a velocity of 11.2 m s^{-1} ignoring drug effects. Is it feasible to run under the shuttlecock and then catch it?

2   A squash player serves from a point 6 m from the front wall. He serves the ball horizontally from a height of 2.1 m with a velocity of only 19 m s^{-1} (a bit of a miss-hit!). Is it a legitimate serve in that, to be so, it must reach the front wall above the service line at 1780 mm, and below the out line at 4570 mm? What if they served the ball with the same poor velocity, but at an angle of 15° above the horizontal?

3   The current world record for the men's hammer was set by Yuriy Sedykh who threw 86.74 m at the European athletics championships held in Stuttgart, West Germany in 1986. He released the hammer from a height of 1.66 m at an angle of 39.9° with a release velocity of 30.7 m s^{-1}. Was this his optimum angle to achieve the maximum range?

4   A tennis court is 39 ft (23.77 m) long. The net is 3 feet 6 inches (1.07 m) high. A child standing on the serve line can only serve underarm and with a maximum velocity of 11.3 m s^{-1}. Is there any angle that the child can launch the ball at which will ensure the ball clears the net? You may assume the ball is launched from

ground level. (See Model 3.1 – Standard trajectory model – Envelope of safety worksheet tab)

5   A soccer center forward takes a shot at the goal from 8.5m from the goalmouth. If he kicks the ball at 25 m s^{-1} at an angle of $20°$ to the pitch, does he score? A soccer goal crossbar is 8 feet (2.44 m) high, and you may assume the player correctly aims between the goalposts. (See Model 3.1 – Standard trajectory model – Standard trajectory equation worksheet tab)

Chapter 4

Impact and bounce

4.1 Introduction

The elastic properties of natural rubber and its ability to deform and snap back to shape have intrigued people for centuries. In particular, ball-sport players have utilized this property since ancient times; indeed, some form of ball game is portrayed on early Egyptian monuments. Of course, in the case of the sports ball, it is not only its elasticity and its ability to rebound that is important, but also good wear and resilience are usually essential properties to consider.

From an energy perspective, the mechanism of bounce involves the transfer of the ball's kinetic energy (by virtue of its velocity immediately before impact) into an elastic potential energy as the ball distorts. In its distorted state, this potential energy is stored within the ball. As the ball starts to reshape to its normal, and usually, spherical condition, it gives up this potential energy to the bouncing surface. Then, by Newton's third law, a 'surface' force is created which reacts back onto the ball imparting kinetic energy to it, and the ball lifts off in the opposite direction to that of the initial impact.

It can be seen that, with all these transfer of energy types occurring in the bouncing process, there is much opportunity for energy loss, by sound, heat and molecular processes. By way of an example, as an elastic medium compresses (or indeed stretches) from its natural state, it will heat up. As it returns to its standard un-stressed state, it will give up that extra heat to the surroundings. So, if a ball were bouncing in an ideal state of perfect thermal insulation, energy would be neither gained nor lost from the bouncing process; the ball would be the same temperature both after and before the bounce, even though it was temporarily hot while in mid-bounce. However, in the real world, as the ball compresses, especially if it is one of the hollow, gas-filled varieties, it will heat up, and this heat will be lost to the surroundings. As the ball returns to its natural state, it will cool down further and, therefore, will be cooler than its pre-bounce state. That heat energy, together with additional heat energy created by a variety of frictional forces, will result in a loss of energy in the bounce and will be one factor in reducing bounce efficiency.

Minimizing these transfer and heat losses will result in balls with optimum bounce potential.

4.2 Coefficient of restitution

The coefficient of restitution (CoR) of an object, symbolized by e, is a fractional value equal to the ratio of velocities before and after an impact. Therefore, a ball which possesses a theoretical value of $e = 1$ would collide with an object perfectly, elastically. If this ball were dropped from a given height, it would, theoretically (but impossibly), bounce back to the same height. By contrast, if a ball had a value of $e = 0$, it would, on contact, stick to the impacting body. As an example, one might like to consider a 1 cm ball of putty rolling and colliding with a 1 cm steel ball bearing. Likewise, if a ball of putty were dropped onto a rigid surface, there would be no bounce; it would merely stick to whatever it landed on; a bit of a cheat but CoR would be zero!

It follows, from a conservation of energy perspective, that no ball can ever achieve a value of $e > 1$. However, note the interesting scenario described in Question 1, at the end of the chapter.

The CoR value is not simply a property of the material in the way that say, elasticity is. Rather, it is a property of the collision process itself; its value depending on the CoR of both bodies in a complex manner, as well as other parameters such as the velocity and angle of impact. One can implicitly understand that a ball's CoR will vary with velocity when one considers that most sporting balls are not of a uniform construction, but are made up of layers, where, often, the innermost layer will be a trapped volume of gas or air. The CoR will depend on just how much each layer is stressed; the higher the velocity of impact, the deeper into the ball the stresses will occur, accounting for a variation in CoR with velocity.

The CoR value is dependent on two entangled processes.

First, there is that of energy loss. Apart from the thermal loss mentioned in Section 4.1, energy loss by vibration into the impacting surface and its environs is a major factor. The nature of this loss varies in scale. At the microscopic end of the scale, molecules will be set into motion by the impact, vibrating with certain resonances, but ultimately damped to negligence with time. At the other, macroscopic, end of the spectrum, whole wall or floor panels may be set into damped resonance. In between these two extremes are such vibrations as the one-dimensional oscillations of racket strings as they oscillate in consonance with their neighbours which form the two dimensional mesh, or the quasi-one-dimensional vibration modes that can be triggered when a ball impacts with a wooden baseball or cricket bat. As a final thought on this subject, guess what the most efficient shape is for the long-term storage of resonant energy. That's right – *the sphere*! The volume of the ball will be the greatest resonator.

The second process that defines the CoR value is that of energy *retention* over the sequence of energy conversions; from the kinetic energy just prior to impact, to the elastic potential energy in the projectile, and thence, finally back to the kinetic energy which fires the projectile off the surface. This succession of energy conversion losses is extremely difficult to quantify for any given projectile. However, it is clearly influenced by such factors as: the elasticity of the bulk material(s) from which the projectile is made, the construction and shape of the projectile, the interplay of the multifarious materials commonly used in modern sporting projectiles and, if one of those materials includes gas, its pressure. Less obviously, it is known that the total impact time (i.e. the time the projectile is in contact with the surface) has a bearing on the CoR. Generally, the shorter the time, the more efficient the energy transfer (e.g. consider the rigidity of a Superball, and its consequential high value of e). However, in seemingly direct disagreement to this, one has to consider that the 'trampoline effect' can also be used to increase the CoR value. This phenomenon works by slowing the energy conversion processes down, controlling them into the best areas for high elasticity and then focusing a large proportion of the energy back to the ball for an efficient launch.

As will be shown in Section 4.4, with regard to a ball bouncing at an angle to a surface, two other properties of the ball become important in modelling the impact process; the moment of inertia of the ball, and the coefficient of kinetic friction between the surface of the ball and the impact surface. In fact, it is a combination of the naturally high values of both CoR and μ_k, which accounts for the beguiling route a Superball takes on its way from launch to its rest state (see the end of Section 4.4 and this chapter's problem no. 4).

4.3 Normal impact

For two bodies colliding with velocities just prior to impact of u_1 and u_2, and velocities immediately following impact of v_1 and v_2, the coefficient of restitution (CoR or e) is given by:

$$e = -\frac{v_2 - v_1}{u_2 - u_1} \tag{4.1}$$

where u_1 is the initial velocity of the first body, u_2 is the initial velocity of the second body, v_1 is the final velocity of the first body and v_2 is the final velocity of the second body.

For a body bouncing off a static surface at 90° (or normally) to it, it is clear that:

$$e = \frac{v}{u} \tag{4.2}$$

where u is the speed of the body immediately prior to impact, and v is the speed immediately following impact.

By the simple application of Newton's laws of motion, it can easily be proved that, for a normal collision, Equation 4.2 may be rewritten as:

$$e = \sqrt{\frac{h_2}{h_1}} \tag{4.3}$$

where h_1 is the drop height and h_2 is the bounce height.

The energy lost in the bounce is $K \propto e^2$.

Table 4.1 lists some common CoR values. The ball is presumed to be bouncing normally against its usual surface.

Table 4.1 Some interesting coefficient of restitutions (values approximate)

Type of ball	CoR
Table tennis ball	0.94
Superball	0.9
Range golf ball	0.858
Ball of rubber bands	0.828
Billiard ball	0.804
Soccer ball	0.76
Hand ball	0.752
Volleyball	0.74
Tennis ball (old)	0.712
Tennis ball (new)	0.67
Hollow, hard plastic ball	0.688
Glass marble	0.658
Lacrosse	0.62
Wooden ball	0.603
Basketball	0.6
Steel ball bearing	0.597
Baseball	0.55
Field hockey	0.5
Softball	0.31
Cricket ball	0.31

Worked Example 4.1

Use the CoR values from Table 4.1 to ascertain the difference in height of bounce between an old and a new tennis ball when they are both dropped from a height of 1.6 m.

From Equation 4.3:

$$e_{old} = 0.712 = \sqrt{\frac{h_{old}}{1.6}} \qquad e_{new} = 0.67 = \sqrt{\frac{h_{new}}{1.6}}$$

$$h_{old} = (0.712)^2 \times 1.6 \qquad h_{new} = (0.67)^2 \times 1.6$$

$$h_{old} = 0.811\,m \qquad h_{new} = 0.718\,m$$

The difference in bounce height between the old and new tennis balls is therefore 9.3 cm.

4.4 Oblique impact

4.4.1 Elastic bounce on a rigid, friction-free surface

Assume a 'smooth-hyper-Superball' impacts with a rigid surface at an angle θ to the normal to that surface, and that it rebounds at an angle ψ to the normal (Figure 4.1). For this hypothetical case of a perfect elastic bounce we can state that $e = 1$.

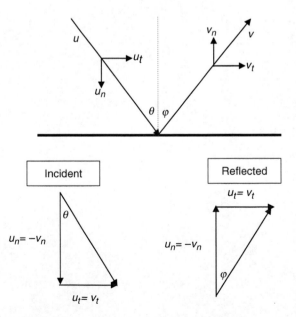

Figure 4.1 Elastic oblique bounce with rigid surface.

We may resolve the incident velocity u into its normal and tangential components, u_n and u_t, respectively. Similarly, we resolve the reflected velocity, v, into its normal and tangential components, v_n and v_t, respectively. Now, in this ideal case, the tangential components do not change before and after the bounce. So u_t is equal to v_t. But, for a perfect elastic collision, the incident normal component reverses direction on reflection. So $u_n = -v_n$.

The vector triangle for the incident and reflected velocities are as shown in Figure 4.1. Now, by inspection, the two right-angled triangles are congruent, and so θ must equal φ and, therefore, the angle of incidence is equal to the angle of reflection.

4.4.2 Inelastic bounce on a rigid, friction-free surface

As before, $u_t = v_t$, so long as there is no friction between the ball and the surface. However, when the inelastic bounce is considered, $v_n = -eu_n$ and the corresponding velocity vector triangles are now as shown in Figure 4.2.

$$u_n = u \cos \theta \quad \text{and} \quad u_t = u \sin \theta$$

So:

$$v_n = -eu_n = -ev \cos \theta \quad \text{and} \quad v_t = u_t = u \sin \theta$$

and

$$\tan \varphi = \text{opp/adj} = u \sin \theta / eu \cos \theta = \tan \theta / e$$

Therefore $\tan \theta = e \tan \varphi$, and the angle of incidence is no longer equal to the angle of reflection, but since e is always less than 1, $\varphi > \theta$.

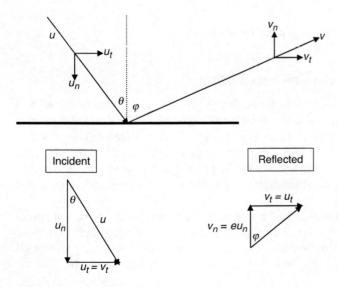

Figure 4.2 Inelastic bounce with a rigid friction-free surface.

4.4.3 Elastic bounce against two surfaces that are 90° to each other

With reference to Figure 4.3, we see that $\angle ABF = \angle FBC = \theta$, and $\angle BCE = \angle ECD = \varphi$. From $\triangle BOC$ we can see that $\angle BCO = \theta$ and $\angle CBO = \varphi$. $\angle GBA$ must equal φ as $\theta + \varphi = 90°$. Now, since OG and CE are obviously parallel to each other and $\angle GBA = \angle ECD = \varphi$ then AB must also be parallel to CD. So, in any collisions involving a double bounce from two surfaces at 90° to each other (consider snooker or squash for instance) the ball will return parallel to the incident shot regardless of the incident angle. Indeed, if the CoR of both surfaces are, more realistically, less than 1, *but are the same*, the parallel rebound rule still applies (see Question 2).

A non-sporting example of this phenomenon can be seen in the radar reflectors found on some small cruisers or fishing boats. These devices are used to aid radio detection and usually hang from a high point on the boat. They are constructed from electrical conducting materials arranged in multiple planes that are orthogonal to each other. Whatever direction an incident radar signal strikes the reflector, it will reflect that signal back in the direction from whence it came; hence achieving a clear 'radio vision' of the boat.

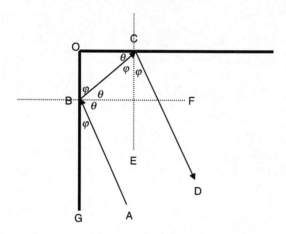

Figure 4.3 Elastic collision with two surfaces at 90° to each other.

4.4.4 Inelastic bounce on a rough rigid surface

Thus far, all the bounce analyses have assumed frictionless interactions between ball and surface. The inclusion of friction adds considerably to the complexity of the analysis since, on the ball's contact with the surface at an oblique angle, there is a tendency for the ball to rotate due to the turning moment of the frictional force on the ball. This rotation takes energy out of the bounce, and the amount of rotation created is dependent on: the incident velocity and angle, the coefficients of restitution and friction between ball and surface, and the moment of inertia of the ball.

Figure 4.4 depicts a ball of mass m and radius a, approaching from the left with an incoming velocity u, rotating at an angular velocity ω in an anti-clockwise direction, and at an angle θ to the normal. It rebounds off the surface (coefficient of restitution $= e$) with a rebound velocity v, an angular velocity ω', maybe in a clockwise direction, and at an angle ψ to the normal.

The tangential component of the incident velocity will be opposed by the frictional force, F, which will reduce the tangential component of the reflected velocity (from u_t to v_t) and reduce

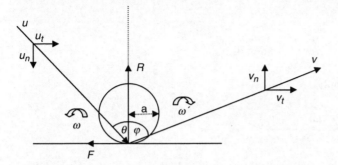

Figure 4.4 The bounce of an inelastic spinning ball against a rigid surface, including friction.

the angular velocity, from ω, to ω', which may result in a reversal of the spin. Finally, there is a reaction force, R, acting normally out of the surface as a consequence of the impact.

By Newton's laws, the change in momentum occurring on the collision must be equal to the impulse of the force producing it. The horizontal and vertical components therefore equate as follows:

$$-\int F\,\mathrm{d}t = m\left(v_t - u_t\right) \tag{4.4}$$

$$\int R\,\mathrm{d}t = m\left(v_n + u_n\right) \tag{4.5}$$

In a similar way, we can equate the change in angular momentum to the impulsive torque produced by F and acting at a distance, a, from the center of the sphere. This is given by $-\int F\,\mathrm{d}t \times a$ and this must equal the difference between the angular momentums before and after the bounce:

$$a\int F\,\mathrm{d}t = I\left(\omega + \omega'\right) = mk^2\left(\omega + \omega'\right) \tag{4.6}$$

where k is the radius of gyration of the ball.

Finally, as in the case of the previous inelastic bounce analysis, if e is the coefficient of restitution, we can state that:

$$v_n = eu_n \tag{4.7}$$

Equations 4.4 to 4.7 are the four equations of motion that must be solved in order to realistically model most cases of ball bounce in sport. Although many different bounce conditions may be analysed, here two of the major ones will be developed in some detail.

Case I – The ball does not acquire enough angular velocity to roll – it slides throughout the duration of the impact process

So, the sliding friction continues throughout the contact with the ground and:

$$F = \mu_k R \quad \text{and therefore} \quad \int F\,\mathrm{d}t = \mu_k \int R\,\mathrm{d}t \tag{4.8}$$

where μ_k is the coefficient of kinetic (i.e. sliding) friction between the surface and the ball.

From Equations 4.4, 4.5 and 4.8 we find:

$$(v_t - u_t) = -\mu_k (u_n + v_n)$$
$$= -\mu_k u_n (1 + e) \quad \text{from Equation 4.7}$$

Then, from Equations 4.6 and 4.8:

$$(\omega + \omega') = \frac{\mu_k a \int R \, dt}{mk^2} = \frac{\mu_k am (u_n + v_n)}{mk^2}$$

Now, it is a fundamental property of all centrally symmetric bodies (i.e. the vast majority of sporting balls, but not including such examples as rugby balls or biased crown-green bowling balls) that the moment of inertia, I, is equal to $\frac{2}{5}ma^2$. Therefore:

$$(\omega + \omega') = \frac{5\mu_k u_n (1 + e)}{2a}$$

The conditions following rebound are given by:

$$v_t = u_t - \mu_k u_n (1 + e) \tag{4.9}$$

$$v_n = eu_n \tag{4.10}$$

$$\omega' = \frac{5\mu_k u_n}{2a} (1 + e) - \omega \tag{4.11}$$

We know that $\tan \theta = u_t/u_n$ and $\tan \psi = v_t/v_n$, so for this case, from Equations 4.9 and 4.10:

$$\tan \psi = \frac{u_t - \mu_k u_n (1 + e)}{eu_n}$$

and so:

$$e \tan \psi = \tan \theta - \mu_k (1 + e) \tag{4.12}$$

Case 2 – The ball manages to grip the surface before completion of impact, and 'rolls off' the surface

The Equations of motion (4.4) to (4.7) must still apply. However, on this occasion F will be zero before the ball leaves the surface.
  From Equations 4.4 to 4.6:

$$(v_t - u_t) = \frac{-\int F \, dt}{m} = \frac{-k^2 (\omega' + \omega)}{a}$$

As before, for a sphere $I = \frac{2}{5}ma^2$ and therefore $k^2 = \frac{2}{5}a^2$ and

$$(u_t - v_t) = \frac{2a}{5} (\omega' + \omega) \tag{4.13}$$

If the ball is completely rolling before it leaves the surface, we can use the standard rotational identity that $\omega' = v_t/a$ and so Equation 4.13 becomes:

$$(u_t - v_t) = \frac{2v_t}{5} + \frac{2a\omega}{5}$$

$$\frac{7v_t}{5} = u_t - \frac{2a\omega}{5}$$

In summary, the conditions immediately following rebound, in the case when the ball is completely rolling before it leaves the surface, are given by:

$$v_t = \frac{5u_t - 2a\omega}{7} \tag{4.14}$$

$$v_n = eu_n \tag{4.15}$$

$$\omega' = \frac{v_t}{a} = \frac{5u_t - 2a\omega}{7a} \tag{4.16}$$

One interesting question which may be considered is: what is the limiting case for slipping to stop, and rolling to *just* begin before the ball leaves the surface? This condition can be found by equating v_t from our Case 2 in Equation 4.14 with that in Case 1 given by its equivalent, Equation 4.9. Obviously, this will be critically dependent on the coefficient of kinetic friction μ_k, but it is also dependent on the coefficient of restitution, e, and the velocity of impact, u.

Combining Equation 4.14 with Equation 4.9 gives:

$$\frac{5u_t - 2a\omega}{7a} = u_t - \mu_k u_n (1 + e)$$

$$\mu_k u_n (1 + e) = \tfrac{2}{7}(u_t + a\omega)$$

So the limiting condition for rolling to occur will be given by:

$$\mu_k u_n (1 + e) \geq \tfrac{2}{7}(u_t + a\omega) \tag{4.17}$$

Finally, for this rolling case, the angles of incidence and rebound can simply be found from Equations 4.14 and 4.15:

$$\tan \psi = \frac{5u_t - 2a\omega}{7eu_n}$$

But:

$$u_n = u \cos\theta$$

Hence:

$$e \tan \psi = \frac{5}{7}\tan\theta - \frac{2a\omega}{7u\cos\theta} \tag{4.18}$$

Worked Example 4.2

Following a tennis serve, the ball hits the grass surface at $50 \, \text{m s}^{-1}$ at an angle of $30°$ to the horizontal (so $\theta = 60°$). If there is a backspin of $100 \, \text{rad s}^{-1}$ just before impact, calculate the angle of rebound. Use the following data: $e = 0.7$, $\mu_k = 0.6$ and $a = 3.25 \, \text{cm}$.

First, we have to decide whether or not the ball has enough angular velocity to roll before breaking contact with the court as that will be the deciding factor in whether Equations 4.12 or 4.18 is used. We use Equation 4.17 to ascertain this. If:

$$\mu_k u_n (1 + e) \geq \tfrac{2}{7} (u_t + a\omega)$$

then rolling will occur:

$$u_n = u \cos\theta = 50 \cos 60 = 25 \, \text{m s}^{-1}$$

$$u_t = u \sin\theta = 50 \sin 60 = 43.3 \, \text{m s}^{-1}$$

For rolling to occur:

$$0.6 \times 25 \, (1 + 0.7) \geq \tfrac{2}{7} (43.3 + 0.0325 \times 100)$$

$$25.5 \geq 13.3$$

So rolling will occur, and therefore, we will use Equation 4.18 in our further calculation:

$$e \tan\psi = \frac{5}{7} \tan\theta - \frac{2a\omega}{7u \cos\theta}$$

$$0.7 \tan\psi = \frac{5}{7} \tan 60 - \frac{2 \times 0.0325 \times 100}{7 \times 50 \times \cos 60}$$

$$= 1.237 - \frac{6.5}{175} = 1.2$$

$$\tan\psi = \frac{1.2}{0.7} = 1.714$$

$$\psi = 59.74°$$

The ball comes off the ground just slightly higher than the approach angle, by $0.26°$. This level of backspin very closely emulates the rebound angle of a theoretically perfect elastic bounce.

However, it is worth considering what the rebound angle would be with these realistic parameters if there was no spin on the ball; in which case, $\omega = 0$.

Equation 4.18 again shows that this condition still allows rolling before the ball leave the court, and following the same procedure as above, we find that $\psi = 60.5°$; a difference of $0.76°$ as a consequence of backspin alone. Furthermore, the effect of spin on the rebound angle is much greater for smaller angles of incidence.

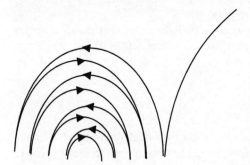

Figure 4.5 The trajectory path of a Superball as a result of oblique impact with a high friction surface.

Final, in this section, consideration is given to an extreme example of Case 2, in which we have a ball with a high value of both e and μ_k; the Superball.

With reference to Figure 4.4, μ_k is so great that F becomes large enough to make $\omega' \lesssim \omega$ and F reduces only slightly on each bounce. For a Superball colliding obliquely with a surface of sufficiently high friction, either with no spin or, better still, with top spin with respect to the surface, the spin will reverse on each bounce and the possibly unanticipated trajectory path shown in Figure 4.5 occurs.

Actually four factors contribute to the Superball behaviour:

1   High value of e.
2   High value of μ_k.
3   High value of I.
4   High value of 'shear e' (i.e. distortions around the surface of the ball are elastic and return energy efficiently as the ball returns to shape; this can only happen for balls that are solid, and are manufactured from homogeneous material throughout.)

4.5 Ball-on-ball impact

4.5.1 The one-dimensional collision

The most common sporting example of the one-dimensional collision is in table-ball games such as snooker and billiards. We know that, when we hit a ball with the cue ball directly straight on, the cue ball more or less stops dead, while the target ball moves off along the same path and with about the same velocity as the cue ball had just prior to impact. We also have a feeling that conservation of energy somehow accounts for this mechanism. However, this is only part of the story. After all, we can indeed assume that, before impact, the cue ball possesses all the kinetic energy of the system, but, following the impact, why should *all* that energy be fully transferred to the other ball? Would not an equally valid solution be for both balls to finish up with a proportion of the available energy, so long as the sum of the energies in the two balls equals the initial cue ball's energy?

The answer lies in the interplay between the conservation of linear momentum and the conservation of energy. For a perfectly elastic collision where $e = 1$, the conservation of linear momentum is mathematically defined as:

$$m_1 u_1 + m_2 u_2 = m_1 v_1 + m_2 v_2$$

where m_1 and m_2 are the masses of the two balls which have velocities before impact of u_1 and u_2, and velocities after impact of v_1 and v_2 respectively.

Now conservation of energy in a two-ball system tells us that the sum of the kinetic energies before impact must equal the sum of the kinetic energies after impact. Mathematically this is defined as:

$$\frac{m_1 u_1^2}{2} + \frac{m_2 u_2^2}{2} = \frac{m_1 v_1^2}{2} + \frac{m_2 v_2^2}{2}$$

Note that, in one-dimensional collisions of this nature, the potential energies of all balls before and after impact are equal and can, therefore, be neglected. In practical terms we are assuming that the playing table is flat and horizontal.

Solving for these equations simultaneously, we obtain:

$$v_1 = \frac{u_1(m_1 - m_2) + 2m_2 u_2}{m_1 + m_2}, \quad v_2 = \frac{u_2(m_2 - m_1) + 2m_1 u_1}{m_1 + m_2} \tag{4.19}$$

$$m_1\left(v_1^2 - u_1^2\right) = m_2\left(u_2^2 - v_2^2\right)$$

$$m_1(v_1 - u_1)(v_1 + u_1) = m_2(u_2 - v_2)(u_2 + v_2) \tag{4.20}$$

Rearranging the momentum equation:

$$m_1(v_1 - u_1) = m_2(u_2 - v_2)$$

and dividing into Equation 4.20 we get:

$$v_1 + u_1 = u_2 + v_2$$

Therefore:

$$v_1 - v_2 = u_2 + u_1 \tag{4.21}$$

Put simply, the speed of separation is equal to the speed of approach. Perhaps the surprising aspect of this rule is that it is true regardless of the masses of the two balls involved in the collision. We might consider that, if a heavy ball hits a much lighter ball, the lighter ball will shoot off comparatively quickly. This is true, but the heavy ball does not lose much of its speed either, so the *difference* in velocities between the two balls, before and after collision, possibly contrary to expectations, does in fact remain the same.

In the case of an elastic collision in which $e < 1$, the energy loss on collision modifies Equation 4.21 to:

$$(v_1 - v_2) = e(u_2 + u_1)$$

This is commonly known as Newton's experimental law of impact.

Now, when, as in the game of billiards, $m_1 = m_2$, Equation 4.19 simplifies right down to $v_1 = u_2$ and $v_2 = u_1$. We can also assume that $u_2 = 0$ during a normal game, so following the impact, v_1 falls to zero and the ball stops dead, while v_2 picks up all the initial velocity of the first ball, as surmised in the first paragraph of this section.

Worked Example 4.3

In a game of crown green bowls the bowling ball strikes the jack head-on with a velocity of $0.8\,\mathrm{m\,s^{-1}}$. If the coefficient of restitution between the two balls is 0.65, what are the velocities of the two balls if the bowling ball weighs 1.15 kg and the jack weighs 0.67 kg?

Applying the law of conservation of momentum $m_1u_1 + m_2u_2 = m_1v_1 + m_2v_2$ where the subscript 1 refers to the bowling ball, and 2 refers to the jack, we can write:

$$(1.15 \times 0.8) + (0.67 \times 0) = 1.15v_1 + 0.67v_2$$
$$0.92 = 1.15v_1 + 0.67v_2$$

Now from Newton's experimental law of impact:

$$v_2 - v_1 = e \times u_1 = 0.65 \times 0.8 = 0.52$$

These two equations may be solved simultaneously to yield:

$$v_1 = 0.31\,\mathrm{m\,s^{-1}}, \quad v_2 = 0.83\,\mathrm{m\,s^{-1}}$$

4.5.2 The two-dimensional collision

The analysis above can successfully be extended to two dimensions by resolving the velocities into two orthogonal directions (say, aligned with the sides of the billiard table as indicated in Figure 4.6). We then apply both the conservation of momentum law, and Newton's experimental impact law to extract the required simultaneous equations for the velocities in the two directions, from which the resultant velocity can then be derived.

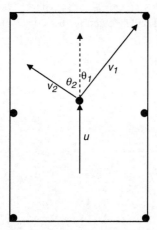

Figure 4.6 Coordinates for snooker/billiard ball collision.

Conservation of the x component and y component of momentum respectively, will be given by:

$$m_1 u_1 = m_1 v_1 \cos\theta_1 + m_2 v_2 \cos\theta_2$$
$$0 = m_1 v_1 \sin\theta_1 - m_2 v_2 \sin\theta_2$$

and, as before, the conservation energy law gives:

$$\tfrac{1}{2} m_1 u_1^2 = \tfrac{1}{2} m_1 v_1^2 + \tfrac{1}{2} m_2 v_2^2$$

Again, these equations simplify for such games as billiards where we can assume that m_1 and m_2 are equal, and so the m terms cancel.

Theoretically, these three equations should be solvable simultaneously. For example, given an incident velocity and one of the ball's angles of deflection, it ought to be possible to derive the other ball's angle of deflection and the two velocities after collision (three equations, three unknowns). Unfortunately, the solutions are generally not trivial and are often solved analytically. However, with just a little more information. We can solve puzzles like the one given below.

Worked Example 4.4

In a game of billiards, a player shoots the cue ball at $4\ \mathrm{m\,s^{-1}}$ down the table parallel to the long side. The target ball is hit with a glancing shot such that its velocity immediately after impact is $3.5\ \mathrm{m\,s^{-1}}$ and it moves at $25°$ to the original path of the cue ball, to land in the corner pocket (obviously!). Assuming the collision to be elastic, calculate the velocity (speed and direction) of the cue ball following its collision.

So, $u = 4\ \mathrm{m\,s^{-1}}$, $\theta = 25°$, $v_1 = 3.5\ \mathrm{m\,s^{-1}}$.

Applying the law of conservation of momentum $m_1 u_1 + m_2 u_2 = m_1 v_1 + m_2 v_2$ where the subscript 1 refers to the cue ball, and 2 refers to the target ball, we can write:

for the y-component:

$$mu = mv_1 \cos\theta_1 + mv_2 \cos\theta_2$$
$$u = v_1 \cos\theta_1 + v_2 \cos\theta_2$$
$$4 = 3.5\cos 25 + v_2 \cos\theta_2$$
$$v_2 \cos\theta_2 = 0.828$$

for the x-component:

$$0 = mv_1 \sin\theta_1 - mv_2 \sin\theta_2$$
$$v_2 \sin\theta_2 = v_1 \sin\theta_1 = 3.5\sin 25 = -1.48$$

Combining by squaring and adding and then using the trigonometric relation: $\sin^2\theta + \cos^2\theta = 1$, we find:

$$\left(v_2\cos\theta_2\right)^2 + \left(v_2\sin\theta_2\right)^2 = v_2^2\left(\cos^2\theta_2 + \sin^2\theta_2\right)$$

$$v_2 = \sqrt{0.828^2 + (-1.48)^2} = \sqrt{2.875}$$

$$v_2 = 1.69\,\mathrm{m\,s^{-1}}$$

To calculate the angle:

$$\tan\theta_2 = \left(\frac{-1.48}{0.828}\right)$$

$$\theta_2 = 60.77°.$$

One interesting aspect concerns the maximum deflection obtainable from the 'billiards scenario'. Figure 4.7 indicates that the condition for maximum deflection occurs when the center of the cue ball is aimed at the periphery of the target ball. Billiard players refer to this as the *half-ball shot*.

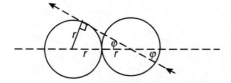

Figure 4.7 Condition for maximum deflection.

Assuming no spin on the cue ball, and all our usual assumptions regarding elastic collisions with smooth-surfaced balls, it is clear that: $\sin\varphi = r/2r = 0.5$ and the maximum deflection angle possible is $\varphi = 30°$. In practical cases, a skilled practitioner may increase this to about 34° by applying appropriate spin.

4.6 Impact of projectile by a striking implement

It is interesting that the term 'coefficient of restitution' entered the common vocabulary when golf club manufacturers began making thin-faced drivers utilizing the 'trampoline effect', which enabled drivers to achieve a greater distance as a result of extra bounce off the clubface. The US Golf Association now places an upper limit of $e = 0.83$ for the club head. As a quantitative indicator, a change in CoR from 0.83 to 0.82 will correspond (all other variables remaining constant) to a reduction in drive distance of approximately 4.2 yards. This reduction in distance will be greater for weaker swings, thereby disadvantaging the less skilled player over the competent. Meanwhile golf balls have a CoR of about 0.78.

The values of CoR stated in Table 4.1 are measured with respect to a hypothetical rigid surface. In the case of the golf driver hitting the ball, a question may be asked: if, for the driver, $e = 0.83$, and for the golf ball, $e = 0.78$, then what is the resulting CoR for that impact? Surprisingly, the answer is difficult to ascertain accurately. To a first approximation, however, the resultant CoR is closer to, and more influenced by, the lower CoR value.

Therefore, to improve the driving range, it is generally better to pay attention to the CoR of the ball, rather than the club.

Combining Equation 4.3 with the conservation of linear momentum leads to:

$$v_2 = \frac{m_1 u_1 (1+e)}{m_1 + m_2} \tag{4.22}$$

Notice that, in the case where v_2 and m_2 are the velocity and mass of the ball respectively, and u_1 and m_1 are those of the striking implement (u_1 being prior to impact), changing the mass of the striking implement will have less of an effect on v_2 than changing the mass of the ball. This is because $v_2 \propto m_1/(m_1 + m_2)$. So, $v_2 \propto 1/m_2$ which is sensitive to m_2 compared with the partial cancellation of variations in m_1 (as m_1 appears both in the numerator and the denominator of the proportionality fraction). In this situation, of course e represents the specific coefficient of restitution between the ball and the particular striking implement.

For completeness, the velocity of the club after impact with the ball will be:

$$v_1 = \frac{u_1 (m_1 - em_2)}{m_1 + m_2} \tag{4.23}$$

CoR and energy loss on impact has been discussed qualitatively in this chapter on two occasions. What follows is a more quantitative approach.

The energy loss on impact will be given by:

$$\tfrac{1}{2}m_1 u_1^2 - \tfrac{1}{2}m_1 v_1^2 - \tfrac{1}{2}m_2 v_2^2$$

Combining with Equations 4.22 and 4.33 and simplifying yields:

$$\text{total impact energy loss} = \frac{m_1 m_2 u_1^2 (1 - e^2)}{2(m_1 + m_2)} \tag{4.24}$$

Worked Example 4.5

A golfer drives a ball of mass 46 g down the fairway with a driver of club head mass equal to 0.25 kg. If the impact velocity of the club is $60 \, \mathrm{m \, s^{-1}}$ and the CoR is taken to be 0.65, calculate both the velocity of the ball and of the club following impact. Also, calculate the energy loss of the impact.

So, $m_2 = 0.046 \, \mathrm{kg}$, $m_1 = 0.25 \, \mathrm{kg}$, $u_1 = 60 \, \mathrm{m \, s^{-1}}$, $e = 0.65$.

From Equation 4.22, the velocity of the ball following impact will be given by:

$$v_2 = \frac{m_1 u_1 (1+e)}{m_1 + m_2} = \frac{0.25 \times 60(1 + 0.65)}{0.25 + 0.046} = \frac{24.75}{0.296} = 83.61 \, \mathrm{m \, s^{-1}}$$

From Equation 4.23, the velocity of the club following impact will be given by:

$$v_1 = \frac{u_1 (m_1 - em_2)}{m_1 + m_2} = \frac{60(0.25 - 0.65 \times 0.046)}{0.296} = \frac{13.206}{0.296} = 44.61 \, \mathrm{m \, s^{-1}}$$

The total energy loss, given by Equation 4.24 is:

$$\frac{m_1 m_2 u_1^2 \left(1 - e^2\right)}{2(m_1 + m_2)} = \frac{0.25 \times 0.046 \times 60^2 \left(1 - 0.65^2\right)}{2 \times 0.296} = \frac{23.9}{0.592} = 40.4\,\text{J}$$

The energy of the club-head just prior to impact is given by:

$$\tfrac{1}{2} m_1 u_1^2 = 0.5 \times 0.25 \times 60^2 = 450\,\text{J}$$

The energy possessed by the ball following impact is given by:

$$\tfrac{1}{2} m_2 v_2^2 = 0.5 \times 0.046 \times 83.61^2 = 160\,\text{J}$$

The efficiency of the impact is defined as:

$$\frac{\text{energy in the ball following impact}}{\text{energy in the club-head prior to impact}} \times 100\% = \frac{160}{450} \times 100 = 35.5\%$$

4.6.1 Impact of ball with a lofted implement

The elastic impact

Consider a striking implement which strikes a static ball horizontally, but presents an angled face of $\theta°$ to the vertical (in golfing terms this is known as the loft of the club) (Figure 4.8).

In the case of a perfectly elastic impact, the horizontal velocity, u_1, with which the head of the implement strikes the ball, will have a velocity component in the R direction, normal to the hitting face, equal to $u_1 \cos\theta$, and a velocity component tangential to the hitting face, in the F direction, equal to $u_1 \sin\theta$. But, in the case of the elastic collision, friction forces are considered to be zero, and so this latter resolved velocity component may also be assumed to be zero. There is no other velocity vector on the ball other than the one normal to the face. Therefore, in the case of an elastic impact, the ball will be launched at an angle to the horizontal equal to θ (i.e. the loft angle of the striking implement), while its launch velocity will be $u_1 \cos\theta$.

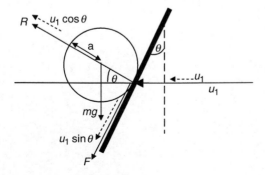

Figure 4.8 Impact of a ball with a lofted implement (implement with an angled hitting face).

The inelastic impact

In the case of the inelastic collision, the frictional force, F, can no longer be assumed to equal zero. So already, we can say that the resultant component vectors of F and R will produce an angle of lift, φ, which will always be less than that of the elastic case, θ. Furthermore, the force, F, will create a backspin on the ball: usually a desirable property.

In the calculations that follow, we neglect the mass of the ball and the effects of gravity on the grounds that the impact forces are so great, that the ball's weight can be safely neglected.

As in the elastic case, and again with reference to Figure 4.8, the horizontal velocity of the club swing can be resolved into the two components, $u_1 \cos\theta$, a reactive force normal to the club face, and $u_1 \sin\theta$, which gives rise to the frictional force F, tangential to the club face. As in the case of a ball suddenly impelled into motion along a flat surface, the ball will start by sliding up the clubface. Then, as the ball's spin increases, the sliding phase ends and the ball starts rolling up the face until such time as it breaks contact with the clubface (Figure 4.9).

During the sliding phase, we can say that $F = \mu_k R$ where μ_k is the coefficient of dynamic friction. The deceleration due to F acting on the ball of mass m_2 is F/m_2, and the deceleration of the short sliding phase, dt is $(F/m_2)/dt$. So, the velocity of the ball up the clubface at the end of the sliding phase will be given by:

$$v'' = u_1 \sin\theta - \int_0^t \frac{F}{m_2}\,dt = u_1 \sin\theta - \frac{\mu}{m_2}\int_0^t R\,dt$$

Now, let us turn our attention to the resistance of the ball to begin rotating. The torque about the ball's center of gravity will be Fa and the angular acceleration will be Fa/I where I is the moment of inertia. The increase in angular velocity of the ball over the short time dt is $(Fa/I)\,dt$. This may be integrated over the whole of the sliding phase:

$$\omega = \int_0^t \frac{Fa}{I}\,dt = \frac{\mu_k a}{I}\int_0^t R\,dt$$

Remembering that $I = mk^2$ where k is the radius of gyration of the ball, gives us:

$$\omega = \frac{\mu_k a}{mk^2}\int_0^t R\,dt$$

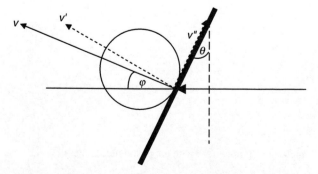

Figure 4.9 Velocity and angle of a ball launch from a lofted implement (inelastic case).

Now, at the point in time, t, when sliding ends and rolling just commences, the angular and linear velocities of the ball must be related by the equation $v = \omega a$, so:

$$u_1 \sin\theta - \frac{\mu_k}{m_2} \int_0^t R \, dt = \frac{\mu_k a^2}{m_2 k^2} \int_0^t R \, dt \tag{4.25}$$

Rearranging, we obtain:

$$u_1 \sin\theta = \frac{\mu_k}{m_2} \left(\frac{a^2}{k^2} + 1 \right) \int_0^t R \, dt$$

But we know (from Table 2.6) that, for a sphere, $k^2 = \frac{2}{5}a^2$. Hence:

$$u_1 \sin\theta = \frac{7\mu_k}{2m_2} \int_0^t R \, dt \tag{4.26}$$

We are now in a position to eliminate $\int_0^t R \, dt$ from Equations 4.25 and 4.26 yielding:

$$v'' = \tfrac{5}{7} u_1 \sin\theta \tag{4.27}$$

Rolling will, therefore, commence when the relative velocity of the ball to the face of the club is 5/7 of its initial value (cf. Equation 4.13 for a ball bouncing off a rigid surface).

Now, since $v'' = \omega a$, we find:

$$\omega = \frac{5u_1 \sin\theta}{7a} \tag{4.28}$$

So, the ball enters the rolling phase with these values of ball spin and relative velocity between the ball and the clubface. In fact, because both the dynamic (rolling) friction and dt are small, these values reduce very little during the rolling phase, and may also be taken as correct at the point of launch.

With reference to Figure 4.9 and from Equation 4.22, we can say:

$$v' = \frac{m_1 u_1 \cos\theta (1 + e)}{m_1 + m_2}$$

While, from Equation 4.27, $v'' = \frac{5}{7} u_1 \sin\theta$. However, the face of the club itself is moving with a component relative to the ground of $u_1 \sin\theta$ in the same direction. Therefore, the vector difference, v'', is given by:

$$v'' = u_1 \sin\theta - \tfrac{5}{7} u_1 \sin\theta = \tfrac{2}{7} u_1 \sin\theta$$

The launch velocity, v, will be the resultant of these two velocities:

$$v^2 = v'^2 + v''^2$$

$$v = \sqrt{\left(\frac{m_1 u_1 \cos\theta (1 + e)}{m_1 + m_2} \right)^2 + \left(\tfrac{2}{7} u_1 \sin\theta \right)^2} \tag{4.29}$$

The ball's vertical velocity will be given by $v' \sin\theta - v'' \cos\theta$ and the ball's horizontal velocity will be given by $v' \cos\theta + v'' \sin\theta$, which makes the launch angle:

$$\tan\varphi = \frac{v' \sin\theta - v'' \cos\theta}{v' \cos\theta + v'' \sin\theta} = \frac{\left(\dfrac{m_1 u_1 \cos\theta(1+e)}{m_1+m_2}\right) \sin\theta - \left(\frac{2}{7}u_1 \sin\theta\right)\cos\theta}{\left(\dfrac{m_1 u_1 \cos\theta(1+e)}{m_1+m_2}\right) \cos\theta + \left(\frac{2}{7}u_1 \sin\theta\right)\sin\theta} \tag{4.30}$$

These equations are modelled in Model 4.1 – Impact with a lofted club.

Finally, we have made the assumption that, in the case of the lofted hit, the ball will go through an initial sliding phase, which will then be followed by a rolling phase before the launch process is completed. However, if the ball were still sliding when it leaves the striking implement, it would not achieve its full backspin angular velocity, and the angle of launch would be somewhat greater than that indicated by Equation 4.30. So, how safe is our assumption?

The impulse of the strike is equal to the change in momentum that produces it (see Equation 2.8). So, if T is the total time of impact:

$$\int_0^T R\,\mathrm{d}t = m_2 v' = \frac{m_1 m_2 v \cos\theta(1+e)}{m_1+m_2}$$

Now, when the time to initiate rolling is less than, or equal to, the impact time, t must equal T; i.e. $T > t$ for rolling to occur. From Equations 4.26 and 4.27, we see that the critical condition occurs when $T = t$ and:

$$\frac{7\mu}{2m_2} \cdot \frac{m_1 m_2 u_1 \cos\theta(1+e)}{m_1+m_2} = v\sin\theta$$

And so provided:

$$\tan\theta \leq \frac{7\mu m_1 (1+e)}{2\left(m_1+m_2\right)} \tag{4.31}$$

rolling will occur before the ball is launched.

Worked Example 4.6

A golfer plays a shot with a 5-iron, with a loft angle of 25°. The horizontal impact velocity is $30\,\mathrm{m\,s^{-1}}$. The radius of the ball is 20.6 mm, its mass is 46 g, while the mass of the club head is 250 g. If the CoR between the ball and 5-iron is found to be 0.7 and the coefficient of friction between them is 0.5:

(a) For a perfectly elastic collision, calculate the velocity and angle of launch.
(b) For the inelastic case, confirm that rolling commences before the ball breaks contact with the clubhead, and thence calculate the launch angle, velocity and rate of backspin of the stroke.

So, $m_1 = 0.25\,\mathrm{kg}$, $m_2 = 0.046\,\mathrm{kg}$, $a = .0206$, $u_1 = 30\,\mathrm{m\,s^{-1}}$, $e = 0.7$, $\mu = 0.5$, $\theta = 25°$.

(a) The elastic case:

$$\varphi = \theta = 25°$$

From Equation 4.22 with $e = 1$ for the elastic case:

$$v_2 = \frac{2m_1 u_1 \cos\theta}{m_1 + m_2} = \frac{2 \times 0.25 \times 30 \times \cos 25}{0.25 + 0.046} = \frac{2 \times 0.25 \times 30 \times \cos 25}{0.25 + 0.046} = \frac{13.6}{0.296}$$

$$v_2 = 46\text{m s}^{-1}$$

(b) The inelastic case:
First, check condition for rolling.
   From Equation 4.31:

$$\tan\theta \le \frac{7\mu m_1 (1+e)}{2(m_1 + m_2)}$$

$$\tan 25 \le \frac{7 \times 0.5 \times 0.25 \times (1+0.7)}{2(0.25 + 0.046)} = \frac{7 \times 0.5 \times 0.25 \times (1+0.7)}{2(0.25 + 0.046)} = \frac{1.487}{0.592} = 2.51$$

$$0.466 \le 2.51$$

So rolling *will* commence before separation of golf head and ball.

Indeed, if we attempt this calculation with any sensible values of club velocity, club loft, CoR and coefficient of friction, this condition will always be met: the ball will always roll off a club head with all types clubs and styles of shots.

$$v' = \frac{m_1 u_1 \cos\theta (1+e)}{m_1 + m_2} = \frac{0.25 \times 30 \times \cos 25(1+0.7)}{0.25 + 0.46} = \frac{11.56}{0.296} = 39$$

$$v'' = \tfrac{2}{7} u_1 \sin\theta = \tfrac{2}{7} \times 30 \times \sin 25 = 3.622$$

From Equation 4.29:

$$v^2 = v'^2 + v''^2$$

$$v = \sqrt{\left(\frac{m_1 u_1 \cos\theta (1+e)}{m_1 + m_2}\right)^2 + \left(\frac{2}{7} u_1 \sin\theta\right)^2} = \sqrt{39^2 + 3.622^2} = 39.17\text{m s}^{-1}$$

To calculate the angle of launch, Equation 4.30 is used:

$$\tan\varphi = \frac{v' \sin\theta - v'' \cos\theta}{v' \cos\theta + v'' \sin\theta} = \frac{39 \times \sin 25 - 3.622 \times \cos 25}{39 \times \cos 25 + 3.622 \times \sin 25} = \frac{16.48 - 3.28}{35.35 + 1.53} = 0.38$$

$$\varphi = 19.7°$$

Finally, Equation 4.28 will give us the rate of backspin:

$$\omega = \frac{5u_1 \sin\theta}{7a} = \frac{5 \times 30 \times \sin 25}{7 \times 0.0206} = 439.61 \,\text{rads/s} = \frac{439.61}{2\pi} \,\text{revs/s} = 70 \,\text{revs/s}$$

Of course, you could have inputted the values into Model 4.1 – Impact with a lofted club, and get these results instantly.

4.7 Collision analysis

Having dealt with balls colliding with surfaces, balls colliding with other balls and balls being struck by implements, we now move into a whole new area; balls 'colliding' with people! To be specific, this section analyses the situation of a ball thrown, kicked or struck such that it moves in its roughly parabolic arch, and then, towards the end of its flight, whether a person can succeed in moving fast enough to catch, trap, kick, strike or otherwise engage with the ball just as it completes its flight.

To commence, let us consider the flight of a ball, say a cricket ball following a cover drive, which unfortunately (for the batsman) heads straight into the hands of the extra-cover fielder, without the fielder having to change position.

We have from Equations 3.3, 3.4, 3.10 and 3.11:

$$s_x = v_0 t \cos\theta_0$$

$$s_y = v_0 t \sin\theta_0 - \tfrac{1}{2}at^2$$

$$t_m = 2\frac{v_0 \sin\theta_0}{a}$$

$$s_m = \frac{v_0^2}{a} \sin 2\theta_0$$

It is clear from Figure 4.10 that, as the ball approaches the fielder, the elevation angle of the ball to fielder is given by:

$$\tan\varphi = \frac{s_y}{s_m - s_x}$$

By substitution of Equations 3.3, 3.4, 3.10 and 3.11 into this expression we obtain the important relation:

$$\tan\varphi = \frac{at}{2v_0 \cos\theta} = (\text{constant})\,t \tag{4.32}$$

The tangent of the approach angle increases in a linear manner with time, until the ball is caught. This statement is quite probably the most important declaration in the book. If, like me, you have reflected on the equations of ball motion, and marvelled at how the human brain can process the data, and then direct the body, so quickly, to catch a ball in mid-parabolic motion, I would suggest that one reason lies in our subconscious awareness that the tangent of the ball's approach angle increases uniformly with time.

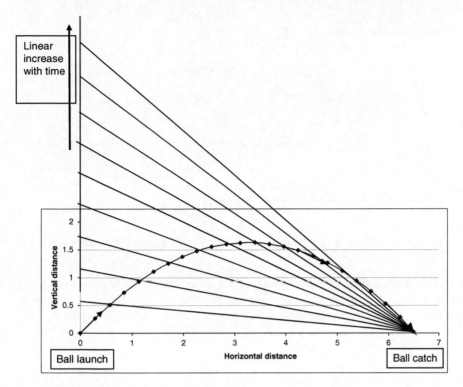

Figure 4.10 Approach angle of incoming ball.

Allow me to explain. Consider, if you will, one of those glass elevators that slide up and down outside of some modern buildings. If it were to rise at a constant velocity without stopping at floors and we were standing some distance away watching it, we could close our eyes for quite some time and when we looked again, we would find the elevator more or less where we might expect it. It is not surprising that our predictive ability in such a scenario is good, and yet Equation 4.32 shows that, as we attempt to catch a ball coming straight into our hands, its elevation at all points in time, from launch to catch, mirrors that of the simple elevator rising with constant velocity. It follows, therefore, that despite all the complex equations of motion discussed so far, at least the elevation of a ball launched accurately towards a fielder is quite predictable for the duration of its journey. This, I contend, must play a major part in the psycho-bio-mechanics of the catch process.

Now let us take this argument further and consider the situation of a ball that is destined to fall short of a fielder who remains static. In this case, $\tan \varphi$ will initially increase from launch, but at a decreasing rate with time, until the ball is at catch height. Thereafter, $\tan \varphi$ falls back to zero in an approximately parabolic curve until it lands on the ground some distance in front of the disappointed catcher. And, perhaps unsurprisingly, for the alternative case of the ball which flies over the fielder's head out of reach, $\tan \varphi$ increases at an exponentially increasing rate. Both these ball position functions are anathema to the catching process. This is indicated in Figure 4.11.

The next stage is to consider the case where the fielder does move, but only along the plane of the ball's flight, either backwards or forwards, to compensate for the difference between, where the fielder is at each point in time, and where the ball might land.

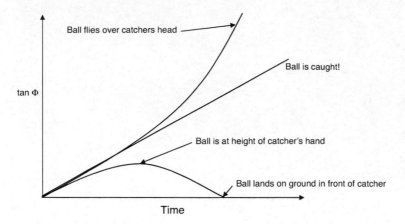

Figure 4.11 Variation of tangent of elevation angle with time, for ball heading directly in towards catcher.

Let us assume that the distance between where the fielder is at the time of launch and where the ball will land is s, and in order to catch the ball he must run for a time τ, at a constant speed of v_f. So, if the fielder just reaches the ball in time, we can state:

$$s = v_f(t_m - \tau)$$

and $\tan \varphi = s_y/s_m - s_x$ becomes:

$$\tan \psi = \frac{s_y}{s_m - s_x + v_f(t - \tau)}$$

where ψ is the new elevation angle between the ball and where the player is standing at any particular time t. The $+$ sign in the denominator signifies that the fielder is standing too close, and the ball will fly overhead if the fielder does not retreat. Following the lines of the previous analysis, we now find that:

$$\tan \psi = \frac{at}{2\left[(v_0 \cos \theta) - v_f\right]} = (\text{constant})\,t \tag{4.33}$$

This time, it is $\tan \psi$ that linearly increases with time. This means that the fielder must run at a constant speed of v_f over the duration of the flight time to reach the catch point at the correct time. Furthermore, they can gauge the correct run speed because, just as in the earlier case of the direct catch, they can watch $\tan \psi$ increasing linearly as they run to the catch point (cf. the predictive aspect of the glass elevator).

Finally, let us consider the more typical case of the ball landing at, not only the wrong range for a simple catch, but also to the left or right of the fielder's initial position. Simple application of Pythagoras on the front/back (axial) and left/right (radial) components of distance yields a similar result to the previous case. Indeed, Equation 4.33 still applies so long as the fielder moves in for the catch along a route known as the path of 'proportional navigation'. This means that they should attempt, at all times, to keep the bearing between the fielder and the ball a constant. The fielder then moves in an arc of approximately parabolic locus to meet the ball. If they can watch $\tan \psi$ increasing linearly then they are maintaining the correct constant velocity of v_f, and the fielder is guaranteed to be in the right position to catch the ball. The rest is down to hand dexterity!

Therefore, the intuitive glass elevator analogy applies to all the catching scenarios, the fielder always moving in the direction and at the constant velocity to maintain that constant elevator ascent.

4.8 Trajectory diversions: a lesson in inverted logic

Sometimes, if we are trying to prove some difficult algebraic relation given some initial expression, after spending some time manipulating equations and making little real progress, it might occur to us to try working backwards from the final expression, juggling terms in an attempt to reach the initial expression. This is known as the 'bottom-up' approach (rather than the initial one, which, not surprisingly, is known as the 'top-down' approach). Obviously, both methods are equally valid from a proof perspective. On occasions, we might even try working both methods together, hoping to meet somewhere in the middle: again, an equally valid method.

What follows is an attempt to solve a seemingly intractable problem by 'thinking outside the box', in this case almost literally. I say 'attempt' because a complete solution is not given, but the line of thinking is novel and we do succeed in placing serious limitations on where the solutions will lie.

The puzzle

A competition billiard table has one white ball positioned somewhere – anywhere – on the baize. Such tables come in a number of officially permitted sizes, but must have an 'aspect ratio' of 2 : 1 ($\pm \frac{1}{8}$ inch), and that is sufficient information for our needs. Let us assume a ball-cushion CoR of 1 (i.e. no loss of energy from the cushion bounce, and the angle of incidence exactly equals the angle of reflection) and, furthermore, assume no other rolling losses, so that, once struck, the ball will go travelling on for ever. Call the angle between the launch direction and the long side of the table (arbitrarily chosen), α. See Figure 4.12a.

The question is this: at what angle(s) should the ball be hit so that it will arrive back at the same point on the table *and* travelling in the same direction, and its motion must, therefore, be cyclic?

It is natural to restrict our attention to a problem that lies on the table; mentally we have placed a restriction on our thinking which proves to be a disadvantage. Thinking this way, it is difficult to envisage possible emerging patterns of motion for varying angles of launch. However, let us think outside the box; or in this case, outside the billiard table! When the ball hits the cushion, the ball reflects back with its angle of incidence equal to its angle of reflection, but, instead of the ball reflecting, *why not reflect the table instead*. The ball carries on across the boundary in a straight line 'onto the new table'. See Figure 4.12b.

The laws of physics have not been violated in this 'thought-experiment', but by manipulating the other 'object' (i.e. the table) instead of the obvious and natural one (the ball), we have broken out of our restrictive framework, as shall be seen.

Now, each time the ball strikes the cushion, instead of reflecting the ball, we will spawn a new table, and the ball continues along its launch path undeterred as shown in Figure 4.12c.

So, we are now in a position to severely limit the allowable values of the angle, α, even if we cannot completely define those angles. The condition for cyclic motion is that the ball travels a certain number of tables up (not including the table the ball starts on), say, p tables, and so many tables to the right, say q tables. Two degenerate cases would be $\alpha = 0$, in which

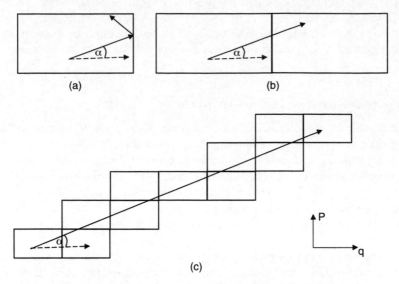

Figure 4.12 (a) Normal ball cushion reflection (b) the corresponding reflection of the billiard table and (c) the cyclic case – the ball returns to its original position following an integer number of reflection.

case $p = 0$, and $\alpha = \pi/2$ when $q = 0$. For the cases in between, $0 < \alpha < \pi/2$ and $p > 0$ and $q > 0$ but they must both be whole integers. In summary, we can now state that the ball will cycle on the billiard table if, and only if:

1    $\alpha = 0°$
2    $\alpha = \pi/2 = 90°$
3    $\tan \alpha = p/q$ (i.e. is a rational number)

As I stated, this is not a complete solution, but an interesting insight into a heuristic method. In justification, even if the perfect solution could be provided, it would be of no practical use to players of the game!

Summary

The coefficient of restitution is the main physical parameter that describes bounce. Its values range $0 < e < 1$, with the hypothetical value of $e = 0$ representing no bounce, and $e = 1$, the perfect, elastic bounce with impact velocity being equal to the rebound velocity. The actual value of e is inextricably linked to the energy loss on impact.

The mathematical analysis of, first normal impact, and then oblique impact are described with the defining equations derived and, in the case of the oblique impact, issues of sliding and rolling of the body on the surface is discussed. The subject of a striking implement hitting a projectile is a particular case of this and the equations governing these dynamic events are derived. We then developed the theory behind ball-on-ball collision dynamics, with particular reference to situations occurring in the game of snooker.

Finally, two simultaneously moving bodies are considered for the case where one body is travelling through the air (the projectile), and the second body (the human player) is moving along the ground in varying directions; the intention being to reduce the distance between the two bodies, in order to enable a 'collision' (e.g. catch or stop the ball).

Problems and questions

1   You place a tennis ball on top of a basketball and hold them 1 m above a rigid surface. If you drop them together, will the tennis ball rebound higher than its drop height? If so, does this imply a CoR > 1 for the tennis ball?

2   Derive an expression that relates the angle of incidence, θ, to the angle of final rebound, ψ, for the case of a double bounce from two surfaces arranged at 90° to each other, if the coefficient of restitution of the first surface is e, and that of the second is e'.

   Hence prove that $\theta = 90° - \psi$ (i.e. the incident path and the final reflected path are parallel as in the case of two perfect elastic bounces) so long as $e = e'$.

3   Calculate the angle between the incident path and the final reflected path for a ball that undergoes a double bounce from two surfaces which are oriented at a general angle $\delta°$ to each other.

4   You have invented a new game involving a Superball, a long table and an opponent. You and your opponent stand at either end of the table and the object is for one player to throw the ball under the table so that it alternately, bounces from floor to under the table and back to the floor again, zig-zagging until it emerges from the other end of the table. With luck, the ball will be caught by the other player, who can then return the ball in a similar manner. The loser would be the first not to catch the ball.

   Would this game be feasible? (Hint: Consider Figure 4.5)

5   If a golf club with a pitch of 15° strikes a golf ball with a velocity of $20 \, \mathrm{m \, s^{-1}}$, assuming a perfectly elastic collision, the theory shows that the ball will launch at 15° to the ground. If, however, we consider a reversal of the process; keep the club static, and fire the ball at the same velocity horizontally at the face of the club, the theory shows that the rebound angle will be 30° to the horizontal. It would appear that these two processes are symmetrical and so is this not a contravention of Newton's third law?

6   For a golf club stroke, the following parameters apply: the mass of the golf ball is 0.046 kg, its radius is 20.6 mm. The weight of the iron head is 0.2 kg and its horizontal impact velocity is $50 \, \mathrm{m \, s^{-1}}$. The CoR is 0.7 and the coefficient of friction is 0.5. Use Model 4.1 – Impact with a lofted club, to suggest an appropriate club that will give a back spin of 200 revs/s. The reference chart on the second tab on the Excel model will help you choose your club once you have found the required loft angle.

Chapter 5

Drag and lift

5.1 Introduction

When a sporting projectile is set into motion by, say, the impact from racket, club or bat, or the kick from a booted foot, it will travel through some fluid (i.e. a gas or liquid), usually air, as it progresses on its trajectory path. The displacement of the medium (air molecules) as it moves out of the path of the trajectory, produces a reactive force on the projectile which acts to slow it down. As the projectile carves its way through the fluid, the molecules will; (a) change direction as they are pushed out of the way of the projectile, or make a detour around it, and, (b) attain some new mean speed, superimposed on the random Brownian motion of the otherwise undisturbed molecules.

The reactive force of the fluid which is impeding the flow of the projectile is known as the *drag* force. Drag is made up of friction forces, which act parallel to the projectile's surface, and pressure forces, which act in a direction perpendicular to the projectile's surface. The analysis that follows shows that the drag force vector always acts in opposition to the projectile's direction of motion, and for most sporting examples, its magnitude, at least to a first approximation, is proportional to the square of its velocity. Now, taking the case of our simple parabolic trajectory, the drag vector will change direction continually over the whole flight duration so as to always act along the tangent to the parabolic curve (or a modified parabolic curve in the case of the more complex trajectory paths) as indicated by Equation 3.8. The drag magnitude will be approximately directly proportional to the right-hand side of Equation 3.7, which is, itself, a function in v^2.

Just as an indication of the impact drag has on a projectile, a soccer ball kicked, without spin, at a launch velocity of $35 \, \text{m s}^{-1}$, and $45°$ to the horizontal, will travel a distance of about 66 m to first bounce. If the player could kick the ball in a vacuum, it would travel 125 m; nearly double the distance.

It might be thought that all aspects of drag are unhelpful in so much as, in sporting applications, it is usually considered that distance and/or speed should, at all times, be maximized. However, the asymmetric shape of a particular projectile, or even the manner in which it is thrown (e.g. the javelin), may result in a component of the drag force which acts at $90°$ to the direction of motion. This vector will have a considerable component acting in opposition to the gravitational force; this is known as the *lift* force. It is, of course, this lift force which keep aircraft flying (see Trajectory diversions at the end of this chapter). In sporting applications, the lift force can extend the period of time the projectile is in the air, thereby, allowing it to travel further. Additionally, increased flight time may be a facet which is important in its own right, such as in the case of American football punts, where flight time is maximized to allow teammates to run downfield as far as possible to prevent a substantial runback.

In summary, over a limited range of velocities faster than, what we might in sporting terms, call 'slow' (!) we can state:

$$F_D = C_D \rho A \frac{v^2}{2} \tag{5.1}$$

$$F_L = C_L \rho A \frac{v^2}{2} \tag{5.2}$$

where F_D and F_L are the drag and lift forces respectively, C_D and C_L are the coefficients of drag and lift respectively, ρ is the density of the fluid, A is the frontal or cross-sectional area exposed to the flow, and v is the velocity of the projectile relative to the fluid. A crude explanation for the second power of velocity in the expression is that, as velocity increases, not only do the molecules sweep past the projectile faster, creating more friction, but correspondingly more 'mass-of-particles-per-second' strike the projectile. Therefore, the change in momentum per second increases as a product of the two, and the resulting force is equivalent to the change of momentum divided by time.

The drag force will cause a retardation of the projectile which is inversely proportional to the mass of the projectile. In simple terms we can say:

$$\text{retardation} \propto \frac{\text{diameter}^2}{\text{mass}}$$

Table 5.1 lists common sporting projectiles in order of, the reciprocal of their retardation values. It is a ballistic league table, with the balls least affected by drag at the top of the table and those which will be expected to slow down considerably by virtue of, either their large size, or their light weight, at the lower end.

The retardation figures explain why it is possible to throw a cricket ball from the boundary to the crease (maybe 75 m), while it is difficult to throw a table tennis ball more than 20 m. Notice that some of the larger sporting balls such as soccer balls and basketballs are not included in the list. The reason for this is that their diameter is such that, in normal play, they will break what is termed the critical Reynold's number. When this happens, drag reduces significantly. This serves to distort the order of the listing in the table. The Reynold's number and the part it plays in drag calculations is explained more fully in Section 5.3.

Table 5.1 Retardation values for common sporting projectiles

Type of ball	Mass m (kg)	Diameter d (m)	m/d^2 (kg m^{-2})
Lacrosse	0.142	0.0635	35.22
Cricket	0.156	0.07	31.84
Hockey	0.156	0.07	31.84
Golf (UK)	0.046	0.041	27.36
Baseball	0.17	0.074	31.04
Golf (US)	0.046	0.043	24.88
Table tennis	0.0255	0.0381	17.57
Squash	0.024	0.04	15.00
Tennis	0.0567	0.0635	14.06

Worked Example 5.1

The International Tennis Federation has, for some time, been concerned that their game is turning into a 'Game of ace serves'. In an effort to slow the game and encourage successful service returns and more prolonged rallies, they have approved a new ball specification which is 6 per cent larger than the traditional ball, and 2 g heavier. If the traditional ball is 6.35 cm and its mass is 58 g, what is the increase in retardation of the new ball over the traditional?

The retardation of the traditional ball is given by:

$$\text{retardation} = \frac{(\text{diameter})^2}{\text{mass}} = \frac{(0.0635)^2}{58 \times 10^{-3}} = 0.07$$

and for the new ball:

$$\text{retardation} = \frac{(0.0635 \times 1.06)^2}{60 \times 10^{-3}} = 0.0755$$

An increase in retardation of about 8 per cent over the old ball.

Simple though Equations 5.1 and 5.2 appear, it is worth noting that, although theoretically precise (in that they define both C_D and C_L), in practice, they are only true for a limited range of velocities. Furthermore, both C_D and C_L depend on, not only the shape of the projectile (the 'blunter' the object, the higher their values), but also on the nature of the airflow around the body, which itself is a function of the velocity, v. And so, the drag and lift forces turn out to be quite complex functions of the velocity of the projectile. The rest of this chapter examines these intricacies in more detail.

5.2 Types of drag

5.2.1 Surface or friction drag

When a projectile cuts through the air, the layer of air molecules which are actually in contact with the surface of the projectile, to some extent, get dragged along with the projectile, and, in so doing, they pick up energy from the moving projectile. These molecules, in turn, cause the next layer of molecules to be dragged along. This process continues for several layers until the energy transfer between layers becomes negligible. Of course, the energy picked up by the molecules can only come from the projectile itself, thereby slowing it down. It is as if the projectile, not only has to carry itself as it travels, but also a certain amount of the surrounding air with it. The moving layers of air in proximity to the projectile are known as the boundary layer and this type of aerodynamic flow is known as *laminar flow*.

As the projectile velocity increases, however, the boundary layer thickness reaches a point where it becomes unstable. It breaks down forming a series of eddy currents which surround the projectile. This reduces the surface drag, usually quite considerably and rapidly, forming what is known as *turbulent* or *chaotic* flow.

The magnitude of the surface drag and the point of onset of chaotic flow are both governed by the velocity of the projectile, its surface area, its shape and surface roughness, and the viscosity of the fluid it is travelling through.

5.2.2 Form drag and interference drag

Form drag, also known as profile or pressure drag, arises because of the shape of the projectile. Projectiles with a larger cross-section area in the direction of the trajectory path will have a higher drag than thinner ones. Sleek, streamlined projectiles such as javelins, where the cross-sectional area changes gradually over its length are also vital for achieving minimum form drag. Any sudden discontinuities, however, such as may occur at the back of the projectile, will lead to vortices containing low-pressure areas which slow the progress of the projectile. This is known as *interference drag*.

Interference drag will be created wherever two surfaces meet at a sharp angle, or even where two parallel surfaces are spaced closely together, even if they happen to be in line with the airflow. In such cases complex vortices can be created in the vicinity of the projectile, which trap both high and low pressure regions in a highly localised manner, disrupting the movement of the projectile through the air. In the case of fast moving vehicles, fairings are added to the structure to smooth out the surfaces, leading to a significant reduction in interference drag.

Note that the form drag of none-circularly symmetric projectiles may vary over the duration of the flight (e.g. javelins or rugby balls) as their orientation changes with respect to the trajectory direction vector.

5.2.3 Wave drag

Wave drag is caused by shock waves building up around the projectile. In air, there is a limit to how fast the molecules can move to accommodate the travelling projectile. When that limit is reached, stability can only be maintained by the generation of longitudinal waves in the air at the front of the projectile. The projectile then has to force its way through those waves which leaches considerable energy from the projectile. Drag will rise considerably and may result in a fourfold increase. That would be enough to cause the projectile to destabilize and tumble to the ground.

We do not see this happen in typical sporting applications because our projectiles travel much less than that critical speed; the speed of sound at $343 \, \text{m s}^{-1}$. There is one common exception. In the case of swimmers, water has a much higher density than air and, as a consequence a much higher velocity of molecular movement (speed of sound in water = $1500 \, \text{m s}^{-1}$). However, the problem arises at the water/air interface. Here shock waves can readily build up on the water surface in front of the swimmer, resisting the motion. For this reason, swimmers try to keep as much of their body underwater for as long a time as possible, reducing the tendency for the buildup of these water-based 'sonic' waves. It is an indication of the importance of wave drag that swimmers would pursue this style to reduce wave drag, even though they would be increasing both their form and their surface drag by swimming submerged. Indeed this has led to a change in the regulations for breast-stroke swimmers who used to attempt to swim the whole of the first length under water. Now, breast-stroke swimmers are only allowed one stroke under water per length.

5.3 Fluid dynamics, laminar and chaotic flow and critical speed

Consider a smooth ball progressing steadily through still air. The air will move around the ball as depicted by Figure 5.1a. The lines depict the flow-lines for a laminar flow (or streamlined) situation.

It can be seen that the lines are spread out at the leading and trailing faces, marked A and C in Figure 5.1a and they are compressed at regions B and D. Since the same amount of air

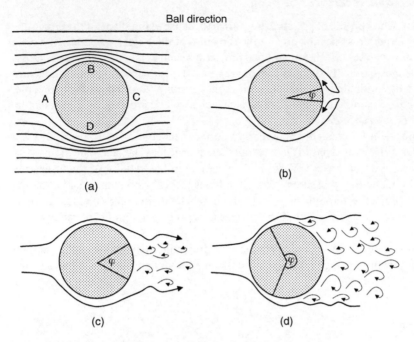

Figure 5.1 Creation of eddies around a travelling ball.

must flow past the ball per second, the velocity of air flow must be greater at the points B and D, where it is being forced along a narrow channel, compared with points A and C where the air splits and rejoins at a comparatively leisurely pace. It follows that the air at points B and D possess more energy than the air at points A and C. Finally we can state that, for this to be the case, the air pressure at A and C must be greater than at B and D in order to sustain the increased energy at B and D. This is known as Bernoulli's Principle.

Bernoulli's Principal states that for an inviscid (i.e. zero viscosity) flow, an increase in the speed of the fluid occurs simultaneously with a decrease in pressure or a decrease in the fluid's gravitational potential energy.

Now, let us consider what happens to the airflow as the ball speeds up. The pressure gradients described above will increase and reach the point where the increased pressure at C is so great that the air can no longer surmount the adverse pressure gradient from B to C and from D to C. The air at C begins to curl back along the direction of the ball's travel creating an eddy (Figure 5.1b).

As the ball's speed increases further, more eddies are formed, which create a wake behind the ball. This becomes thicker with increasing speed, gradually moving forward in the direction of the ball's travel. In Figures 5.1b,c the 'break' angle, φ, increases with increasing speed. The air in this wake is no longer laminar and energy is drawn from the progressing ball to sustain the wake; over this range of velocities, the drag force increases with the ball's velocity as expected. The eddying flow of air in the wake region is known as chaotic flow, as opposed to the smooth laminar air flow the ball creates at lower speed.

If we want to design the shape of a projectile in order to minimize drag, it is clear that as much consideration should be given to the rear contour, as reducing the frontal surface area. In fact, judicious design of the rear of the object can considerably narrow the wake and improve drag, as well as in-flight stability.

Finally, in the case of balls and other projectiles with convex rear surfaces, if the projectile speed is allowed to increase still further, a point is reached where the laminar flow completely breaks down and a chaotic boundary layer is created which completely surrounds the projectile, as shown in Figure 5.1d. As stated in Section 5.2, this can occur quite suddenly and, in the case of spherical forms, when $\varphi \approx 90°$. The result is a turbulent air flow which, rather than drawing energy from the ball, surprisingly draws it from the free stream of air further away from the ball's surface. At this point, known as the critical speed, the mixing of free speed fluid with boundary layer fluid allows the air to overcome the adverse pressure gradient more easily. The wake, therefore, narrows and the drag reduces quite considerably.

Obviously, the critical speed value, in the case of most sporting projectiles, is crucial and players would like its value to be as low as possible. Roughly speaking, it is inversely proportional to the diameter of the ball but it can also be reduced by increasing surface roughness. As a consequence, footballs kicked with some force will usually overcome the critical speed quite easily by virtue of their large diameter; on the other hand, golf balls, with their much smaller diameter, have the dimples impressed into the surface in order to hasten the onset of chaotic flow and reduce drag.

5.4 The Reynold's number and drag-to-weight ratio

It is clear from the preceding treatment that the magnitude of fluid drag on sporting projectiles is a complex affair, which behaves in a non-linear, though somewhat predictable, fashion, dependent on such variables as projectile shape, velocity and surface roughness.

One way of normalising these dynamic phenomena is to define a non-dimensional parameter, the Reynold's number R_e, which describes, precisely the flow of a fluid around a body.

The Reynold's number, R_e, is a measure of the ratio of inertial forces, given by $v_s \rho$ (where v_s is the mean fluid velocity and ρ is the density of the fluid) to viscous forces, μ/L, (where μ is the absolute dynamic fluid viscosity and L is known as the characteristic length: in the case of a sports ball this would be the diameter) and, consequently, it quantifies the relative importance of these two types of forces for given flow conditions.

$$R_e = \frac{\text{dynamic pressure}}{\text{shear stress}} = \frac{\rho v_s L}{\mu} = \frac{v_s L}{\upsilon} = \frac{\text{inertial forces}}{\text{viscous forces}} \qquad (5.3)$$

where $\upsilon =$ kinematic fluid viscosity given by μ/ρ.

When two geometrically similar flow lines, in perhaps different fluids with possibly different flow rates, have the same Reynold's number, they are said to be in dynamic similitude, and they will have similar flow geometry. Notice that, in common with customary fluid dynamic notation, the projectile is now considered to be static and focus shifts to the movement of fluid as it passes and accommodates the obstruction.

Laminar flow, which is distinguished by smooth, constant fluid motion occurs at low Reynold's numbers where viscous forces are dominant. Turbulent flow, on the other hand, occurs at high Reynold's numbers and is dominated by inertial forces, random eddies, vortices and chaotic turbulence. The cross-over point between laminar and chaotic flow, which can be quite abrupt in the case of smooth-surfaced bodies, occurs at what is known as the critical Reynold's number, R_{ecrt}. All other factors being equal, a rough sphere will have a R_{ecrt} which may be as much as three times lower than a smooth one.

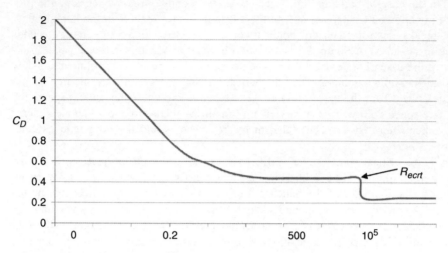

Figure 5.2 Variation of C_D with R_e.

Figure 5.2 shows the typical variation of the coefficient of drag, C_D, with the Reynold's number R_e for a smooth sphere. With reference to Equation 5.1 a more accurate refinement which covers a wider range of velocities would be:

$$F_D = C_D \rho A \frac{v^n}{2}$$ (5.4)

where now n is an integer which is dependent on the projectile's velocity, v. At low speeds and correspondingly low R_e, $n = 1$ and the drag, F_D, is proportional to v. In Figure 5.2, we see in this region, C_D is inversely proportional to R_e. However, at high speeds and high R_e, $n = 2$, and in this region, C_D is independent of v. Finally, the value R_{ecrt} is clearly seen occurring at a much higher value of R_e.

Model 5.1 – Calculation of R_e for range of projectiles, gives a more complete table, with the added option of inserting your own values of velocity, v and representative length, l, allowing you to quickly and easily calculate R_e for any projectile of your choosing.

Comparing Table 5.2 with the chart of Figure 5.2 it would appear that many games are played around the region of R_{ecrt} which, as a general rule, may be taken as $\approx 1.4 \times 10^5$. Very approximately, games played with balls below this Reynold's number will exhibit a drag coefficient, C_D, in the region of 0.45. Games played with balls above the critical value will exhibit the lower C_D value of 0.2.

Table 5.2 R_e values for common sporting projectiles (see Appendix I for more complete list)

Projectile	R_e
American rules football	3.60E+05
Cricket ball	1.63E+05
Golf ball	1.87E+05
Soccer football	4.40E+05
Squash ball	1.33E+05
Tennis ball	1.60E+05

The final column on the Model 5.1 – Calculation of R_e for range of projectiles, is the value of the 'drag-to-weight' ratio, ε, defined as:

$$\varepsilon = F_D/mg = \frac{\rho A C_D v^2}{2mg}$$

For the purposes of this model, ρ is the density of air, taken as $1.23\ \text{kg m}^{-3}$, g is the acceleration due to gravity at $9.81\ \text{m s}^{-2}$, and C_D is either 0.2 or 0.45 as appropriate, dependent on the Reynold's number.

A small value of ε serves to indicate that the drag effect is much smaller than the gravitational effect. A large value of ε would suggest that drag forces will dominate over the gravitational force. The result of large ε is a ball which typically begins its trajectory by flying in a straight line, until such time that the drag forces slow the projectile down to a point where the gravitational force takes over and pulls it to the ground. It is interesting to note the wide range of values of ε; from the almost negligible figures of the shot-put and the hammer, right up to the values for the shuttlecock and table tennis balls where, arguably, the nature of the game is characterized by the drag properties of the projectiles.

Worked Example 5.2

Calculate the drag-to-weight ratio for an arrow fired with a velocity of $70\ \text{m s}^{-1}$. By comparing your answer with the values given in Model 5.1 – Calculation of R_e for range of projectiles, would you say that the arrow's trajectory is 'dominated by its drag'?

The specifications for the arrow are:

Length $= 0.6\ \text{m}$, mass $= 30\ \text{g}$, diameter $= 9.3\ \text{mm}$, $\rho = 1.23\ \text{kg m}^{-3}$. As in the case of the javelin, assume the arrow does not reach its critical velocity, so take $C_D = 0.45$.

The drag-to-weight ratio is given by:

$$\varepsilon = \frac{\rho A C_D v^2}{2mg} = \frac{1.23\left[\pi \times \left(\dfrac{9.3 \times 10^{-3}}{2}\right)^2\right] \times 0.45 \times 70^2}{2 \times 30 \times 10^{-3} \times 9.81} = \frac{0.184}{0.59} = 0.312$$

The drag-to-weight ratio is on a par with a cricket ball or basket ball. So, I would suggest that, although not insignificant, the trajectory is not really dominated by drag.

5.5 The projectile flight equations for a resistive medium

Up to this point in the book, we have assumed that the equations of motion for projectile flights have been projectiles which are moving freely under gravity. In this section, we will consider derivations for flight equations for projectiles which are moving through a resistive medium which obeys Equation 5.4.

5.5.1 The linear resistive medium

In this, the simplest of drag cases, Equation 5.4 will be utilised with $n = 1$, implying that the resistive force is proportional to velocity. In fact, this is only true for very small objects moving very slowly through a highly viscous medium. However, the analysis will form the basis for the more realistic solutions derived in the next section. As stated, the drag will always act in opposition to the direction of motion throughout the flight. So the drag value will vary in both magnitude and direction, over the whole of the flight duration. However, only in the *linear* resistive case are the governing equations both linear and coupled, leading to solutions which can be mathematically derived.

Assume a sporting projectile is fired from ground level with a launch velocity v_0 at an angle θ_0 to the horizontal. Further, let us define a constant, k, the resistance coefficient per unit mass. With reference to Figure 5.3 and using the standard vector notation with the unit vector $\hat{\mathbf{i}}$, lying along the horizontal x direction, and the unit vector $\hat{\mathbf{j}}$ along the vertical y direction.

The equation of motion for the projectile is obtained by equating forces in conjunction with Newton's second law:

$$F = ma = m\frac{\mathrm{d}^2 r}{\mathrm{d}t^2} = -mkv\hat{\mathbf{v}} - mg\mathbf{j}$$

$$\ddot{\mathbf{r}} = \begin{pmatrix} -kv_x \\ -g - kv_y \end{pmatrix} \tag{5.5}$$

We solve Equation 5.5 for x and y components respectively, with values at launch of:

$$\theta = \theta_0, \quad v = v_0, \quad v_x = v_{x0}, \quad v_y = v_{y0}, \quad \mathbf{r} = 0 = s_0 = s_{x0} = s_{y0}$$

Horizontally:

$$\frac{\mathrm{d}v_x}{\mathrm{d}t} = -kv_x$$

$$\int_{v_{x0}}^{v_x} \frac{\mathrm{d}v_x}{v_x} = -k \int_0^t \mathrm{d}t$$

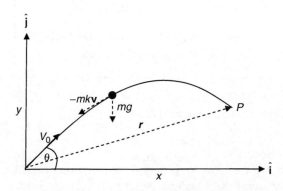

Figure 5.3 Coordinate system for projectile with linear drag.

therefore:

$$\log_e \frac{v_x}{v_{x0}} = -kt$$

So:

$$v_x = v_{x0}\, e^{-kt} = \frac{ds_x}{dt}$$

$$\int_0^{s_x} ds_x = v_{x0} \int_0^t e^{-kt}\,dt$$

$$\therefore \qquad s_x = \frac{v_{x0}}{k}(1 - e^{-kt}) \tag{5.6}$$

Vertically:

$$\frac{dv_y}{dt} = -g - kv_y$$

$$\int_{v_{y0}}^{v_y} \frac{dv_y}{g + kv_y} = -\int_0^t dt$$

therefore:

$$\frac{1}{k}\log_e \left(\frac{g + kv_y}{g + kv_{y0}}\right) = -t$$

So:

$$v_y = \frac{1}{k}\left(g + kv_{y0}\right) e^{-kt} - \frac{g}{k} = \frac{ds_y}{dt} \tag{5.7}$$

$$\int_0^{s_y} ds_y = \int_0^t \frac{1}{k}\left[\left(g + kv_{y0}\right) e^{-kt} - \frac{g}{k}\right] dt$$

therefore:

$$s_y = \frac{1}{k^2}\left(kv_{y0} + g\right)(1 - e^{-kt}) - \frac{gt}{k} \tag{5.8}$$

We now combine the parametric Equations 5.6 and 5.8 by equating t to obtain the equation of the trajectory. From Equation 5.6:

$$t = -\frac{1}{k}\log_e \left(1 - \frac{ks_x}{v_{x0}}\right)$$

and substitute into Equation 5.8:

$$s_y = \left(v_{y0} + \frac{g}{k}\right)\frac{s_x}{v_{x0}} + \frac{g}{k^2}\log_e \left(1 - \frac{ks_x}{v_{x0}}\right)$$

But $v_{x0} = v_0 \cos \theta_0$ and $v_{y0} = v_0 \sin \theta_0$. Therefore:

$$s_y = s_x \tan \theta_0 + \frac{gs_x \sec \theta_0}{kv_0} + \frac{g}{k^2} \log_e \left(1 - \frac{ks_x \sec \theta_0}{v_0}\right) \tag{5.9}$$

which is the equation of the trajectory. It is interesting to compare this equation with that of the dragless case given by Equation 3.6, restated below:

$$s_y = s_x \tan \theta_0 - \frac{as_x^2 \left(1 + \tan^2 \theta_0\right)}{2v_0^2}$$

And, as expected, Equation 5.9 approaches Equation 3.6 as $k \to 0$.

Worked Example 5.3

In Jan Zelezny's world record javelin throw in 1996 (see Worked Example 3.1), we calculated that if the javelin was thrown at 36° to the horizontal, his throw velocity was 31.87 m s^{-1}. Calculate the height of the javelin when it is 20 m from the thrower (a) in the dragless case and (b) when there is a drag coefficient, k, of 0.07

For the dragless case, from Equation 3.6:

$$s_y = s_x \tan \theta_0 - \frac{as_x^2 \left(1 + \tan^2 \theta_0\right)}{2v_0^2}$$

$$= 20 \times \tan 36 - \frac{9.81 * 20^2 \left(1 + \tan^2 36\right)}{2 \times 31.87^2}$$

$$= 14.53 - \frac{6000}{1993} = 11.52 \text{ m}$$

For the case where the drag coefficient is 0.07, from Equation 5.9:

$$s_y = s_x \tan \theta_0 + \frac{gs_x \sec \theta_0}{kv_0} + \frac{g}{k^2} \log_e \left(1 - \frac{ks_x \sec \theta_0}{v_0}\right)$$

$$= 20 \times \tan 36 + \frac{9.81 \times 20 \times \sec 36}{0.07 \times 31.87} + \frac{9.81}{0.07^2} \log_e \left(1 - \frac{0.07 \times 20 \times \sec 36}{31.87}\right)$$

$$= 14.53 + \frac{242.51}{2.23} + 2002 \times \log_e (0.946)$$

$$= 14.53 + 108.75 - 111.77$$

$$= 11.51 \text{ m}$$

A reduction due to the inclusion of this low drag factor of 1 cm only.

An expression based on k for the terminal velocity v_T yields an interesting result. Equation 5.5 may be rewritten as:

$$\frac{dv}{dt} = -kv\hat{\mathbf{v}} - g\mathbf{j}$$

Terminal velocity is reached when $dv/dt = 0$ and $v_T = -(g/k)\mathbf{j}$. There are a number of experimental methods for obtaining values for v_T from which accurate calculated values of k can be derived for use in the equations above. Furthermore, we can see from Equation 5.5 and 5.7 that, as $t \to \infty$, $v_x \to 0$ and $v_y \to -g/k$; if the flight is long enough, the trajectory will degenerate into a vertical fall at the terminal velocity of $-g/k$. The curve of the trajectory is asymptotic with a vertical line at $s_x = (v_0 \cos\theta_0)/k$.

Model 5.2 – General trajectory model, has a worksheet tab 'Trajectory model – linear drag' which models Equation 5.9 with v_0, θ_0 and k as variables. The asymptotic line is also plotted as a series.

Range, height and enveloping parabola for the linear resistive medium

The range is found by calculating the flight time, τ, which will be the value of t when $v_y = 0$ in Equation 5.8:

$$\frac{g\tau}{k} = \frac{\left(kv_{y0} + g\right)\left(1 - e^{-k\tau}\right)}{k^2}$$

$$\tau - \frac{\left(kv_{y0} + g\right)\left(1 - e^{-k\tau}\right)}{kg} = 0$$

Unfortunately this equation cannot be solved analytically. However, given values of k, v_0 and θ_0, a value for τ can be obtained by means of an iterative process such as the Newton–Raphson's method. This value can then simply be inserted into Equation 5.6 to obtain a value of s_x for the known flight duration.

By way of example, if we allow $k = 0.1$, $v_0 = 10\,\mathrm{m\,s^{-1}}$, $\theta_0 = 45°$, τ works out at $1.382\,\mathrm{s}$ which gives a range of $9.13\,\mathrm{m}$. This is compared to $10\,\mathrm{m}$ if the calculation is performed with no air resistance. It should be noted, however, that this is not the maximum range, as $45°$ will not be the optimum launch angle for flight through a resistive medium. De Mestre (1991, p. 31) has shown that for small values of $\varepsilon gt/v_0$ (i.e. small drag) the angle for maximum range, θ_m, is given by:

$$\theta_m = \frac{\pi}{4} - \frac{\sqrt{2}\varepsilon}{6} + \text{smaller terms in } \varepsilon$$

So the optimum angle for maximum range is the dragless optimum angle ($45°$) $-\sqrt{2}\varepsilon/6$. With this as the launch angle, and under the same conditions of small drag, the optimum range is reduced over the dragless case by an amount: $2\sqrt{2}kv_0^3/3g^2$.

Unlike the range calculation, the calculation for maximum height, H, can be accomplished analytically. The maximum height will occur when $v_y = 0$ and this can be put into Equation 5.7 to find the time to the apex of the flight, T:

$$v_y = \frac{1}{k}\left(g + kv_{y0}\right)e^{-kT} - \frac{g}{k} = 0$$

$$\Rightarrow \left(g + kv_{y0}\right) e^{-kT} = g$$

$$e^{-kT} = \frac{g}{g + kv_{y0}}$$

$$T = \frac{1}{k} \log_e \frac{\left(g + kv_{y0}\right)}{g}$$

Substituting this value of T into Equation 5.8 yields:

$$H = \frac{1}{k^2} \left(kv_{y0} - g \log_e \frac{\left(g + kv_{y0}\right)}{g} \right)$$

For the values given above for k, v_0 and θ_0, we note that $T = 0.683$ s which is less than $\tau/2$; it takes longer for the projectile to reach its maximum height than it does to fall back to earth. For comparison, H calculates out to be 2.4 m compared with 2.5 m for the drag free case.

Although it is clear that the trajectory for the projectile subject to drag is no longer a parabola, it is interesting to investigate whether there is still a 'parabola of safety' associated with the resisted motion case, or indeed, any bounding function.

To study the family of curves associated with equal launch velocity at differing angles, we should eliminate θ_0 from the two function equations:

$$f(x, y, \theta) = 0 \quad \text{and} \quad \frac{\partial f}{\partial \theta}(x, y, \theta) = 0$$

From Equation 5.9:

$$f(x, y, \theta) = s_x \tan \theta_0 + \frac{gs_x \sec \theta_0}{kv_0} + \frac{g}{k^2} \log_e \left(1 - \frac{ks_x \sec \theta_0}{v_0}\right) - s_y = 0$$

Differentiating with respect to θ_0 gives:

$$\frac{\partial f}{\partial \theta_0} = s_x \sec^2 \theta_0 + \frac{gs_x \sec \theta_0 \tan \theta_0}{kv_0} - \frac{gs_x \sec \theta_0 \tan \theta_0}{k\left(v_0 - ks_x \sec \theta_0\right)} = 0$$

therefore:

$$s_x \sec \theta_0 + \frac{gs_x \tan \theta_0}{kv_0} - \frac{gs_x \tan \theta_0}{k\left(v_0 - ks_x \sec \theta_0\right)} = 0$$

Solving for s_x gives:

$$s_x = \frac{v_0^2}{v_0 k \sec \theta_0 + g \tan \theta_0} \tag{5.10}$$

Substituting Equation 5.10 into Equation 5.9 yields:

$$s_y = \frac{v_0^2}{v_0 k \csc \theta_0 + g} + \frac{g}{k\left(v_0 k + g \sin \theta_0\right)} + \frac{g}{k^2} \log_e \left[1 - \frac{v_0 k}{v_0 k + g \sin \theta_0}\right] \tag{5.11}$$

Equations 5.10–5.11 are the parametric equations for the bounding curved. They are solved for values of v_0, k and θ_0 in Model 5.3 – Bounding curve for linear drag.

With reference to Equation 5.10, and holding v_0, k and g all constant, we see that the bounding curve for s_x is, in fact, *not* a parabola as θ_0 varies, although the curve does look similar. Also, just as in the drag-free case, the velocity vector of the projectile where it grazes the bounding curve is perpendicular to the launch direction.

For completeness, Murphy (1979) derived the bounding curve for the linear drag case to be:

$$y^* = \frac{\sqrt{\left[1 + x^{*2}\left(1 - \varepsilon^2\right)\right]}}{\varepsilon} + \frac{1}{\varepsilon^2}\log_e\left[\frac{1 - \varepsilon\sqrt{\left[1 + x^{*2}\left(1 - \varepsilon^2\right)\right]}}{1 - \varepsilon^2}\right]$$

where:

$$y^* = \frac{s_y}{v_0^2/g}, \qquad x^* = \frac{s_x}{v_0^2/g}, \qquad \text{and} \qquad \varepsilon = \frac{kv_0}{g}$$

Projectile trajectory up an inclined plane for the linear resistive medium

With reference to Figure 3.4, and with the plane of inclination given by ψ, we find: that $\tan\psi = s_{ym}/s_{xm}$ where s_{ym} and s_{xm} relate to the point of impact of the projectile with the inclined plane. If this expression is substituted into Equations 5.10 and 5.11, we find:

$$\tan\psi = \frac{1 + \varepsilon\sin\theta_m}{\varepsilon\cos\theta_m} + \frac{\varepsilon + \sin\theta_m}{\varepsilon^2\cos\theta_m}\log_e\left(\frac{\sin\theta_m}{\sin\theta_m + \varepsilon}\right)$$

To solve this we make the transformation $W_m = 1 + \varepsilon\,\mathrm{cosec}\,\theta_m$ such that the equation simplifies to:

$$W_m - 1 + \varepsilon^2 - W_m\log_e W_m - \varepsilon\tan\psi\sqrt{\left(W_m - 1\right)^2 - \varepsilon^2} = 0$$

This equation can be solved graphically or iteratively using Newton's successive approximation algorithm to obtain the value for W_m, and, once this is known, we can obtain a value for θ_m:

$$\theta_m = \arcsin\left[\frac{\varepsilon}{W_m - 1}\right] \tag{5.12}$$

Thence, the maximum range can be obtained by substitution into Equation 5.10:

$$s_{xm} = \frac{v_0^2}{v_0 k\sec\theta_m + g\tan\theta_m} \tag{5.13}$$

Worked Example 5.4

Calculate the optimum launch angle for a projectile which is launched at an angle above a horizontal surface with the special case where the initial drag equals the force of gravity. Compare the maximum distance attainable with the dragless case.

So $\varepsilon = 1$ and $\psi = 0$.

Under such conditions the expression:

$$W_m - 1 + \varepsilon^2 - W_m \log_e W_m - \varepsilon \tan \psi \sqrt{(W_m - 1)^2 - \varepsilon^2} = 0$$

simplifies to:

$$W_m - W_m \log_e = 0$$

therefore:

$$W_m = e$$

From Equation 5.12:

$$\theta_m = \arcsin \left[\frac{1}{e - 1} \right] \approx 35.59°$$

Now, from Equation 5.13:

$$s_{xm} = \frac{v_0^2}{v_0 k \sec \theta_m + g \tan \theta_m} = \frac{v_0^2}{1.23 v_0 k + 7}$$

compared with the dragless case where:

$$s_{xm} = \frac{v_0^2}{9.81}$$

5.5.2 Trajectory equations for resistive medium proportional to velocity-squared

Most sporting projectiles more accurately obey this 'drag proportional to v^2' law although several of the derived equations do not have analytic solutions. Notwithstanding, many useful derivations can be obtained and, as will be seen, our Excel models are still capable of plotting trajectories using mathematically discrete techniques.

For this model, and with reference to Figure 5.3, we can write:

$$F = ma = m \frac{d^2 r}{dt^2} = -mkv^2 \hat{v} - mg\mathbf{j}$$

$$= -mk |\mathbf{v}|^2 \frac{\mathbf{v}}{|\mathbf{v}|} - mg\mathbf{j} = -mk |\mathbf{v}| \mathbf{v} - mg\mathbf{j}$$

As in the linear case we have again assumed the sporting projectile to be fired from ground level with a launch velocity v_0 at an angle θ_0 to the horizontal. We have taken the constant, k, as the resistance coefficient per unit mass. Further, the unit vector $\hat{\mathbf{i}}$ lies along the horizontal x direction, and the unit vector $\hat{\mathbf{j}}$ along the vertical y direction.

Given that, at any point in the trajectory: $v = |\mathbf{v}| = \sqrt{(v_x^2 + v_y^2)}$, the equation of motion for the projectile is given by:

$$\ddot{\mathbf{r}} = \begin{pmatrix} -k\sqrt{\left(v_x^2 + v_y^2\right)}v_x \\ -g - k\sqrt{\left(v_x^2 + v_y^2\right)}v_y \end{pmatrix}$$

So:

$$\frac{d^2 s_x}{dt^2} = \frac{dv_x}{dt} = -k\sqrt{\left(v_x^2 + v_y^2\right)}v_x$$

$$\frac{d^2 s_y}{dt^2} = \frac{dv_y}{dt} = -g - k\sqrt{\left(v_x^2 + v_y^2\right)}v_y$$

(5.14)

Equation 5.14 represent the parametric equations of motion for a projectile under the influence of drag that is proportional to the square of the velocity. The equations are both coupled (i.e. they are dependent on each other) and are highly non-linear; two properties which account for the fact that a truly analytic solution has yet to be found. In fact, this is true for all models of drag which are proportional to any power of the velocity other than 1 (i.e. the linear case described in the previous section). A Runge–Kutta method may be employed, but a time-discrete model is provided in Model 5.2 – General trajectory model on the third worksheet tab, named: 'Trajectory model – Drag prop v^2' which is capable of good accuracy.

Equation 5.14 does allow terminal velocity calculations to be performed. For a projectile moving in a purely vertical direction where $v_x = 0$, Equation 5.14 reduces to:

$$\frac{d^2 s}{dt^2} = \frac{dv}{dt} = -g - k|v|v$$

which, at terminal velocity, when dv/dt reduces to zero, becomes:

$$-g = -k|v_T|v_T$$

$$v_T = \sqrt{\frac{g}{k}}$$

where v_T is the terminal velocity.

Various mathematical techniques have been pursued in order to obtain more useful values from Equation 5.14. One such study involves taking 'intrinsic coordinates' rather than Cartesian coordinates. These are moving coordinates where the origin maintains alignment with the projectile as it traverses its path. By this technique, an expression can be derived which relates the velocity of the projectile at any point in its flight with the angle of the projectile to the horizontal at that point. The equation is:

$$v = \left\{ \frac{\cos^2\theta}{v_0^2 \cos^2\theta_0} - \frac{k\cos^2\theta}{g}\left[\ln\left(\frac{\sec\theta + \tan\theta}{\sec\theta_0 + \tan\theta_0}\right) + \frac{\sin\theta}{\cos^2\theta} - \frac{\sin\theta_0}{\cos^2\theta_0}\right]\right\}^{-1/2}$$

From this, three further equations may be derived which, given the velocity and direction of the projectile at any point in the flight, allows the position and time into the flight to be deduced:

$$s_x = \frac{-1}{g} \int_{\theta_0}^{\theta} v^2 \, d\theta$$

$$s_y = \frac{-1}{g} \int_{\theta_0}^{\theta} v^2 \tan\theta \, d\theta$$

$$t = \frac{-1}{g} \int_{\theta_0}^{\theta} v^2 \sec\theta \, d\theta$$

where θ and v are the velocity and direction of the projectile at the time t.

These three equations have been found to be most useful in modelling projectiles subject to the v^2 drag law.

5.5.3 The ballistic coefficient and ballistic tables

As we have seen, if we assume linear drag, equations may be derived analytically for most flight parameters, albeit with some difficulty. For the 'drag $\propto v^2$' relationship, equation derivations are no longer possible, although some useful properties can be gleaned from some of the derived expressions. Matters become even worse for higher orders of non-linearity where $n > 2$ in Equation 5.4.

However, if one is only trying to obtain accurate values of range and time of flight, help is at hand. Ballistic tables have been available since the end of the eighteenth century when the concept of 'quadratic ballistics' as it was called, was developed out of earlier work by Huygens and Newton. The tables were refined and used extensively throughout the Great War. They were also used during the Second World War but, by then, their application was supported by the earliest computers. The legendary ENIAC in particular was fully equipped to handle ballistic tables and predict shell trajectories with good accuracy.

Table 5.3 is a modified version which allows predictions for spherical projectiles (rather than the elongated and pointed shape of typical ordnance). The values in the tables are derived from a mix of complex mathematical analysis and empirical data. Note that the values given are for non-spinning balls only.

The application of the table to obtain values of range and time is presented by means of a numbered 'recipe' for clarity.

1 First, calculate the ballistic coefficient, C, which, for non-spinning spherical projectiles only, is given by:

$$C = \frac{1}{C_D} \frac{m}{d^2} \tag{5.15}$$

noting that m/d^2 for a given type of ball can be found from Table 5.1, while C_D is the drag coefficient.

Table 5.3 Ballistic tables for range and time of flight

Z	A	T'	Z	A	T'
0.00	0.0	0.00			
0.20	2.2	0.22	1.00	15.0	1.39
0.25	2.8	0.28	1.05	16.1	1.47
0.30	3.5	0.35	1.10	17.2	1.57
0.35	4.1	0.41	1.15	18.4	1.66
0.40	4.8	0.48	1.20	19.6	1.76
0.45	5.6	0.55	1.25	20.8	1.86
0.50	6.3	0.62	1.30	22.1	1.96
0.55	7.1	0.69	1.35	23.4	2.06
0.60	7.8	0.76	1.40	24.7	2.17
0.65	8.7	0.83	1.45	26.1	2.27
0.70	9.5	0.90	1.50	27.6	2.38
0.75	10.3	0.98	1.55	29.0	2.49
0.80	11.2	1.06	1.60	30.4	2.60
0.85	12.1	1.14	1.65	31.9	2.71
0.90	13.0	1.22	1.70	33.5	2.82
0.95	14.0	1.30	1.75	34.9	2.94
			1.80	36.4	3.06
			1.85	37.9	3.18
			1.90	39.4	3.29
			1.95	41.6	3.41

Extension to basic table

Z	A	T'
2.0	44	3.6
2.2	56	4.3
2.4	68	5.0
2.6	80	5.8
2.8	94	6.6
3.0	112	7.5
3.2	131	8.4

2   Calculate the function A from:

$$A = v_0^2 \frac{\sin 2\theta_0}{C} \tag{5.16}$$

where, as usual, v_0 and θ_0 are the velocity and angle of launch, respectively.

3   With reference to the ballistic table (Table 5.3) read off the corresponding values of Z and T' relating to the value of A derived from Equation 5.16.

4   The range of the ball is derived from:

$$R = CZ \tag{5.17}$$

5   The time of flight of the ball is derived from:

$$T = \frac{CT'}{v_0 \cos\theta_0} \tag{5.18}$$

Worked Example 5.5

Use the ballistic tables to calculate the range and time of flight of a cricket ball thrown at a speed of $35\,\mathrm{m\,s^{-1}}$ at an angle of $40°$ with the ground. Take the value of m/d^2 as $32\,\mathrm{kg\,m^{-2}}$ and assume the speed is below the critical Reynold's number and can be taken as 0.45.

So from Equation 5.15:

$$C = \frac{1}{C_D}\frac{m}{d^2} = \frac{32}{0.45} = 71$$

From Equation 5.16:

$$A = v_0^2\frac{\sin 2\theta_0}{C} = 35^2\frac{\sin 80}{71} = 17$$

Using the table, the values of Z and T' are:

$$Z = 1.08 \quad \text{and} \quad T' = 1.5$$

So from Equation 5.17, the range:

$$R = CZ = 71 \times 1.08 = 76.68\,\mathrm{m}$$

From Equation 5.18, the time of flight, T is given by:

$$T = \frac{CT'}{v_0\cos\theta_0} = \frac{71 \times 1.55}{35 \times \cos 40} = \frac{110.05}{26.81} = 4.1\,\mathrm{s}$$

5.6 Lift effects

We noted in Section 5.1 the similarity between the expressions for a projectile's drag force and its lift force, as indicated by Equations 5.1 and 5.2. We stated that the direction of the drag vector is always aligned with the projectile's changing direction as it progresses on its flight, but acts against that direction. By contrast, the lift force is always orthogonal to the drag force and, with the exception of the vertical throw, it will always have a component that opposes the gravitational force which will align when the projectile is at its peak.

However, for a non-spinning, smooth ball travelling through air, the lift coefficient C_L (and consequently the lift force) is negligibly small compared to the drag. This is not true, however, for a back-spun ball which will have a significant value of C_L, resulting in a noticeably altered flight. Similarly, top-spin reduces the lift coefficient to the point where the lift force acts downwards (!), adding to the gravitational force, again altering the flight trajectory significantly. Ball spin analysis of this nature is investigated in Chapter 6.

However, projectiles do not have to be spun to create lift. Lift coefficient values can also be significant for projectiles which are asymmetric in form either side of the projectile's direction vector. An obvious example is the aerofoil/plane wing which creates considerable lift because of its aerodynamic shape, which is clearly not symmetric about its direction of travel. Although not a sporting issue, it is worth noting that Equation 5.2 indicates that the lift generated is directly proportional to the air density, ρ. The density parameter reduces approximately linearly with altitude, which explains why every aircraft has an 'altitude ceiling'; the height at which the air density is such that the vehicle can no longer create enough lift to counteract its weight. The aerofoil shape is considered in some detail in Section 5.8.

Projectile asymmetry can even be produced in objects which actually have symmetric form. Consider a crude balsa wood toy plane for instance. Its wing is not carved or moulded. It is a simple sheet of wood but it is held at a small angle to the direction of travel. This is known as the yaw angle of the projectile. In such cases, the lift is proportional to the sine of the yaw angle. So, for small angles, we can say that the lift is proportional to the yaw angle. Unfortunately, the pay-off is that the drag will also increase with yaw angle. As the angle increases, the lift increases. But, once a critical value is reached, the lift force drops suddenly and the projectile stalls, usually resulting in it plummeting to the ground.

The application of yaw to create lift is utilized in the throwing of the javelin. The javelin is thrown so that its point is angled upwards by a few degrees above its throw angle. For a good throw, that difference between angle of the javelin axis and trajectory angle should be maintained for the duration of the flight, thereby maintaining a constant lift. It should be noted that this technique creates greater drag than throwing the javelin with its axis along the launch angle (zero yaw), but the added lift keeps the javelin in the air longer which, under certain preferential circumstances, can result in the greater range.

5.7 Headwind, tailwind and crosswind

Dealing with headwind and tailwind mathematically can be quite simple. To a first approximation, wind flows horizontally only, and therefore only the x-component of an appropriate parametric equation is affected. Taking the simplest case of no drag other than a headwind or tailwind, we can modify Equation 3.3 to:

$$s_x = v_x t = v_0 t \cos \theta_0 - \varsigma\, t^2$$

where ς is a parameter representing a constant wind force over the whole of the flight. Simple vector analysis for this basic case leads to the resulting throw vector relative to the headwind, v_w, given by:

$$v_w^2 = v_0^2 + 2v_0 \varsigma \cos \theta_0 + \varsigma^2$$

while:

$$\tan \phi = \frac{v_0 \sin \theta_0}{v_0 \cos \theta_0 + \varsigma}$$

where $\tan \phi$ is the angle of the ball relative to the headwind.

More accurately, however, the wind can be modelled by what meteorologists refer to as a laminar flow. A laminar wind flow means that the wind increases in a linear manner with the altitude, s_y. Again, in a parametric piece-wise manner, this is easy to model. Finally, the most accurate model is the $\log_e$ model, in which the wind increases as a function $\log_e(s_y)$. The x-component of the parametric equation of motion then becomes:

$$s_x = v_x t = v_0 t \cos \theta_0 - \varsigma \ln (s_y) t^2$$

Of course, the y-component remains as Equation 3.4:

$$s_y = v_y t - \tfrac{1}{2}at^2 = v_0 t \sin \theta_0 - \tfrac{1}{2}at^2$$

These parametric equations may be plotted using Model 5.2 – General trajectory model, in which trajectories can be studied for all three wind functions. It may be seen that, as the variables are adjusted, the trajectory profiles quickly become very oddly shaped, for all but lightest, breezy conditions. In practice, although it is safe to assume wind vectors are always horizontal to the ground, accurate models cannot be constructed due to variations in wind with time (short gusts) and over distances that are small compared with most sports trajectory ranges. This may be demonstrated by studying the corner flags at soccer matches. Not only does the degree of flapping vary constantly and over short time durations, but, while one flag may lie flaccid for a while, others may be extended to their fullest, caught in a short blast of air.

Crosswinds of moderate strength or less do not affect the range of most balls very much (their drag-to weight ratio is too small). However, they can deflect their line quite considerably, especially in the case of large balls such as soccer balls. In the case of a crosswind which does not vary over the duration of the flight and is constant with height (unlikely!), the horizontal deflection would be a straight line at an angle to the intended target. However, because the horizontal component usually reduces over the duration of the flight due to natural drag effects, the deviation traces out the curved path indicated in Figure 5.4b.

With reference to Figure 5.4a, assume a ball is propelled along OA with a velocity of v_0 and at a launch angle of θ_0 to the horizontal, with the wind travelling from left to right, of value ς.

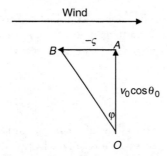

(a) Direction in which ball is propelled

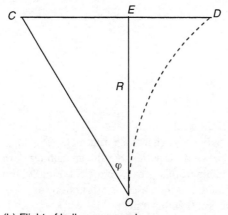

(b) Flight of ball over ground

Figure 5.4 (a, b) Crosswind vector calculations.

The forward horizontal velocity vector OA will have a magnitude $v_0 \cos \theta_0$. Vector OB will be the velocity relative to the airflow, and the angle φ will be given by:

$$\tan \varphi = \frac{\zeta}{v_0 \cos \theta_0}$$

Now, with reference to Figure 5.4b the path of the ball relative to the air is represented by the vector OC at an angle φ with the direction OE in which it is actually propelled. However, the ball will be carried over to D; the sideways deflection being given by the vector ED. For small deflections we can assume that the range R is the same as the deflected path. Now, if the time of flight is given as t, wind speed is ζ then CD is equal to ζt. So the deflection $ED = CD - CE = \zeta t - R \tan \varphi_0$:

$$= \zeta t - \frac{R \zeta}{v_0 \cos \theta_0} = \zeta \left(t - \frac{R}{v_0 \cos \theta_0} \right) \tag{5.19}$$

Worked Example 5.6

In an American football punt a ball is kicked at $60°$ to the horizontal, at $35\,\mathrm{m\,s^{-1}}$. The range is $22\,\mathrm{m}$ and the time of flight is $0.6\,\mathrm{s}$. If it is kicked in a constant crosswind of $8\,\mathrm{m\,s^{-1}}$, calculate its deflection.

So from Equation 5.19: the deflection, ED is given by:

$$ED = \zeta \left(t - \frac{R}{v_0 \cos \theta_0} \right) = 8 \left(0.6 - \frac{22}{35 \cos 60} \right)$$

$$= -8 \times 0.66 = -5.26\,\mathrm{m}$$

5.8 Trajectory diversions

5.8.1 The Bernoulli fallacy

In the light of what we have learnt in this chapter, we may think we can easily understand the workings of a plane wing. It's all down to our Bernoulli expression; isn't it? Our understanding may go something like this.

The airflow splits at the leading edge of the wing; some follows the lower edge, while the rest follows the upper edge which travels the greater distance as it has to negotiate the curved profile of the upper edge. The air must meet again at the trailing edge of the wing at the same time to allow a smooth continuance of flow as indicated by Figure 5.5.

We can, therefore, safely assume the air flowing over the upper surface must be travelling faster than the air travelling along the lower surface. So far so good?

Right, now we know from Bernoulli's principle that air pressure is lower where it travels faster. There is, therefore, a higher pressure on the lower surface of the wing than on the upper surface. The result? The plane flies.

… er … no!

Although, in principle, there is some lift provided by this method, as we will see, it is nowhere near enough. This is known as the 'hump' principle of lift and is still wrongly

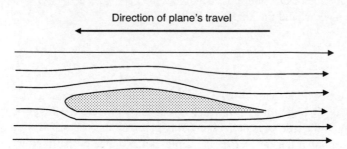

Direction of plane's travel

Figure 5.5 Airflow over an aerofoil/wing.

taught as *the* principle of flight in many quarters. There are some fundamental difficulties with the theory.

First, it violates the law of conservation of momentum. For lift to occur, there must be a net flow of air displaced downwards by the wing. Consider a helicopter on the deck of a carrier. Would the carrier be lighter if the pilot set the rotors in motion? Of course not! The lift force is matched by the downward force against the deck. In fact, the pilot of the carrier is only aware that his ship is lighter when the chopper has taken off and flown quite high or clear of the flight deck. Another example: would a racing pigeon transporter van be lighter if all the pigeons left their perch and hovered in the middle of their cages? I rest my case.

A second point; cheap balsa wood planes and folded paper planes seem to have pretty good flight characteristics without the need to involve the traditional aerofoil shape of Figure 5.5.

Finally, let me prove once and for all that the hump theory doesn't hold water by way of a (fairly) realistic calculation involving the largest airliner now flying; the Airbus A380. It's a fabulous machine without doubt, but can Bernoulli, and Bernoulli alone, keep it flying? First, some Airbus statistics (data is approximate and variable, dependent on various flight conditions such as altitude and fuel load, and also on the particular model type):

Mass = 560 tonnes (= 560×10^3 kg)
Wing area = 845 m^2
Flight speed = mach 0.89 (=$\sim$ 300 m s^{-1})
Difference in surface path between upper surface and lower surface of wing = 2%

Now a simple form of the Bernoulli expression of particular use to aerodynamic engineers states:

$$\Delta p = \tfrac{1}{2}\rho\left(v^2 - u^2\right)$$

where Δp is the difference in pressure between the upper and lower surface of the wing; ρ is the density of air which varies inversely with altitude and with temperature but for the purposes of this calculation we will take it to be 1.2 kg m^{-3}; v is the velocity of air over the lower surface of the wing (= the air speed); and u is the velocity of air over the upper surface of the wing.
So:

$$\Delta p = \tfrac{1}{2} \times 1.2 \times \left(300^2 - (300 \times 1.02)^2\right) = 2.18 \times 10^3 \text{Pa}$$

The lift force will be 2.18×10^3 Pa times the total area of the wing, 845 m^2.

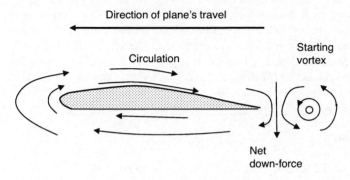

Direction of plane's travel

Circulation

Starting
vortex

Net
down-force

Figure 5.6 Airflow over an aerofoil/wing with vortices.

The lift force due to the hump principle alone is 1.84×10^6 N. Quite a force!

But, the Airbus A380 pays a price for its double-decker design with banqueting suite, gymnasia and the like; its weight is $560 \times 10^3 \times 9.81$ N $= 5.5 \times 10^6$ N; a force three times greater than the lift provided by the Bernoulli method. In fact, for the Airbus to gain enough lift to fly by hump principle alone, it would need to attain a speed of $518\,\mathrm{m\,s^{-1}}$. Not only is this getting on for twice the speed it is capable of, but sadly, at mach 1.5, it would need to fly beyond the sound barrier, and good though the Airbus is, it isn't Concorde!

So enough of the fallacy; what of the truth of flight? Well, first, in the preceding argument we stated that the air splits at the wing's leading edge and recombines at the trailing edge smoothly – this is known as the theory of equal transit time and, for aero-foils, this *cannot* be assumed. Yes, the air does travel over the upper surface of the wing faster than the lower surface, but not fast enough to recombine with the air it split from initially – there is a lag between the two airstreams which produces a vortex beyond the wing's trailing edge as shown in Figure 5.6.

This vortex, known as a 'starting vortex' causes a contra-flow circulation of air which travels right around the wing, upwards at the leading edge and downwards at the trailing edge. Figure 5.6 shows this circulation tendency, but note that it is not a complete and true circulation as simplistically represented in the figure; rather it is a small circulatory airflow which is superimposed on the straight airflow from left to right which actually dominates the flow.

Two points are relevant here. First, air certainly does move faster on the leading edge of the wing than the trailing edge, so the Bernoulli principle leads to the wing being 'sucked' though the airstream, which considerably reduces its drag. Second, the two contra-rotating vortices do actually lead to a net downward deviation of air flow; this is lift which does not violate the conservation of energy, and of a magnitude enough to raise the behemoth from the ground.

Summary

In this chapter we look at how the basic projectile flight trajectory is modified by the effects of both drag and lift. We see how drag and lift forces are generally dependent on the square of the velocity, as well as other parameters associated with the projectile and the fluid it is flowing through. We look, in qualitative terms, at the different types of drag and show that, beyond a certain critical velocity, chaotic drag can take over from laminar drag, resulting in an almost instantaneous drop in drag value.

Mathematically, we derive the modified equations of motion for projectiles subject to drag from, first a linear resisting medium, and then the more complex case of the resistance proportional to the square of the velocity. In the latter case, analytic solutions are not possible, and so alternative, numeric and graphic solutions are presented. The effects of headwind, tailwind and crosswind are considered, and a surprising result is stated that, a projectile travelling into a headwind and subject to lift may, under certain circumstances, fly further than one which cannot create a lift force.

Problems and questions

1   In the worked Example 5.2 and in Model 5.1 – Calculation of R_e for range of projectiles, we note that both the javelin and the archer's arrow typically do not achieve critical velocity even though, in the case of the arrow, its velocity is around $70 \, \text{m s}^{-1}$. In both cases we assumed a value of C_D of 0.45, rather than the lower value of 0.2, indicating that both projectiles' drag is of a laminar rather than a chaotic nature. Why is this when the velocities, are so high?

2   Use Model 5.1 – Calculation of R_e for range of projectiles, to compare the Reynold's number and the drag-to-weight ratio for a rugby ball kicked along its long axis with a ball kicked with the same velocity broadside. Assume ellipse area of the ball to be 0.045 m and the representative length to be 0.286 m.

3   Using the linear resistive model find the point (in x, y coordinates) where a projectile is moving in a direction 90° to its initial launch direction, if the launch velocity is $30 \, \text{m s}^{-1}$, the launch angle is 38° and $k = 1.1$. (Hint: This is the point where the projectile grazes the bounding curve, so use Equations 5.10 and 5.11.)

4   Assuming travel though a linear resistive medium, calculate the resistance coefficient per unit mass, k, for, (i) a raindrop whose terminal velocity is $6.7 \, \text{m s}^{-1}$, (ii) a golf ball whose terminal velocity is $40 \, \text{m s}^{-1}$, and (iii) a table-tennis ball whose terminal velocity is $9 \, \text{m s}^{-1}$.

5   Repeat the calculations of Question 4 but for the resistive medium which is proportional to v^2.

6   Use the ballistic table (Table 5.3) to calculate the initial velocity required to kick a football 40 m if it is launched at 45° to the horizontal. Assume the critical Reynold's number is exceeded throughout the flight, so $C_D = 0.2$ and the m/d^2 value for a soccer ball is $8.8 \, \text{kg m}^{-2}$.

Chapter 6

The effects of spin

6.1 Introduction

Having investigated flight deviations of projectiles due to motion through viscous media (both still air, and air flowing in the different directions relating to headwind, tailwind and crosswind), we now turn our attention to flight deviations caused by projectile rotations.

Deliberate spin deviations in sport are usually applied in order to deceive an opponent by creating a degree of unpredictability in the flight of the projectile. To this end, the bowler or pitcher in the games of cricket and baseball respectively will attempt, where possible, to hide both the direction and the degree of applied spin from the facing opponent. Specifically in cricket, the batsman may be further confused by an unpredictable bounce off the pitch, due to ball spin (which may be further complemented by any surface undulations at the point of bounce).

In many ball sports, topspin or backspin may be applied with the intention of altering the time of flight. Backspin will allow the ball to 'hang' longer in the air, and so travel further; a technique used in golf drives to increase the ball's range. Topspin, on the other hand, will reduce flight time and cause the parabolic curve to deviate in a manner that emphasizes and tightens the radius of the curve. In tennis, this allows the ball to traverse the court faster while both clearing the net, and landing within the base line. The result is, hopefully, a faster legal shot than may be attained from an otherwise un-spun ball.

One other common application of spin is in obstacle avoidance. Snooker players will apply spin to the cue ball, causing it to swerve around, and not touch, an illegal ball. The cue ball should then continue onward to hit the intended ball, thereby avoiding the ubiquitous 'snooker'. A similar strategy, which may involve a deviation from the simple parabolic trajectory in all three dimensions, may be applied in the case of a soccer free kick being taken close to the attacking goal. The defenders might form a wall of bodies as a barrier between the point of launch and the goal. The talented soccer player may attempt to spin the ball so that it curves around the side of the wall, or, alternatively, with judicious application of topspin, cause the ball to dip sharply once it has cleared the top of the wall, confounding the goalkeeper into a saving misjudgement.

Of course, not all ball spins are intentional. A misjudged hit with a racket or golf club will not only cause the ball to shoot off in an unexpected direction; it will typically also spin the ball about a vertical axis, causing greater embarrassment as the ball veers even further from target. However, note the scenario discussed in Chapter 10. Here, the convex face of golf club-heads in current use, impart a corrective influence on the slice shot. A ball which is accidentally sliced off to the right, for instance, will be caused to spin in a counter-clockwise direction (looking from above), which will produce an inward curving ball. This, to a greater or lesser extent, will result in a correction of the miss-hit ball.

It was in 1672 that Sir Isaac Newton at the youthful age of 23 noted the deviating effect on the flight of a tennis ball when launched with a spin about an axis perpendicular to its direction of motion. He surmised:

> For, a circular as well as a progressive motion ..., its parts on that side, where the motions conspire, must press and beat the contiguous air more violently than the other, and there excite a reluctancy and reaction of the air proportionably greater.

Later, in 1742, Benjamin Robins experimentally showed that a rotating sphere could create a force at 90° to the axis of rotation. However, it is interesting to note that the distinguished polymath, Euler, completely rejected the possibility of such a spin force. However, in a paper by Lord Rayleigh in 1877, entitled, 'On the irregular flight of a tennis ball', Magnus is first credited with an analytical explanation of the effect. It has henceforth become known as the Magnus effect. Magnus, rather than using a rotating sphere, mounted a cylinder such that it was perpendicular to the airflow. When he set the cylinder into rotation about its central axis, he could accurately measure the force produced in a direction that was both perpendicular to the airflow direction, and to the spin axis direction.

Rayleigh's analysis for a 'frictionless fluid' stated that the spin force was both proportional to the velocity of the fluid flow passing the cylinder *and* the angular speed of the cylinder, thereby confirming that Euler, for once, did indeed get something wrong!

6.2 Basic principles

Figure 6.1 shows how air flows over a ball which is moving from right to left at a velocity v, while, at the same time, spinning in a clockwise direction at an angular velocity ω. The resulting lift force, F_L, is shown acting orthogonally to the ball's direction vector.

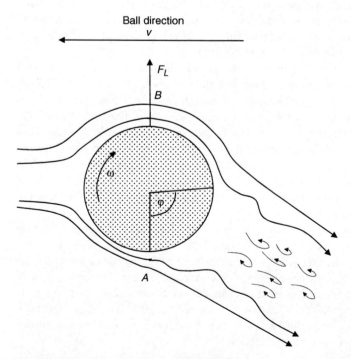

Figure 6.1 Flow lines around a spinning ball.

In a manner similar to the explanation of the aerofoil discussed in Section 5.8, it is conceptually possible to partially account for the lift force created by spin as an application of the Bernoulli effect.

The argument would be as follows. When the ball spins, the layer of air closest to the surface of the ball will be dragged around with it. Below the ball (marked 'A' in Figure 6.1), the thin layer of air will act in contra-flow with the surrounding air as the ball advances from right to left, and the resulting airflow will slow down. Conversely, above the ball (marked 'B' in Figure 6.1), the air layer will be swept along in the airflow and so its velocity will increase. Now, the Bernoulli effect states that the pressure is greater where the velocity is less, so a net pressure differential is created between the top and bottom of the ball, with the higher pressure present below it. This is how it was thought that the lift force is generated.

To be fair, under certain circumstances, such as in the case of slow moving, smooth(ish) spinning spheres, a small lift force may be created by the Bernoulli effect as described. However, for most practical sporting cases, the Bernoulli effect alone cannot account for the major deviations witnessed in some games. Furthermore, the simple argument, as stated, contravenes the law of conservation of momentum, in as much as, for the ball to rise, something (i.e. the airflow) must be forced downwards, and the Bernoulli explanation alone does not allow for this.

It should be stated that there are many other shortcomings in the Bernoulli explanation. Notwithstanding, for many years, it remained the accepted explanation for the spin effect and was quoted by fluid dynamicists and mechanical engineers the world over.

The modern, more accurate explanation of lift due to spin is ascribed to an asymmetric boundary layer separation phenomenon and has become known as the Magnus effect.

We have learnt from the previous chapter that airflow past an object broadly fall into one of two categories; laminar, or turbulent (chaotic) flow. The type of flow that manifests is very dependent on the flow speed. The critical Reynold's number defines the point at which the flow switches from laminar to turbulent. We also know that, although drag increases with projectile velocity, once the critical Reynold's point is reached and the flow changes from laminar to chaotic, the drag may reduce suddenly and significantly only to start rising again as the projectile velocity continues to increase. Attributed to each of these flow types is a pressure drag factor: a parameter which describes how the airflow 'sucks' at the surface of the projectile.

It must be stated that this is a rather over-simplified explanation of the actual state of affairs, but it will suffice to explain and justify the existence of the Magnus force. It will not, unfortunately provide sufficient quantitative detail to aid the development of an accurate analytical model of the spin trajectory.

The creation of the Magnus force is explained as follows. As the spinning ball moves through the air with a velocity such that both sides ('A' and 'B' in Figure 6.1) are beyond the critical velocity (i.e. both sides are in turbulent flow mode), the boundary layer separation point will advance forward on the side of the ball that is spinning in opposition to the airflow ('A' in Figure 6.1). This is due to a reduction in pressure drag in this region. Conversely, the boundary layer will retreat further to the back of the ball on the side where the spin is moving with the airflow ('B' in Figure 6.1), as the increased pressure drag holds the boundary layer close to the ball for longer.

The asymmetric boundary layer separation due to the ball spin leads to a deflection of the wake behind the ball. Now Newton's third law tells us that there must be a reactive force acting upwards as a consequence of the downward pointing deflected wake. That is, the momentum change due to the wake deflection is balanced by a momentum change in the ball causing it to move upwards. This is the Magnus force.

At smaller Reynold's numbers (possibly corresponding to a slower speed, a smaller ball, a higher viscosity fluid or a smooth ball, which is rotating slowly) a small reverse Magnus effect may occur. Here the boundary layer on the side moving with the flow is laminar ('B' in Figure 6.1) since the net-airflow velocity is below the Reynold's critical number. However, the boundary layer on the side moving against the flow ('A' in Figure 6.1) may have a relative velocity that is greater than the Reynold's critical number and is turbulent. The turbulent boundary layer now has the higher pressure drag and will separate later, causing the wake to deflect upwards (in Figure 6.1), resulting in a small Magnus force in the opposite direction. This effect can be seen over a range of golf drive velocities when a *smooth* (dimple-less) golf ball is used. If the usual backspin is applied to such a drive, the ball will be seen rising quickly and then diving into the ground resulting in a considerable reduction in expected range.

In reality, practical balls are seldom perfectly spherical and usually possess various stitching patterns or dimples. As a consequence, varying and complex combinations of laminar and turbulent boundary layers and eddies can be created around the ball which, under certain circumstances, can produce some unusual and impressive spin effects. Although the details of the airflow and the consequence on ball trajectories are seldom appreciated by the sports practitioner, elite players do possess a unique feel for how to produce optimal spin, and the most beneficial use of spin induced ball deviations in their game.

6.3 A quasi-empirical mathematical approximation of ball spin analysis

The analysis outlined below utilizes a combination of empirical values obtained from measuring forces on a variety of spinning balls and then incorporating those values into a modification of the trajectory equations. The result is a model that shows good accuracy over a range of ball types and launch angles and velocities.

The experimental data is taken from work performed by MacColl (1928), who investigated smooth spheres spinning in a wind tunnel, Tait (1893) and, later, Davies (1949) who measured lift parameters on golf balls, and finally, Briggs (1959) who measured the lift parameter on baseballs. All results bear a close resemblance to each other and the graph, Figure 6.2, represents a consolidation of the values obtained.

The following mathematical analysis is a modification of the work presented by Tait (1893) and cited in Daish (1972), and was intended to analyse the flight of a back-spinning golf ball

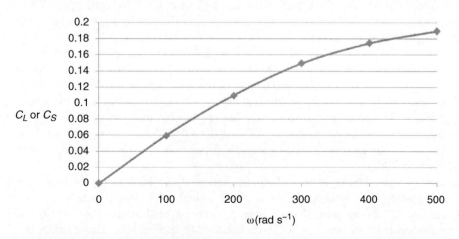

Figure 6.2 Lift or sideways coefficient versus spin (ω).

driven by a club. However, so long as the stated assumptions are adhered to, the model maintains good accuracy for many other flight trajectories involving spin deviations.

Consider the lift and drag forces on a spinning ball to be L and D respectively. As before, we assume the drag to be proportional to v^2, but the lift to be directly proportional to v.

So:

$$D = k_1 v^2 = \tfrac{1}{2} C_D \rho A v^2 \quad \text{and} \quad L = k_2 v = \tfrac{1}{2} C_L \rho A v \tag{6.1}$$

where C_D and C_L are the drag and lift coefficients (values obtained from Figures 5.2 and Figure 6.2), respectively, ρ is the density of the fluid, A is the cross-sectional area of the projectile and v is the projectile's velocity.

Let us define two 'Tait' coefficients, κ_1 and κ_2 as:

$$\kappa_1 = \frac{m}{k_1} \quad \text{and} \quad \kappa_2 = \frac{k_2}{m}$$

The retardation due to the drag force will then be:

$$\frac{D}{m} = \frac{v^2}{\kappa_1} \tag{6.2}$$

and the acceleration due to the lift force is:

$$\frac{L}{m} = \kappa_2 v \tag{6.3}$$

With reference to Figure 6.3, the force along the projectile direction vector equates to:

$$-m\frac{dv}{dt} = m\frac{v^2}{\kappa_1} + mg\sin\theta$$

$$-\frac{dv}{dt} = \frac{v^2}{\kappa_1} + g\sin\theta \tag{6.4}$$

The net acceleration orthogonal to the trajectory will be $(\kappa_2 v - g\cos\theta)$. If this acceleration is assumed constant then the projectile is destined to travel on a perpetual circular path.

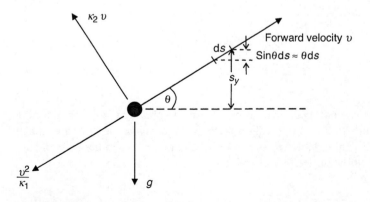

Figure 6.3 Forces on a projectile subject to both lift and drag.

However, a closer approximation is for the lift force to reduce over the duration of the flight which results in a path of increasing radius; a sector of a spiral. Keeping with the more simple case of the constant acceleration and the circular path, the force turning the projectile through the circle must be balanced by a centripetal force given by $mr\,d\theta/dt$. The centripetal acceleration will then be given by:

$$\left(\kappa_2 v - g\cos\theta\right) = v\frac{d\theta}{dt}$$

So:

$$\frac{d\theta}{dt} = \kappa_2 - \frac{g\cos\theta}{v} \tag{6.5}$$

In order to obtain the equations of motion for this spinning projectile it is necessary to integrate Equations 6.4 and 6.5. This may be carried out by a number of computational methods, but for our purposes, we will assume that θ is small, thereby allowing an analytical solution to be extracted. Although this assumption may appear rather limiting, it is worth remembering that, although many trajectories in sport may have started their flight with launch angles of greater than, say 30°, the projectile will still be less than 10° for much of its flight. If θ is assumed small then we may say that $\sin\theta \approx 0$ and $\cos\theta \approx 1$. Equations 6.4 and 6.5 now become:

$$\frac{dv}{dt} = -\frac{v^2}{\kappa_1} \tag{6.6}$$

and:

$$\frac{d\theta}{dt} = \kappa_2 - \frac{g}{v} \tag{6.7}$$

Rearranging Equation 6.6 we obtain:

$$\frac{dv}{v^2} = -\frac{dt}{\kappa_1}$$

which, when integrated with respect to time with the boundary condition that, at launch time, $t = 0$ and $v = v_0$ we obtain:

$$\frac{1}{v} - v_0 = \frac{t}{\kappa_1} \tag{6.8}$$

Combining Equation 6.8 with 6.7 by substituting for v gives:

$$\frac{d\theta}{dt} = \kappa_2 - \frac{gt}{\kappa_1} - \frac{g}{v_0}$$

This can now be integrated once again with respect to time to obtain an expression for the angle of the projectile at any time t into the flight for a given launch angle θ_0 and a launch velocity v_0. Values for the Tait variables κ_1 and κ_2 may be obtained experimentally and are based on the drag and lift of the projectile respectively:

$$\theta = \theta_0 + \left(\kappa_2 - \frac{g}{v_0}\right)t - \frac{gt^2}{2\kappa_1} \tag{6.9}$$

If the projectile moves an incremental distance ds in flight, then its corresponding incremental height will increase by an amount $\sin\theta\, ds$, which equals $\theta\, ds$ if we apply our assumption that θ is small. Its height s_y at any time t will then be given by:

$$s_y = \int_0^s \theta\, ds = \int_0^t \theta\frac{ds}{dt}dt = \int_0^t \theta v\, dt \tag{6.10}$$

From Equations 6.8 and 6.9 we obtain an expression for θv:

$$\theta v = \frac{\theta_0 \kappa_1 v_0}{v_0 t + \kappa_1} + \frac{(\kappa_1 \kappa_2 v_0 - g\kappa_1)}{v_0 t + \kappa_1}t - \frac{gt^2}{2\kappa_1}\frac{\kappa_1 v_0}{(v_0 t + \kappa_1)} \tag{6.11}$$

So, if Equation 6.11 is integrated with respect to t, we will obtain our expression for s_y. Each term will be considered separately. First:

$$\int_0^t \frac{\theta_0\kappa_1 v_0}{v_0 t + \kappa_1}dt = v_0\kappa_1\left[\ln\left(v_0 t + \kappa_1\right)\right]_0^t = v_0\kappa_1\ln\left(1 + \frac{v_0 t}{\kappa_1}\right)$$

The second term:

$$\int_0^t \frac{\kappa_1\left(\kappa_2 v_0 - g\right)}{v_0 t + \kappa_1}t\, dt = \int_0^t \frac{\kappa_1}{v_0}\left(v_0\kappa_2 - g\right)\left(1 - \frac{\kappa_1}{v_0 t + \kappa_1}\right)dt$$

$$= \frac{\kappa_1}{v_0}\left(v_0\kappa_2 - g\right)\left[t - \frac{\kappa_1}{v_0}\ln\left(v_0 t + \kappa_1\right)\right]_0^t$$

$$= \frac{\kappa_1}{v_0}\left(v_0\kappa_2 - g\right)t - \frac{\kappa_1^2}{v_0^2}\left(v_0\kappa_2 - g\right)\ln\left(1 + \frac{v_0 t}{\kappa_1}\right)$$

And the final term:

$$\int_0^t \frac{gt^2 v_0}{2\left(v_0 t + \kappa_1\right)}dt = \frac{g}{2}\int_0^t \left[t - \frac{\kappa_1}{v_0}\left(1 - \frac{\kappa_1}{v_0 t + \kappa_1}\right)\right]dt$$

$$= \frac{g}{2}\left[\frac{t^2}{2} - \frac{\kappa_1 t}{v_0} + \frac{\kappa_1^2}{v_0^2}\ln\left(v_0 t + \kappa_1\right)\right]_0^t$$

$$= \frac{g}{2}\left[\frac{t^2}{2} - \frac{\kappa_1 t}{v_0} + \frac{\kappa_1^2}{v_0^2}\ln\left(1 + \frac{v_0 t}{\kappa_1}\right)\right]$$

Combining the three terms and from Equation 6.10 we obtain an expression for s_y:

$$s_y = \left[\kappa_1\theta_0 - \frac{\kappa_1^2}{v_0^2}\left(v_0\kappa_2 - g\right) - \frac{\kappa_1^2 g}{2v_0^2}\right]\ln\left(1 + \frac{v_0 t}{\kappa_1}\right) + \frac{\kappa_1\left(2v_0\kappa_2 - g\right)}{2v_0}t - \frac{g}{4}t^2 \tag{6.12}$$

and by a similar (but much simpler!) means, and again assuming a small trajectory angle such that $\cos\theta = 1$, the horizontal range at any point in time t, is given by:

$$s_x = \int_0^t v\,dt = \int_0^t \frac{\kappa_1 v_0}{v_0 t + \kappa_1}\,dt = \left[\kappa_1 \ln\left(v_0 t + \kappa_1\right)\right]_0^t = \kappa_1 \ln\left(1 + \frac{v_0 t}{\kappa_1}\right) \qquad (6.13)$$

Notice how only the y coordinate is actually dependent on the amount of lift caused by top or back spin: the variation of horizontal distance with time is unaffected by spin.

Model 6.1 – Drag and lift trajectory, evaluates Equations 6.12 and 6.13 and presents the results as a trajectory graph with launch height and angle, as well as the two Tait variables as adjustable parameters.

Worked Example 6.1

A golf ball is driven at a launch angle of $10°$ to the horizontal with a launch velocity of $70\,\mathrm{m\,s^{-1}}$. Calculate the horizontal and vertical position 2 seconds into the flight and 4 seconds into the flight, taking Tait variables, $\kappa_1 = 150$ and $\kappa_2 = 0.18$. How do these values compare with the 'dragless' and 'spinless' case?

For 2 s:

$$s_x = \kappa_1 \ln\left(1 + \frac{v_0 t}{\kappa_1}\right) = 150 \times \ln\left(1 + \frac{70 \times 2}{150}\right) = 98.89\,\mathrm{m}$$

$$s_y = \left[\kappa_1 \theta_0 - \frac{\kappa_1^2}{v_0^2}\left(v_0\kappa_2 - g\right) - \frac{\kappa_1^2 g}{2v_0^2}\right]\ln\left(1 + \frac{v_0 t}{\kappa_1}\right) + \frac{\kappa_1\left(2v_0\kappa_2 - g\right)}{2v_0}t - \frac{g}{4}t^2$$

$$= \left[150 \times 0.175 - \frac{150^2}{70^2}(70 \times 0.18 - 9.81) - \frac{150^2 \times 9.81}{2 \times 70^2}\right]\ln\left(1 + \frac{70 \times 2}{150}\right)$$

$$+ \frac{150(2 \times 70 \times 0.18 - 9.81)}{2 \times 70} \times 2 - \frac{9.81}{4} \times 2^2$$

$$= -9.155 \times 0.66 + 33.0 - 9.81 = 17.15\,\mathrm{m}$$

For 4 s:

$$s_x = \kappa_1 \ln\left(1 + \frac{v_0 t}{\kappa_1}\right) = 150 \times \ln\left(1 + \frac{70 \times 4}{150}\right) = 158\,\mathrm{m}$$

$$s_y = \left[\kappa_1 \theta_0 - \frac{\kappa_1^2}{v_0^2}\left(v_0\kappa_2 - g\right) - \frac{\kappa_1^2 g}{2v_0^2}\right]\ln\left(1 + \frac{v_0 t}{\kappa_1}\right) + \frac{\kappa_1\left(2v_0\kappa_2 - g\right)}{2v_0}t - \frac{g}{4}t^2$$

$$= \left[150 \times 0.175 - \frac{150^2}{70^2}(70 \times 0.18 - 9.81) - \frac{150^2 \times 9.81}{2 \times 70^2}\right]\ln\left(1 + \frac{70 \times 4}{150}\right)$$

$$+ \frac{150(2 \times 70 \times 0.18 - 9.81)}{2 \times 70} \times 4 - \frac{9.81}{4} \times 4^2$$

$$= -9.155 \times 1.05 + 66.0 - 39.24 = 17.08\,\mathrm{m}$$

Notice that the heights at 2 seconds and 4 seconds are similar – the height peaks at about 3 seconds and it is on its way down at 4 seconds. In addition, notice that there is clearly a drag effect since, if there was not, the range at 4 seconds would be double that at 2 seconds.

For the 'dragless' and 'liftless' case:

For $t = 2$ s, from Equation 3.3:

$$s_x = v_0 t \cos\theta_0 = 70 \times 2 \times \cos 10° = 137.88\,\text{m}$$

from Equation 3.4

$$s_y = v_0 t \sin\theta_0 - \tfrac{1}{2}at^2 = 70 \times 2 \times \sin 10° - \tfrac{1}{2} \times 9.81 \times 2^2 = 24.31 - 19.62$$

$$= 4.7\,\text{m}$$

and for $t = 4$ s:

$$s_x = v_0 t \cos\theta_0 = 70 \times 4 \times \cos 10° = 275.75\,\text{m}$$

$$s_y = v_0 t \sin\theta_0 - \tfrac{1}{2}at^2 = 70 \times 4 \times \sin 10° - \tfrac{1}{2} \times 9.81 \times 4^2 = 48.62 - 78.48$$

$$= -29.86\,\text{m}$$

So, after 4 seconds, the 'lossless', ' dragless' projectile has burrowed itself nearly 30 m into the ground, compared with the spun ball which, at 4 seconds, is still hanging in flight 17 m above ground level; quite a difference!

The values for the Tait variables in the above example are typical of back-spun sporting projectiles including tennis balls, table-tennis balls and golf balls. They are characterized by an initial approximately straight line path for about half the flight. The projectile then begins to curve and reach an apex about two-thirds into the flight and finally falling to the ground in only the final third to a quarter of the flight.

Using the more detailed, 0.1 second interval worksheet of Model 6.1 – Drag and lift trajectory, we can study the range of a projectile for varying launch velocities, all other factors held constant. Figure 6.4 shows this plot and, unlike the simpler, non-spinning 'dragless'

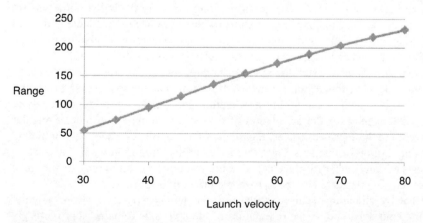

Figure 6.4 Plot of range versus launch velocity.

model in which the range is proportional to the square of the launch velocity (Equation 3.11), we now see that the range is an approximate linear function of launch velocity.

6.4 Gyroscopic stabilization

6.4.1 The basics

There is a further and completely separate spin phenomenon to consider which may be crucial in dictating the flight path of certain sporting projectiles, and, with correct understanding and appreciation, can be utilized to yield improved performance on the field.

Consider the launch of a discus into flight; a matter dealt with in some detail in Chapter 8. The projectile leaves the athlete's hand at speeds up to $30 \, \mathrm{m \, s^{-1}}$ with typical release angles lying somewhere between 35° and 40°. It is also launched such that the major plane of the discus lies a little above the launch angle. This allows further range to be extracted by utilizing the lift created by the asymmetric form shape on either side of the launch direction, a technique described in Section 5.6.

Now, as the discus progresses, it remains approximately in the same orientation over the whole of the flight path; this despite having to cope with a number of potential destabilizing forces. Such forces, which would tend to flip the discus into a somersaulting motion or chaotic disturbance, may be caused by factors such as: slight manufacturing variations in weight distribution around the central axis of the discus, wind variations striking from any direction, or a slight mis-throwing action that may be created by the athlete on release by a 'lazy' trailing finger. Indeed, the aerodynamic strategy of maintaining an angle between the rotating plane and the direction vector would itself, create a turning couple (or moment) which, in the absence of some opposing stabilizing force, would set the discus into a continuous back-flipping roll. In the case of an airplane, the tail-wing assembly prevents the back-rolling action of the aerofoil wing; obviously, a discus does not possess such a stabilizing assembly.

Let us now turn our attention to the firing of a bullet from a gun. The bullet is set into motion by a short, massive impulse to its rear end. This momentary but enormous force would tend to flip the bullet over if something did not prevent it. The tendency for the bullet to flip is not helped by the fact that, by virtue of the bullet being pointed at the front, its center of gravity is set somewhat backward of its mid-point. As the bullet is propelled forward, there would be a strong propensity for it to flip over as it attempts to bring its center of gravity further towards the front and into, what would be, a more stable flying position (c.f. throwing a dart flight-forwards). Finally, we may again consider manufacturing tolerances. If the bullet is not manufactured perfectly symmetrically about the long axis (say the bullet point was set slightly off-center) the strong asymmetric airstream produced by this lop-sidedness will again tend to flip the bullet into unwanted motion. Unlike darts, archer's arrows, or some missiles, the use of stabilizing flights or fins are not a practical solution to stabilizing a bullet's flight.

In both the examples above, the stabilizing force that prevents disastrous projectile motion is perhaps the most peculiar force found in nature. It is known as the gyroscopic force, and its controlling effect is known as gyroscopic stabilization. Put simply, if a body rotates about an axis, there is a tendency for the orientation of that axis to remain constant of angular movement in all of the three principle axes, regardless of any applied torque that might tend to cause a twisting motion of that axis. Note that it is only twisting motions of the rotating axis that are resisted: any linear displacements of the rotating axis along any of the three orthogonal axes remain unaffected by the gyroscopic effect of a spinning body.

We now see that, by launching the discus with spin, there will be a resistance to any angular movement of the spin axis, and so the orientation of the spin axis remains approximately constant over the duration of the flight. In the case of the bullet, the gun barrel has spiral grooves cut or etched into the inside tube; a practice known as rifling. This causes the bullet to exit the barrel spinning around its long axis, which again stabilizes any potential pitch (rotation about a horizontal axis perpendicular to the bullet's direction) or yaw (rotation about a vertical axis perpendicular to the bullet's direction). Obviously, the bullet's spin can be built up more efficiently and to a much greater extent in the case of a rifle with its long barrel, compared with that of a hand gun.

Further examples of gyroscopic forces in action include the obvious case of the spinning top and, more importantly from a sporting perspective, the balance of bicycles and motorbikes on two wheels; a feat which is much easier to accomplish when the wheels are spinning. In this latter case, it is worth emphasizing that the gyroscopic forces offer no resistance to forward motion, as the bike's travel only results in a linear displacement of the two wheels' axes. However, any tendency for the bike to tip over will be restricted by the gyroscopic force, as that motion would result in the wheels' axes tipping downward through an angle. However, the gyroscopic force *will* be noticeable as the racing biker leans into a corner, or even turns the front wheel in order to steer. Both these actions would have to take place *against* the resisting gyroscopic force.

6.4.2 Gyroscopic precession

The simplest and most obvious example of gyroscopic stabilization in action is a child's spinning top: it spins and so it stays upright. A more efficient variant of this apparatus is, not surprisingly, the gyroscope; a device capable of spinning much faster and storing much greater angular momentum in a disc of higher moment of inertia than the child's toy. The gyroscope can be used to demonstrate the full range of strange behaviours and tricks (after all, they are used by magicians) that make up the enigma that is the gyroscopic action.

Fundamental to the gyroscopic action is the motion known as precession. Gyroscopic precession naturally falls into two distinct types. The first, torque-induced precession, occurs when a vertically balanced spinning gyroscope, resting on a horizontal base, is given a slight knock at 90° to the upper part of the gyroscope's axle. Rather than fall over, the axle of the gyroscope will execute a circular motion around the vertical. If the gyroscope is fixed at the base, the axis will trace out an inverted conical figure with a certain period of rotation. See Figure 6.5a. The second type of precession is known as torque-free precession and can be witnessed when a badly thrown discus wobbles its way along its trajectory. In this case, the lower axle is not fixed, so the discus's 'virtual' spin axis will trace out a roughly double conical locus in its precessing motion. See Figure 6.5b.

To understand the phenomenon of torque induced precession, consider the gyroscope's flywheel to be constructed, not from a continuous disc of material, but from four point source masses on the end of four light spokes set at 90° to each other around the central axis. Referring to Figure 6.6 in which, this time, the structure is rotating around a horizontal axis, consider what happens as a torque is applied which tends to pull the axle down on the left and up on the right.

When such a torque is applied, the point mass labelled 'A' in Figure 6.6 will tend to move to the left, while the point mass labelled 'C' will tend to move to the right. If the flywheel were not rotating, the gyroscope would simply tip over. However, the flywheel is in constant motion so, at some later stage (in the second diagram of Figure 6.6), the masses will have moved through a 90° rotation. Now, Newton's first law of motion states that a body in motion continues to move at a constant speed along a straight line unless acted upon by an unbalanced force. Masses 'A' and 'C' have indeed been acted on by a force and so still want to move to the left and right respectively, *even though they have now moved through the 90° angle*.

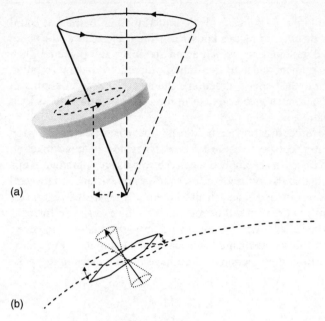

(a)

(b)

Figure 6.5 (a) Torque-induced precession: gyroscope in precessing motion. (b) Torque-free precession: the wobbly discus.

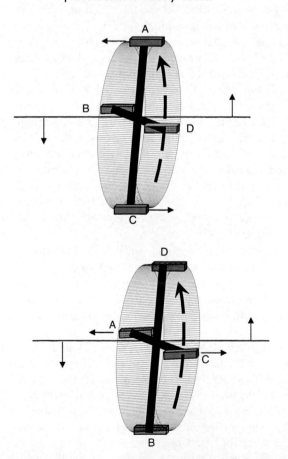

Figure 6.6 Forces on a gyroscope as a torque is applied to the axis.

So now there is a torque on the flywheel tending to turn it through an axis vertical to the page (in an anti-clockwise sense, looking from the top).

Indeed, at some time later, when the flywheel has turned through another 90°, the forces at masses 'A' and 'C' will still be acting in the same direction ('A' to the left, 'B' to the right), but their positions have fully swapped over. Now the torque is in the opposite sense to the original one, and so their motions cancel. If we now consider the mass to be distributed evenly around the flywheel again, over the period of one complete rotation, all the torques on the small elemental masses around the circumference of the wheel will cancel. The gyroscope will therefore not tip over, but remain hanging horizontally in the air, assuming one end of the axis is fixed, say, on a stand of some sort.

If there is a temporary mis-balance of applied forces, the gyroscope attempts to compensate by executing the precessing motion as described above. A full mathematical treatment of gyroscopic motion shows that there is also a minor, low period, low amplitude oscillation of the precession angle superimposed on the circular path, similar to that found in planetary motion. This motion is known as *nutation*, but for the purposes of sports projectile rotations, it can be safely ignored as being too small to affect trajectory motion.

6.4.3 Precession frequency

An expression for the gyroscopic precession frequency will reveal the factors which produce regular oscillations or wobbles in slightly destabilized discuses and Frisbees as they progress on their trajectory flight.

Torque-induced precession

We commence obtaining an expression for the torque-induced precession frequency by considering the rotational form of Newton's second law. The linear displacement form (for bodies of fixed mass) is, of course:

$$\mathbf{F} = \frac{d\mathbf{p}}{dt}$$

where $\mathbf{F}$ is the force vector and $\mathbf{p}$ the momentum vector. With reference to the conversion table, Table 2.5, this translates to the rotational form:

$$T = \frac{d\mathbf{L}}{dt}$$

where T is the torque, and $\mathbf{L}$ is the angular momentum vector.

Now, the torque, T, will be given by:

$$T = rmg$$

where m is the mass of the spinning body, g is the acceleration due to gravity and r is the horizontal component of the distance between the pivot point and the center of mass of the body. So:

$$d\mathbf{L} = rmg \, dt$$

The precessing motion of the body is caused by a created torque, which, at all times in the motion, acts both perpendicular to the weight vector, w, and to r (Figure 6.7a). It is caused

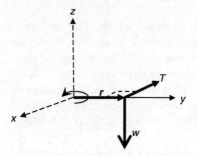

(a) The torque *T* is perpendicular to *r* and *w*

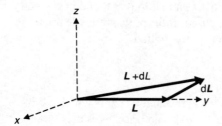

(b) Hence d**L** is perpendicular to **L**

Figure 6.7 (a, b) Momentum vectors associated with a precessing body.

by the rate of change of momentum, and the vector addition triangle indicated in Figure 6.7b shows that, during the time d*t*, the incremental angle the body precesses through, must equal the change in momentum divided by the old momentum.

The angle dθ through which the axis swings in a time d*t* is therefore given by:

$$d\theta = \frac{d\mathbf{L}}{\mathbf{L}} = \frac{rmg \, dt}{\mathbf{L}}$$

so:

$$\frac{d\theta}{dt} = \frac{rmg}{\mathbf{L}}$$

This is the angular velocity with which the axle rotates around the horizontal plane. It follows that the precession frequency is given by:

$$\omega_p = \frac{rmg}{\mathbf{L}} = \frac{rmg}{I\omega} \tag{6.14}$$

The period of precession is then given by:

$$\tau_p = \frac{2\pi}{\omega_p} = \frac{2\pi I\omega}{rmg} \tag{6.15}$$

So, the period of precession is proportional to the angular velocity of the flywheel's spin and its moment of inertia, while being inversely proportional to the flywheel's mass and the cosine of the angle of precession. If the flywheel spin-rate is great enough, precession can occur with a precession angle in the region of 90° (i.e. the axis of spin is around the horizontal), so long as one end of the axis is supported. See Section 6.5 for an interesting example of this. In such a circumstance, the period of precession is independent of small variations in precessing angle around the horizontal. However, in sporting applications, such spins are generally unattainable.

Torque-free precession

In order to derive an expression for the torque-free precession frequency, it is necessary to use Euler's equations, which relate torque and angular velocity for irregular three-dimensional bodies in three principle axes. For the purposes of this treatment, the equations are quoted, but proofs can be found in most analytical mechanics books.

If an irregular object is thrown through the air at an arbitrary angle and spin direction, its angular velocities in the three principle axes will be given by these equations. One may care to think of the complex spin and displacement motion of a tennis racket thrown with some vigour across the court following perhaps a poor service return!

If τ_1, τ_2 and τ_3, ω_1, ω_2 and ω_3, and I_1, I_2 and I_3 are the torques, angular velocities and moments of inertia, respectively, of a general three-dimensional body moving with general spin and displacement, then Euler's equations state:

$$\tau_1 = I_1 \frac{d\omega_1}{dt} + \omega_2\omega_3 \left(I_3 - I_2\right)$$

$$\tau_2 = I_2 \frac{d\omega_2}{dt} + \omega_1\omega_3 \left(I_1 - I_3\right)$$

$$\tau_3 = I_3 \frac{d\omega_3}{dt} + \omega_1\omega_2 \left(I_2 - I_1\right)$$

Now, in the case of a discus or Frisbee, we can say that the moment of inertia about the two principle axes that are at 90° to the axis of spin will be equal so we can re-label our moments of inertia: $I_s = I_3$, while $I_\perp = I_1 = I_2$, where I_s is the inertia around the normal spin axis and $I_\perp$ is the inertia around the other two orthogonal axes.

For torque-free rotation, $\tau_1 = \tau_2 = \tau_3 = 0$, so Euler's equations become:

$$I_\perp \frac{d\omega_1}{dt} + \omega_2\omega_3 \left(I_s - I_\perp\right) = 0 \tag{6.16}$$

$$I_\perp \frac{d\omega_2}{dt} + \omega_1\omega_3 \left(I_\perp - I_s\right) = 0 \tag{6.17}$$

$$I_s \frac{d\omega_3}{dt} = 0 \tag{6.18}$$

Integrating Equation 6.18 yields:

$$\omega_3 = \text{constant}$$

We may now define a new constant:

$$\omega_p = \frac{\omega_3 \left(I_s - I_\perp\right)}{I_\perp}$$

(This is one of those cases in mathematics where you begin to suspect someone has been here before, but please bear with it!)

Equations 6.16 and 6.17 may now be rewritten as:

$$\frac{d\omega_1}{dt} + \omega_p\omega_2 = 0 \tag{6.19}$$

$$\frac{d\omega_2}{dt} - \omega_p\omega_1 = 0 \tag{6.20}$$

Differentiating Equation 6.19 with respect to time:

$$\frac{d^2\omega_1}{dt^2} = -\omega_p\frac{d\omega_2}{dt}$$

and substituting into Equation 6.20 yields:

$$\frac{d^2\omega_1}{dt^2} = -\omega_p^2\omega_1 \tag{6.21}$$

Equation 6.21 is the standard equation for simple harmonic motion in which ω_1 has the solution:

$$\omega_1(t) = a\cos\omega_p t \tag{6.22}$$

where a is a constant representing the amplitude of the oscillation.

A similar treatment may be applied to ω_2 and by substituting Equation 6.22 into Equation 6.20 we obtain the complementary equation:

$$\omega_2(t) = a\sin\omega_p t \tag{6.23}$$

The 90° phase difference between Equations 6.22 and 6.23 indicate that the motion is circular in the ω_1/ω_2 plane with a radius, a, and it has an angular precession frequency ω_p (hence the earlier choice of symbol).

The angle α between the ω spin vector and the ω_3 direction (the symmetry axis) represents the angle of discus wobble and is given by:

$$\omega_3 = \omega\cos\alpha$$

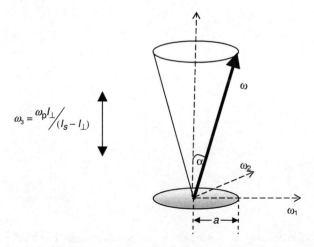

Figure 6.8 Torque free rotation.

and from the equation which defines ω_p:

$$\omega_p = \frac{\omega_3 \left(I_s - I_\perp \right)}{I_\perp}$$

we find:

$$\omega_p = \omega \cos\alpha \left[\left(\frac{I_s}{I_\perp} \right) - 1 \right] \tag{6.24}$$

Finally, as in the case of the torque-induced gyroscope the period of precession, τ_p is given by:

$$\tau_p = \frac{2\pi}{\omega_p}$$

so:

$$\tau_p = \frac{2\pi}{\omega \cos\alpha \left[\left(\frac{I_s}{I_\perp} \right) - 1 \right]} \tag{6.25}$$

Note that, in this torque-free case, the period of precession is inversely proportional to the spin rate of the disc. By contrast, in the case of the torque-induced gyroscope spin, Equation 6.15 indicates that the precession period increases with increasing flywheel spin.

Worked Example 6.2

Calculate the rate of wobble (i.e. the period of precession) for a discus which weighs 2 kg, thrown with a spin rate of 7 rev/s, when it is precessing at an angle of 10° to its spin axis.

From Equation 6.25:

$$\tau_p = \frac{2\pi}{\omega \cos\alpha \left[\left(\frac{I_s}{I_\perp} \right) - 1 \right]}$$

where, in this case, $\omega = 7 \times 2\pi \text{ rad s}^{-1}$ and $\alpha = 10°$

However, it would appear that the main complexity in this problem is in obtaining values for the moment of inertias, I_s and $I_\perp$ for the discus. Worked Example 6.3 includes the calculation for the derivation of I_s, and, as can be seen, it is not a trivial exercise. However, by utilizing the 'perpendicular axis theorem' we can neatly bypass the process of deriving the moment of inertias in calculating the precession period for thin bodies rotating about an axis at 90° to that plane.

The perpendicular axis theorem states that, for three dimensional, yet thin bodies, with their principle plane lying in the x, y direction, the moment of inertia in the z direction is given by:

$$I_z = I_x + I_y$$

Applying this to the discus, we can state:

$$I_s = I_1 + I_2 = 2I_\perp$$

since it is circularly symmetric about 1, 2. Now Equation 6.25 may be simplified to:

$$\tau_p = \frac{2\pi}{\omega \cos\alpha}$$

Therefore, if our projectile is both thin and planar (e.g. discus or Frisbee), its period of precession is independent of its three moments of inertia. So:

$$\tau_p = \frac{2\pi}{7 \times 2\pi \times \cos 10} = \frac{1}{6.9} = 0.15 \text{ precesses per second}$$

Worked Example 6.3

A discus of mass 2 kg is thrown with a release velocity of 35 m s^{-1} and a spin rate of 7 revs/s. Calculate the total kinetic energy on release.

The total kinetic energy is made up of the linear kinetic energy given by $\frac{1}{2}mv^2$ and the rotational kinetic energy, given by $\frac{1}{2}I\omega^2$.

Linear kinetic energy $= \frac{1}{2} \times 2 \times 35^2 = 1225$ J.

We next need to calculate the moment of inertia around the spin axis, I_s.

The approximate dimensions of a men's competition discus are shown (see Figure 6.9).

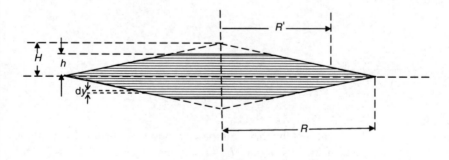

$m = 2\,\text{kg}$

$R = 0.11\,\text{m}$

$h = 0.023\,\text{m}$

$H = 0.032\,\text{m}$ (calculated by geometry)

$\rho = 2511\,\text{kg m}^{-3}$ (calculated by m/v where v is calculated from the volume of cones)

Figure 6.9 Men's competition discus.

Consider the discus to be constructed from two truncated cones joined at their bases. With regard to the upper cone only, we may consider this to be made up of a series of horizontally sliced discs mounted vertically, each of radius R' and with an incremental width given by dy.

The moment of inertia of the upper half of the discus will be the sum of the inertias of each slice:

$$I = \int I_{\text{slice}} \, dR'$$

The moment of inertia of each disc is given from Table 2.6 as:

$$I = \tfrac{1}{2} m_{\text{slice}} R'^2$$

So, for the upper half of the discus:

$$I = \int \tfrac{1}{2} m_{\text{slice}} R'^2 \, dR'$$

$$m = \rho v$$

where ρ is the density and v is the volume, which, for each disc is $\pi R'^2 \, dy$.

The moment of inertia for the upper half of the discus is, therefore:

$$I = \int_0^h \tfrac{1}{2} \pi \rho \left[\left(\frac{R}{H} \right) y \right]^4 dy$$

where h is the half-height of the discus, H is the height of the upper cone and R is the radius of the discus.

$$I = \frac{\pi \rho R^4}{2H^4} \left[\frac{y^5}{5} \right]_0^h = \frac{\pi \rho R^4 h^5}{10 H^4}$$

The moment of inertia of the whole discus is twice this and is, therefore, given by:

$$I = \frac{\pi \rho R^4 h^5}{10 H^4} \times 2$$

$$I = \frac{\pi \rho R^4 h^5}{5 H^4}$$

For a men's competition discus, and with reference to Figure 6.9, $R = 0.11$ m, $h = 0.023$ m, $H = 0.032$ m and $\rho = 2511 \, \text{kg m}^{-3}$ (Found from $m = 2$ kg and $v = 8 \times 10^{-4}$ m).

The moment of inertia is, therefore, equal to:

$$I = \frac{\pi \rho R^4 h^5}{5 H^4} = \frac{\pi \times 2511 \times 0.11^4 \times 0.023^5}{5 \times 0.032^4} = 1.142 \times 10^{-3}$$

The rotational kinetic energy $= \frac{1}{2} I \omega^2 = \frac{1}{2} \times 1.142 \times 10^{-3} \times (7 \times 2\pi)^2 = 1.1$ J.

Total kinetic energy $= 1225 + 1.1 = 1226.1$ J

So little of the kinetic energy that is transferred to a discus on throwing goes into the spin of the discus.

6.5 Trajectory diversions: the impossible lift

I have made several allusions to the peculiar properties of the gyroscope and how some of these may be applied to spinning sporting projectiles. The applications of gyroscopic action are numerous and varied, including transport stabilization, navigation and many conjuring tricks. Obviously, in this last application, the Magic Circle rules prevent me from giving away details, but you might like to consider how gyroscopy may be used in relation to, say, those 'solid' spheres that magicians manage to balance and travel along the edge of a cloth or rod.

The 'Levitron' is an executive toy with a hidden battery-powered gyroscope within it which, with the aid of opposing magnetic fields, produces a stable hovering platform producing a quite amazing visual impact.

But, to bring us back, albeit loosely, to a sporting application, consider, if you will, a motorbike wheel; say one of those 'mean' Harley-Davison sized bikes with tyres of almost dragster proportions. They can weigh up to about 20 kg. They can be lifted, but with care. Now imagine it to be attached to an axle of 2 m in length, and you are holding the far end of the axle so that it is horizontal. At the other end, the wheel is vertical and its tyre is resting on the ground. Could you lift the wheel now, from that position? The answer is most decidedly no, even using all your strength and both hands.

Now let us turn the system into a gyroscope with the wheel as a spinning flywheel and you as the pivot. The wheel will now have to rest on a roller-bed to allow it to spin freely without it rapidly moving off in a circular motion along the ground, with you spinning around with it! The wheel is spinning fast enough for the gyroscopic action to take affect (don't ask how it's powered!). What happens now when you try to raise the axle and wheel?

Amazingly, though not surprisingly, if you have followed all the discussion of Section 6.4, you *can* now lift the wheel, even from the far end of the axis. It still, of course, weighs the full 20 kg and this is what you would feel as you lift it. However, the gyroscopic action stabilizes the applied torque caused by the couple, which comprises of the lifting force and the weight of the wheel displaced by the 2 m axle.

Obviously, what is happening is that the gyroscopic effect resists any attempt to move the axle through a vertical angle, although it does allow for the vertical displacement of the wheel/axle/hands system. So, whatever force is applied to the axle, it stays horizontal, giving you the impression, or *feeling*, that you are actually lifting the wheel from directly above it. However, be warned: as soon as the wheel leaves the roller-bed and is in free air, precession commences and the wheel will rotate around you – the pivot – at the precession period, ω_p. So, as you lift the wheel higher, you and the wheel will rotate and the wheel actually traces out the path of a coiled spring as it rises, and also, when you begin to weaken, as it lowers back down to the ground. Weird, or what?

Summary

This chapter looks at the way in which the spin of a projectile can affect its flight. The aerodynamic flow lines around a spinning body are diagrammatically shown and, from this, the creation of the Magnus effect is explained. The value of the Magnus force is often measured experimentally, but the way in which this force modifies the flight of a given projectile is mathematically described in some detail in this chapter.

A further purpose of spin is to stabilize the projectile in flight by means of gyroscopic forces. Gyroscopic theory is described and mathematically expounded with reference to typical spinning sports projectiles such as the discus. Spin rates, stabilizing forces and precession (wobble) rates are all justified in quantitative terms.

Problems and questions

1   Use the equations of motion involving lift and drag (Equation 6.12 and 6.13) to compare the trajectory of a British golf ball with the larger American golf ball which has a cross-sectional area $7\frac{1}{2}$ per cent larger than its British counterpart. Calculate the range and height of both balls at 2 and 4 second intervals if they are both driven at $12°$ to the horizontal, and with a launch velocity of $65\,\mathrm{m\,s^{-1}}$. The Tait variables for the British ball are $\kappa_1 = 150$, $\kappa_2 = 0.18$, and for the American ball are, $\kappa_1 = 139$, $\kappa_2 = 0.193$.

2   A Frisbee is thrown with a launch velocity of $30\,\mathrm{m\,s^{-1}}$ and a launch angle of $1.2°$ to the horizontal. If the Tait values are $\kappa_1 = 100$, $\kappa_2 = 0.5$, use Model 6.1 – Drag and lift trajectory, to plot its trajectory. What happens on the rising part of the trajectory?

3   Two identical tennis balls are returned at an angle of $5°$ above the horizontal and a return speed of $50\,\mathrm{m\,s^{-1}}$. The first Tait variable for the balls have a common value κ_1 equal to 150. However, the first ball is returned with a topspin that results in a second Tait value, κ_2, of 0.15. The second ball is returned with a backspin resulting in a value of κ_2 equal to 0.21. Calculate the range and height at 1 second and 2 second intervals for both balls.

Part II

Practical applications

Chapter 7

Shot put and hammer

7.1 History of the events

7.1.1 The shot put

Although it might be assumed that most throwing events originated during the ancient Greek Olympic era, there is no record of any projectiles similar to the shot put being thrown in Greek competition. It would appear that Greek contestants preferred the superior grace and style that accompanied the throwing of discuses and javelins.

The first evidence of competitively throwing heavy 'dead' weights was about 2,000 years ago in the Scottish Highlands, where undoubtedly the requirement for brute explosive strength was a matter of shear survival. In Great Britain, in the Middle Ages, records show that the competitive hurling of cannon balls by soldiers was a popular sport, and this spread to certain areas of Europe during the seventeenth century. Shot put competitions using stones weighing 18 pounds (8.16 kg) took place in early nineteenth Century Scotland and were a part of the British Amateur Championships which began in 1866. The International Amateur Athletics Federation recognized the first world record in 1876, when the American, J. M. Mann putted the shot almost 9.4 m.

At around this time, the throwing area was a rectangular, slightly raised board, which precluded any throwing style other than standing still and heaving the shot across the field, with little or no contribution from lower body movement. In the modern event, the athlete must project the shot from the interior of a 2.15 m circle, with the range measured from the inside edge of the stop-board. This launch pad allows a variety of styles of throw, some of which will be discussed in Section 7.2.

The shot put, together with the more traditional Greek throwing activities of discus and javelin, was one of the sporting events of the first modern Olympics of 1896 in Athens, when the American, Robert Garrett, won the event with a put of 11.22 m. In fact, male Americans have dominated this sport up to the present day, with athletes such as Ralph Rose (a giant of a man at 6 ft 6 in and 250 lb), who won Olympic Golds in both the 1904 and the 1908 games, and Leo Sexton who was the first to throw 16 m in the 1932, Los Angeles Games. Rather analogous to Roger Bannister's perceived four-minute mile barrier, the 60 foot (18.3 m) barrier was thought to be the shot putter's 'challenge of the day'. Strangely, this barrier was broken by the American, Parry O'Brien, only two days after Bannister's historic achievement on 8th May, 1954. O'Brien had, in fact, previously broken the world record twice, in both May and June of 1953. The current male world record holder is again an American, Randy Barnes, with a 23.12 m put, unsurpassed since 1990. He also holds the indoor shot put record of 22.6 m.

Women's shot put entered the Summer Olympics in 1948 but, by contrast, this event has been dominated by the East European and Soviet Union competitors, such as Marianne Adam (East Germany) who threw 21.67 m in 1976, and the current world record holder, Natalya Lisovskaya (USSR) who threw 22.63 m in 1987. The women's current indoor world record holder is a Czechoslovakian, Helena Fibingerová, with a throw of 22.5 m, attained in 1977.

As a point of interest, the shot put thrower can realize only the shortest of throwing field event distances. Probably next higher up the range table is the discus with typical throwing distances of around 70 m, then comes the hammer throw at around 80 m and, finally, the javelin around 90 m. However, outside the Olympic field events, the world record for the 'longest throw of any object without any velocity-aiding feature' belongs to the flying ring which has been thrown the massive distance of 406.3 m. It was in flight for 30 seconds, or thereabouts (but note Problem 1 of Chapter 8).

7.1.2 The hammer

Arguably, the hammer throw event represents one of the most complex athletic contests of the field sporting canon. The game is thought to be Irish in origin, and legend has it that, at around the time of Christ, chariot wheels were competitively thrown in the Tailteann Games; an annual competition held in honor of Queen Tailtiu, in an area which is now in County Meath, in The Republic of Ireland. Legend even records a mythological early champion of the game; a heroic warrior who went by the name of Cuchulainn. Somewhat later, the Irish tribes adopted the sport for religious festivals, whirling a boulder attached to a wooden shaft around their heads and then releasing the projectile (which now began to resemble a hammer) in honor of the God, Thor.

During the Middle Ages the sport was adopted, first in Scotland, and then in England; the projectile most commonly used being a blacksmith's hammer, from which the modern event takes its name. A sixteenth century drawing shows King Henry VIII, a keen 'all-rounder', competitively throwing just such an implement.

The throwing of the 'sledge' was a popular event in the Scottish Games of the 1800s and throwing a rounded sledgehammer is still considered the purist form of the sport by participants and followers of the Scottish Highland Games.

Throwing the hammer has been an established sport in track and field events of Great Britain and Ireland since 1866. The hammers were manufactured from forged iron (some still are), but weights varied between 9 and 15 pounds (4 and 7 kg) and handle lengths could be anything between 3 and $3\frac{1}{2}$ feet. The variance was due to each Games possessing and using their own implements. The hammer was subsequently standardized by the English in 1875 at 16 pounds, and with a handle length of exactly $3\frac{1}{2}$ feet.

Up until this time, the only throwing technique used was to stand still and swing the hammer around the head before releasing. The distance was measured between the nearest foot and the point of landing. Typical recorded distances at the time were in the region of 40 m. Later the hammer was thrown from a line marked on the field, allowing more opportunity for differing throw techniques. Donald Dinnie is reported to be the first to rotate his body (usually twice) prior to the hammer's release in order to gain throw velocity. In 1887, the Amateur Athletic Union of the United States adopted the seven-foot (now 2.15 m) circle and the sixteen-pound hammer with 4 foot handle. In 1895, A. J. Flanagan of Ireland adopted the three-turn technique and achieved many records as a consequence.

The hammer throw was first included in the Olympics as an exclusively male event in 1904. In those early days, the medals were won, predominantly by Irishmen, or Irish-Americans, such as Flanagan himself, Matt McGrath and Pat Ryan. At around this time, the wooden handle was replaced by a wire, terminated by a pair of steel grips. By 1910, the world record had levelled off at around the 55 m mark and, for the next 20 years or so, little improvement occurred. In the 1930–40s, no country dominated the sport, and the record hovered around 58–59 m.

By contrast, the latter half of the twentieth Century saw some remarkable increases in throwing distances, as the Russians and Eastern Europeans, such as Mikhail Krivonosov of the Soviet Union and Gyula Zsivotzky of Hungary, took control. In particular, Gyula Zsivotzky of Hungary took the world record beyond 73 m. From the late 1960s to the present day, the Soviet Union has dominated the sport. In fact, since 1969, of the 19 record breakers, 14 were from the Soviet Union and the remaining 5 were either East or West German. The current record lies with the Russian, Iouri Sedykh, who threw a remarkable 86.74 m at Stuttgart, unbroken since 1986.

The phenomenal increase in attained distances may be explained by a number of factors which include: a greater knowledge of the physics and mechanics of the throw, the ability to spin faster using smooth-soled shoes on cement throwing rings instead of the spiked shoes used on dirt rings in the mid-1950s,[1] the implementation of the single grip handle design and precision manufactured hammers, better training methods, improved nutrition and, finally, one has to mention the part played by the legal and illegal use of performance enhancing substances and pharmaceuticals.

It was not until 2000 that the women's hammer was added to the Olympic programme. The first Olympic record holder was the seventeen-year-old Kamila Skolimowska of Poland with a throw of 71.16 m. Sadly, she died in February 2009 at the age of 26, as a result of a pulmonary embolism during a training session. The current world record holder is Anita Wlodarczyk of Poland, with a distance of 77.96 m.

One aspect of the rules of the game which seems to have gone through a series of regular revisions is the reduction in the angle of the throwing sector. Over the years, the sector marked on the field for valid throws has reduced from $90°$ to $60°$ in the 1960s, $45°$ and $40°$ in the 1970s, and now down to its present sector angle of $34.92°$. Maybe this serves to indicate the increasing popularity of field events, and the resultant decrease in 'real estate' available on the in-field for these sports as they become condensed into ever-smaller areas.

7.2 Throw techniques

7.2.1 The shot put

There are two main types of shot put throw classification in use today; the *glide method* which was developed by Parry O'Brian in 1951, and the *turn method*, developed by many exponents, but most famously attributed to both Aleksandr Baryshnikov and Brian Oldfield.

The glide technique is considered less technically complex and, although the current world record is held by the 'spinner', Randy Barnes, there are some creditable distances recorded by 'gliders' including the second longest recorded throw by Ulf Timmerman and also by Michael Carter. Additionally, nearly all female shot put athletes use the glide method.

---

1 The modern hammer shoe has a sole which runs up the outer side of the shoe. This allows the thrower to roll on the outer edge of the foot.

It has proved difficult to compare directly the two techniques: partly because it is believed by many that key throws have only been achieved under the influence of enhancing substances, and partly because throwing styles can seldom be clearly defined as purely glide or spin, but are usually a blend of the two methods, albeit with a heavy emphasis on one or the other. In summary though, most athletes begin throwing by learning the glide method; they may then gradually evolve their style into the spin method. Many feel that the glide technique favours the tall, strong throwers, while the spin technique is preferred by shorter, more powerfully built throwers. (Both techniques described below assume a right-handed, male thrower.)

The glide technique

The shot is placed on the base of the fingers – not in the palm – and the fingers are slightly spread. The shot is then pushed firmly into the neck, under the chin. The thumb should be under the shot with the throwing elbow pointing outwards, away from the body. The throwing arm should be at about a 45-degree angle to the ground.

The athlete stands at the back of the circle, facing away from the direction of the throw. The right foot should be near the back edge of the circle, with the left leg extended forward. Now, with most of the weight on the right foot, the knees should be bent as if moving back into a seated position, while the left leg is drawn back so that the toes of the left foot line up with the heel of the right. This is known as the 'unseating'.

The left leg is then extended toward the target area and the athlete pushes off with the right foot, 'gliding' to the front of the circle while keeping his center of mass low. This is known as the 'shoulder lift'. The athlete's feet should land simultaneously; the left foot in the front of the circle, just behind the toe-board and slightly left of center, and the right foot in the middle of the circle. The weight should now be on the right leg and the right knee bent to approximately $75°$. This is known as the 'power position', with the feet shoulder width apart, the left arm extended from the body and the knees bent.

The right elbow is kept up as the athlete's weight is shifted to the left. The left leg is straightened (the 'left leg drive') as the hips are rotated so they are square on to the target. Keeping the left side firm, the shot-arm is punched upwards and the throw is completed with a flip of the wrist and a strong follow through.

By this means the body builds a combination of linear and angular momentum from the legs, through the hips, the chest, the shoulders, the arms, wrist, hand and fingers, *in that order*. Key to this efficient energy creation and transfer is the respective moments of inertias of these various body elements; to an extent, the moment of inertias reduce with each respective body motion, as the velocities – both linear and rotational – increase.

The body's movement may, therefore, be considered analogous to a bullwhip action which is capable of making a cracking noise as the whip's velocity increases until it breaks the sound barrier. This is only possible as the kinetic energy builds because, as the wave travels along the length of the whip, its mass per unit length reduces and so the wave velocity increases, in an effort to hold the momentum constant.

The turn technique

The turn technique offers many advantages over the glide technique, the main one being that both the thrower and the shot travel a much greater distance in the shot ring, so the shot can be accelerated to a greater velocity than the glide. Furthermore, the body levers can be used to greater effect over this increased movement. However, it is not an easy technique to master as the throwing ring is much smaller than a discus ring, and so the turn performed with this weight must be tight. Also, on release, the discus is thrown with the non-throwing arm extended, which counterbalances the body, whereas the shot must remain close to the center of rotation and, therefore, it offers little help in balancing the athlete.

Details of this complex method vary in detail, but briefly, the method is as follows. The athlete prepares with the shot tucked into the neck and held, as before. This time, he faces directly away from the target on the far side of the ring, with the left foot on the line (actually a concrete-filled steel ring), the feet set apart wider than shoulder-width and in a semi-sitting position, but balanced with trunk and left arm directed out of the ring. The motion commences with the torso turning anti-clockwise (looking from above). As the body rotates on the left foot it leans inwards towards the center of the ring, counterbalanced by a wide sweeping right leg motion through the air as the totality of the weight is on the left leg.

Having completed one full rotation, the athlete should be again facing backwards with his weight on his right foot and the left foot now performing counterbalancing duties. His position should be slightly forward of the center of the ring, leaning upright (away from target), in a 'seated' position, now balanced by both the left leg, and the centripetal force created as a consequence of the rotation.

Finally, as the rotation goes into the last half turn, the left foot is planted, hopefully, on or near the launch line, the right foot slightly behind it, separated by more than a shoulder width (i.e. the power position again). The shot is dislodged from the neck as the athlete begins his final push action, with the fine-tuning of the counterbalance performed by the left arm on the shot's release and follow-through.

Key to the process is the rotate and the lift. The rotate function is counterbalanced at all times by alternate trailing legs. Lift is achieved by commencing with the lowest 'seating' posture possible commensurate with maintaining an easy balance, as well as finishing with the strongest stretch possible culminating in a high release point.

7.2.2 The hammer

The technique for the hammer throw naturally divides into four phases which, for a successful throw, must be performed with smooth fluidity and with the phases joined seamlessly together: the whole routine creates a constant acceleration of the hammer to the point of release. If one section of the routine is performed badly, there is little opportunity of compensation or correction later in the routine.

The four phases are commonly termed; *the winds, the transition, the turns* and *the delivery* or *release*. Again, the technique description outlined below is for the right handed, male thrower.

The winds

The athlete begins with his back to the throwing area, feet set apart by about a shoulder's width and the hammer dangling to his front. He begins swinging the hammer in front of him and then over his head. Two, or sometimes three winds, are enough, and on each wind, the angular velocity is increased and the arms are stretched away from the body a little further, thereby increasing the hammer's rotating radius and resulting in an increasing centripetal force. At this stage, the force is counterbalanced by a rotating hip movement extending in a direction diametrically opposing the hammer position. The circle traced out by the rotating hammer is not horizontal, but lies in an inclined plane with the lowest point at $90°$ to the left of the direction of release (looking from above).

The transition

This is the change over from the winds to the turns. Towards the end of the winds, the centripetal forced is too great to be balanced by mere hip motion. The athlete must now begin to sit back against the weight of the spinning hammer, while pivoting on the heel of his left foot and the ball of his right foot. In such a balanced and pivoted position, our athlete will be ready to commence the turns.

The turns

If the transition is fully accomplished, the athlete should 'fall naturally into' his turns. The accelerating angular velocity is controlled by the legs, hips and back; the arms are relatively passive at this point with regard to control. The number of turns varies with the athlete's ability. Beginners can maintain hammer control and achieve adequate release velocity with only three turns. Elite performers, on the other hand, will spin four or even five times, increasing their velocity steadily and smoothly, since any jerkiness will result in a potentially catastrophic imbalance.

The force that increases the angular velocity of the hammer comes from the hips, whose rotations at all times lead the hammer, thereby dragging it around ever-faster. The foot movement involves pivoting on the left foot and rolling on the right. The legs are held quite close together and the right foot lags behind the body; its main purpose is to maintain overall balance. Over the duration of the spins, the feet move the body across the ring from the initial starting position at the back of the ring to the launch point, as close to the front of the ring as possible.

The angle between the plane of the circle the hammer travels through and the horizontal plane increases with each successive spin until, on the last spin, it matches the required release angle. However, the low point of the incline remains unchanged; being 90° anti-clockwise of the launch direction (looking from above), and the high point is obviously opposite this, at 90° to the left of the launch direction.

The turns phase is completed with the athlete planting both feet firmly, as close to the front rim as possible and facing away from the launch direction, in readiness for the delivery or release phase.

Ideally, the athlete's center of gravity travels in a straight line from the center back of the circle to the center front.

The delivery or release

Once the feet have gripped the ring surface, the hammer is pulled down to the low point, and then pulled almost vertically by driving upwards with both legs and back. The left leg is then straightened to create a 'block'. The final acceleration is created by swinging the hammer up around this block set by the leg and hips. The hammer is released at, or above, shoulder height without any additional impulse, as the acceleration continues to remain constant throughout the process.

Figure 7.1 illustrates the *linear* velocity of the hammer as it progresses from the last wind/transition at about 0.1 s, through the three turns and the release at 1.6 s.

Panoutsakopoulos (2006) analysed the hammer throw at the 2006 Olympic Games in Athens and discovered a clear negative correlation between the total time of the four phases that make up the throw and the distance achieved ($r = -0.82$; $p < 0.01$). Specifically, Murofushi won the contest with a throw of 82.01 m and a total throwing time of 1.96 s.

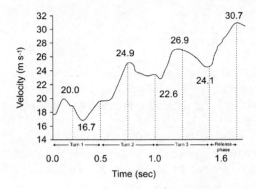

Figure 7.1 Velocity curve of the hammer during the three turns and the release phase of a throw by two-time Olympic champion Yuriy Sedykh. (Reprinted by permission from R. M. Otto, 1992, 'NSA photo sequence 22-hammer throw', *New Studies in Athletes* 7(3): 51–64.)

While the ninth ranked contestant, Epalle, managed a mere (!) 71.43 m, and it took him 2.42 s to complete his throw.

7.3 Notes on shot put and hammer manufacture

7.3.1 International Association of Athletics Federations Rules Extract from IAAF Rule Book 2008

THE SHOT
RULE 188.4

Construction. The shot shall be of solid iron, brass or any metal not softer than brass, or a shell of such metal filled with lead or other solid material. It shall be spherical in shape and its surface finish shall be smooth. To be smooth, the surface average height must be less than 1.6 μm, i.e. a roughness number N7 or less.

RULE 188.5

It shall conform to the following specifications:

Table 7.1 IAAF shot put regulations

Shot				
Minimum weight for admission to competition and acceptance of a record:				
	4.000 kg	5.000 kg	6.000 kg	7.260 kg
Information for manufactures.				
Range for supply of implement	4.005 kg	5.005 kg	6.005 kg	7.265 kg
for competition:	4.025 kg	5.025 kg	6.025 kg	7.285 kg
Minimum diameter	95 mm	100 mm	105 mm	110 mm
Maximum diameter	110 mm	120 mm	125 mm	130 mm

THE HAMMER
RULE 191

4   *Construction*. The hammer shall consist of three main parts: a metal head, a wire and a handle.

5   *Head*. The head shall be of solid iron, brass or other metal not softer than brass or a shell of such metal filled with lead or other solid material. If a filling is used, this shall be inserted in such manner that it is immovable and that the center of gravity shall not be more than 6 mm from the center of the sphere.

6   *Wire*. The wire shall be a single unbroken and straight length of spring steel wire not less than 3 mm in diameter and shall be such that it cannot stretch appreciably while the hammer is being thrown. The wire may be looped at one or both ends as a means of attachment.

7   *Handle*. The handle shall be rigid and without hinging joints of any kind. The total deformation of the handle under a tension load of 3.8 kN shall not exceed 3 mm. It shall be attached to the wire in such a manner that it cannot be turned within the loop of the wire to increase the overall length of the hammer. The handle shall have a symmetric design and may have a curved or straight grip and/or brace with a maximum length inside of 110 mm. The minimum handle breaking strength shall be 8 kN. Note: Other designs complying with the specifications are acceptable.

8   *Connections for wire*. The wire shall be connected to the head by means of a swivel, which may be either plain or ball bearing. The handle shall be connected to the wire by means of a loop. A swivel may not be used.

9   The hammer shall conform to the following specifications:

Table 7.2 IAAF hammer regulations

Hammer			
Minimum weight for admission to competition and for acceptance of a record:			
4.000 kg	5.005 kg	6.005 kg	7.265 kg
Information for manufactures: Range for supply of implement for competition:			
4.005 kg	5.005 kg	6.005 kg	7.265 kg
4.025 kg	5.025 kg	6.025 kg	7.285 kg
Length of hammer measured from inside of handle:			
Minimum 1160 mm	1165 mm	1175 mm	1175 mm
Maximum 1195 mm	1200 mm	1215 mm	1215 mm
Diameter of head:			
Minimum 95 mm	100 mm	105 mm	110 mm
Maximum 110 mm	120 mm	125 mm	130 mm

Center of gravity of head

It shall be not more than 6 mm from the center of the sphere, i.e. it must be possible to balance the head, less handle and wire, on a horizontal sharp-edged circular orifice 12 mm in diameter.

7.3.2 Shot and hammer manufacture

The shot and hammer balls are manufactured in much the same way. There is a choice of materials available; cast iron, turned iron, steel, stainless steel or brass.

Cast iron is used for the most basic of shots, being cast in moulds at between 900°C and 1200°C. For the hammer, a swivel thread is actually cast into the ball. Once manufactured, they are tested for weight and diameter; those that fall outside the allowable range are rejected or used for practice equipment.

Turned iron balls are the better quality alternative. They are cast at similar temperatures to the cast iron shots, but are made in a two-piece mould that produces a hollow-centered shot which is over-size. More silicon and magnesium are applied to the molten mix to allow for easier turning of the ball in a lathe to the required size and weight. Athletes may have a preference of shot diameter according to the size of their hands or, indeed, personal whims.

Steel shot are harder and denser, allowing for smaller diameter balls (within the allowed regulations). The melt temperature lies around 1400°C. The two-piece mould forms a hollow shell with a threaded access point. Again, the ball is turned to size and the smooth shell is weighed and then molten lead is poured into the hole to achieve the regulation weight. Once cooled and solid, the lead is covered with an epoxy resin to fill the remaining cavity.

The manufacture of the stainless steel shot is identical to that of steel but with a slightly higher melt temperature, in the region of 1500°C.

Casting is simpler for brass, with its lower melting point which is below 1000°C. Some athletes like the feel of brass, but it is the softest of the materials, and is therefore, more prone to wear than the alternatives.

All the shot materials, apart from the cast iron variant, may be cast with a hollow central cavity with a threaded hole leading to the outer world, which is then filled by screwing in a plug. The advantage of this is that the weight can be 'fine-tuned' by inserting lead shot, maybe to compensate for the loss in weight through natural wear and tear.

The balls of the hammer are manufactured in exactly the same way, but the threaded section is deeper in order to accommodate the threaded swivel. The swivel is made from steel and ball bearings to enable free movement. It has a 3 mm steel wire attached, and to that is attached a wire form handle of dimensions governed by IAAF Rule 191.7.

7.4 Trajectory dynamics

7.4.1 Basic dragless case

If initially we assume the drag-to-weight ratio for a shot, ε, is so low (it is about 0.01, compared to $\varepsilon = 1.23$ for a golf ball) as to render drag forces inconsequential, then the simplest useable equations might be considered to be Equation 3.15 and 3.16, in which all modifying factors to the most basic Equations 3.6 are neglected, apart from the launch height:

$$s_m = \frac{v_0\sqrt{v_0^2 - 2ah}}{a} \tag{3.15}$$

$$\theta_m = \arctan\left[\frac{v_0}{\sqrt{v_0^2 - 2ah}}\right] \tag{3.16}$$

Worked Example 7.1

Compare the optimum release angle for a shot put released with a launch velocity of $20\,\mathrm{m\,s^{-1}}$ at a height of 1.98 m with that of the simplest mathematical model of trajectory motion which assumes launch at ground level.

We know that if the projectile is launched at ground level, the optimum launch angle is $45°$ regardless of launch velocity. For the dragless shot put case, using Equation 3.16:

$$\theta_m = \arctan\left[\frac{v_0}{\sqrt{v_0^2 - 2ah}}\right]$$

where the launch velocity, $v_0 = 20\,\mathrm{m\,s^{-1}}$ and the launch height, $h = -1.98\,\mathrm{m}$. (Note that h is negative as the ground is below the launch height.)
So:

$$\theta_m = \arctan\left[\frac{20}{\sqrt{20^2 + 2 \times 9.81 \times 1.98}}\right] = \arctan\left[\frac{20}{\sqrt{400 + 38.85}}\right]$$

$$= \arctan\left[\frac{20}{20.95}\right] = 43.67°.$$

For biomechanical reasons, a greater launch velocity can be achieved with launch angles nearer $40°$ with the shot put. By contrast, the hammer may be launched near to the ideal optimum launch angle.

We may rewrite Equation 3.15 as:

$$s_m = \frac{v_0^2\sqrt{1 - 2(ah/v_0^2)}}{a}$$

If we now make the assumption that $v_0^2 > ah$ which is equivalent to assuming $s_m \gg h$, then the equation simplifies to:

$$s_m = \frac{v_0^2}{a} - h \tag{7.1}$$

Again we see how the range is so very dependent on the launch velocity; a 1 per cent increase in launch velocity will increase the range by about 2 per cent, regardless of the angle. Perhaps more surprisingly, note the linear relationship between range and height of release indicated by Equation 7.1. In the vicinity of the ballistically optimal launch angles of around $45°$, any increase in launch height will be matched by a similar increase in range (note $h = 0$ at ground level and upwards is positive).

We can also combine Equation 3.16 and 3.17 by eliminating v_0. We now obtain the surprisingly simple result:

$$s_m = h \tan 2\theta_m$$

This can be a useful equation since both h and s_m are easy to measure and, from these values, the optimum release angle can be derived and then compared with that measured from an experimental throw. The extent of this discrepancy may be an important statistic for coaches.

For the sake of completeness, we may eliminate the height, h, from Equation 3.16 and 3.17 to obtain:

$$s_m = \frac{v_0^2 \cot \theta_m}{a}$$

It is interesting to write s_m in terms of the kinetic energy, K, and the potential energy, P, where, at the point of release, $K = \frac{1}{2}mv_0^2$ and $P = mah$.

Equation 7.1 then becomes.

$$s_m = \frac{(2K + P)}{ma}$$

So, the kinetic energy has twice the effect on the maximum range as that of the potential energy. If there was a fair trade off between these two energies transferred to the shot (and there is not!), then primary consideration should be given to increasing the transfer of kinetic energy. However, to increase either v_0 or h, the putter must impart extra energy to the shot. This will, in turn cost him additional energy, although the result on the maximum range will be greater if the velocity is increased in preference to the height, by an approximate factor of two.

7.4.2 Equations including drag

As mentioned, the effects of drag are small as indicated by the minimal drag to weight ratio (the lowest of all sporting projectiles), and its relatively low projectile velocity. But, let us see whether the drag effect is truly negligible in our mathematical model of the trajectory. If we neglect any spin effects on the shot and only consider the drag under varying wind conditions, the equation of motion in vector terms will be:

$$m\frac{d\mathbf{v}}{dt} = -mg - \frac{F_D(\mathbf{v} - \mathbf{w})}{|\mathbf{v} - \mathbf{w}|} \tag{7.2}$$

where $\mathbf{w}$ is the wind vector, F_D is the drag force given by Equation 5.1:

$$F_D = C_D \rho A \frac{v^2}{2} \tag{7.3}$$

and the other terms have their usual meaning. The value of C_D, the drag coefficient, lies just below 0.5 for a shot put launched at normal velocities, as can be seen from Figure 5.2 with the Reynold's number around 1×10^5. Note that, it is just less than the Reynold's critical number.

Combining Equation 7.2 with Equation 7.3 and separating out the vector components, we obtain:

$$\left.\begin{array}{l} dv_x/dt = -\left(\frac{1}{2}\right)C_D A\rho |\mathbf{v} - \mathbf{w}| \left(v_x - w_x\right)/m \\[2mm] dv_y/dt = -\left(\frac{1}{2}\right)C_D A\rho |\mathbf{v} - \mathbf{w}| \left(v_y - w_y\right)/m \\[2mm] dv_z/dt = -g - \left(\frac{1}{2}\right)C_D A\rho |\mathbf{v} - \mathbf{w}| \left(v_z - w_z\right)/m \end{array}\right\} \tag{7.4}$$

where x is the horizontal vector, z the vertical vector and y lies horizontally perpendicular to the velocity vector. This group of coupled differential equations is difficult to solve analytically, but if we assume the drag is small, we can make certain assumptions which make a numerical solution possible.

Lichtenberg and Wills (1978) evaluated this family of equations by making the assumption that:

$$|\mathbf{v} - \mathbf{w}| \cong \left[(v_x - w_x)^2 + w_y^2 + v_z^2 + w_z^2 \right] \tag{7.5}$$

on the basis that the term v_y can be ignored; it being, at all times, perpendicular to the line of flight. We have also dropped the term $-2v_z w_z$ because wind seldom blows upwards. Even if it did, the shot would spend half the time travelling 'with the wind' on the upward trajectory, and half the time 'against the wind' on the downward path, so the effect of the small value of w_z would be cancelled on the overall flight.

Let us now assume that, if the wind is constant over the duration of the flight we can replace Equation 7.5 with the constant value, γ, which will have a maximum value when the velocity is greatest (on launch and landing when $v = v_0$) and a minimum when the shot reaches the apex of the flight (when $v = v_0 \cos \theta_0$). So we can now state:

$$v_0^2 \cos^2 \theta_0 + w^2 - 2v_0 w_x \cos \theta < \gamma^2 < v_0^2 + w^2 - 2v_0 w_x \cos \theta$$

$$v_0 \cos \theta_0 < \gamma < v_0 \tag{7.6}$$

What we are saying here is that the drag will vary over the duration of the flight as the velocity varies with the progression of the shot; it will be a maximum at the launch and landing points, where $\gamma = v_0$, and it will be a minimum when the velocity is lowest, at the peak of the trajectory when $\gamma = v_0 \cos \theta_0$.

With this approximation, the group of Equations 7.4 now becomes:

$$\left. \begin{array}{l} dv_x/dt = -\alpha \left(v_x - w_x \right) \\[2mm] dv_y/dt = -\alpha \left(v_y - w_y \right) \\[2mm] dv_z/dt = -g - \alpha \left(v_z - w_z \right) \end{array} \right\} \tag{7.7}$$

where:

$$\alpha = \left(\tfrac{1}{2} \right) C_D A \rho \gamma / m \tag{7.8}$$

The group of Equations 7.7 can now be readily integrated utilizing the initial launch conditions:

$$\left. \begin{array}{l} x = \left[(v_0 \cos \theta - w_x)(1 - e^{-\alpha t})/\alpha \right] + w_x t \\[2mm] y = \left[-w_y (1 - e^{-\alpha t})/\alpha \right] + w_y t \\[2mm] z = h + \left[(v_0 \sin \theta + g - w_x/\alpha - w_z)(1 - e^{-\alpha t})/\alpha \right] + (w_z - g/\alpha) t \end{array} \right\} \tag{7.9}$$

Remembering that α is small and so higher order terms can be neglected, we can expand the x and z functions of Equation 7.9:

$$x = v_0 t \cos \theta + \left(\tfrac{1}{2} \right) \alpha t^2 \left(w_x - v_0 \cos \theta \right) \tag{7.10}$$

$$z = h + v_0 t \sin \theta - \left(\tfrac{1}{2} \right) g t^2 + \left(\tfrac{1}{2} \right) \alpha t^2 \left[w_x - v_0 \sin \theta + \left(\tfrac{1}{3} \right) g t \right] \tag{7.11}$$

We may now set boundary conditions for the landing; $x = R$ and $z = 0$, when $t = T$, where T is the total time of flight and R is the range.

Equation 7.10 now becomes:

$$T = R \sec \theta / v_0 + \left(\tfrac{1}{2}\right) \alpha T^2 \left(1 - w_x \sec \theta / v_0\right) \tag{7.12}$$

Now, for the dragless case, we see from Equation 3.3:

$$T = R \sec \theta / v_0$$

and this can be substituted into the right-hand side of Equation 7.12. The resulting expression is then substituted into Equation 7.11, with $z = 0$ at $t = T$, and, ignoring the higher orders of α, we obtain:

$$0 = h + R \tan \theta - \left(\tfrac{1}{2}\right) \frac{g R^2 \sec^2 \theta}{v_0^2}$$

$$+ \alpha R^2 \sec^2 \theta \frac{\left[\dfrac{\left(\tfrac{1}{2}\right) g w_x R \sec^2 \theta}{v_0^2} - \left(\tfrac{1}{2}\right) w_x \tan \theta + \left(\tfrac{1}{2}\right) w_z - \dfrac{\left(\tfrac{1}{3}\right) g R \sec \theta}{v_0} \right]}{v_0^2} \tag{7.13}$$

Let us now focus on, not simply the range, but the *change in range as a consequence of introducing the drag terms*. We now take the derivative of the range, R, over small changes in α and evaluate for $\alpha = 0$:

$$\frac{dR}{d\alpha} = \frac{R^2 \left(2 g R v_0 \cos \theta + 3 v_0^2 w_x \sin \theta \cos \theta - 3 g w_x R - 3 v_0^2 w_z \cos^2 \theta\right)}{6 v_0^2 \cos^2 \theta \left(v^2 \cos \theta \sin \theta - g R\right)} \tag{7.14}$$

Now we are interested in the change in R as a consequence of the drag factor at the optimum launch angle, θ_m. In this case, R_m has been previously given by Equation 3.14:

$$R_m = \frac{v_0^2}{a} \cot \theta_m \tag{3.14}$$

Substituting Equation 3.14 into Equation 7.14 and setting $d\alpha = \alpha$, we obtain the final expression for the change in maximum range as a result of introducing the drag term. This will not produce a single answer as α varies with velocity (because γ varies as Equation 7.6). We will shortly see how we can work around this issue:

$$dR_m = \frac{-\alpha v_0^3}{g^2 \sin \theta_m} \left(\frac{1}{3 \sin \theta_m \cos \theta_m} - \frac{w_x}{2 v_0 \sin \theta_m} - \frac{w_z}{2 v_0 \cos \theta_m} \right) \tag{7.15}$$

Note that, since α varies with v_0, dR_m varies as v_0^4. For shot puts, we can safely neglect wind effects, but let us see if drag is significant for a standard-range put.

Worked Example 7.2

On a windless day, a shot put is thrown with a launch velocity of $14 \, \mathrm{m \, s^{-1}}$ at the optimum angle of $\theta_m = 42°$. Calculate the effects of drag on the range for a standard men's shot put with; cross-sectional area, $A = 0.011\,\text{m}^2$, $m = 7.26\,\text{kg}$, with a drag coefficient, $C_D = 0.5$, air density, $\rho = 1.2\,\text{kg m}^{-3}$

From Equation 7.15 with w_x and $w_z = 0$:

$$dR_m = \frac{-\alpha v_0^3}{g^2\,\sin\theta_m}\left(\frac{1}{3\,\sin\theta_m\,\cos\theta_m}\right)$$

α is given by Equation 7.8:

$$\alpha = \left(\tfrac{1}{2}\right)C_D A\rho\gamma/m$$

With γ lying between the extremes of v_0 and $v_0\cos\theta_m$.
So, $10.4 < \gamma < 14$.

The upper value of α, α^+ is given by:

$$\alpha^+ = \left(\tfrac{1}{2}\right)C_D A\rho\gamma^+/m = 0.5 \times 0.5 \times 0.011 \times 1.2 \times 14/7.26 = 6.36 \times 10^{-3}$$

The lower value of α, α^- is given by:

$$\alpha^- = \left(\tfrac{1}{2}\right)C_D A\rho\gamma^-/m = 0.5 \times 0.5 \times 0.011 \times 1.2 \times 10.4/7.26 = 4.73 \times 10^{-3}$$

The greater reduction in range will be given by dR_{m^+} as:

$$dR_m{}^+ = \frac{-\alpha^+ v_0^3}{g^2\,\sin\theta_m}\left(\frac{1}{3\,\sin\theta_m\,\cos\theta_m}\right) = \frac{-6.36\times10^{-3}\times14^3}{9.81^2\times0.67}\left(\frac{1}{3\times0.67\times0.74}\right)$$

$$= -0.18\,\text{m}$$

The lesser reduction in range will be given by dR_{m^-} as:

$$dR_m{}^- = \frac{-\alpha^- v_0^3}{g^2\,\sin\theta_m}\left(\frac{1}{3\,\sin\theta_m\,\cos\theta_m}\right) = \frac{-4.73\times10^{-3}\times14^3}{9.81^2\times0.67}\left(\frac{1}{3\times0.67\times0.74}\right)$$

$$= -0.13\,\text{m}$$

Now, if a full analytical solution had been carried out, it would be established that the drag actually lies mid way between these two values at just under $-0.15\,\text{m}$. This also ties in closely with shot put ranges found in practice using video analysis to obtain the release velocity and angle.

The question may be asked, if drag affects the range, and our assumptions have been based on a launch of the shot at the optimum angle, what happens if the drag itself alters the optimum angle?

In Equation 7.13, and with $w = 0$ again, we take the derivatives of R with respect to θ, and set it equal to zero. This gives the condition:

$$R_m = \frac{v_0^2\,\cot\theta_m\left(1 - \alpha v_0\,\operatorname{cosec}\theta_m/g\right)}{g} \tag{7.16}$$

We can now substitute Equation 7.16 back into Equation 7.13 to eliminate R_m and obtain an expression for θ_m. This we can then differentiate with respect to α, and, as before, set $d\alpha = \alpha$, to obtain:

$$d\theta_m = \frac{-\alpha v_0 \left(\cos \theta_m - \sec \theta_m/3\right)}{g} \tag{7.17}$$

with $d\theta_m$ expressed in radians. If we now put the values from Worked Example 7.2 into Equation 7.17 and convert the difference-angle back to degrees, we find:

$$d\theta_m = -0.132°$$

So, the optimum launch angle does not alter significantly as a consequence of introducing the drag term.

The first tab of Model 7.1 – Shot put range reduction through drag, executes both Equation 7.15 and 7.17. Wind vector values can be included and the mid-value of γ and α is used, although strictly this would only apply to the windless case.

Let us now consider the drag on a hammer. One might assume the drag of a hammer to be significantly greater than that of the shot. After all, drag is so dependent on velocity, and a hammer is released at roughly twice the velocity of a shot. Furthermore, the combined effect of the wire and handle will increase the comparative drag over the shot. Indeed, the drag-to-weight ratio of the hammer would seem to be twice that of the shot.

However, experimental measurements seem to indicate similar drags between the hammer and the shot. The reason for this is that the value of the drag *coefficient* for the hammer is actually lower than that of the shot because it travels much of its flight at a velocity which takes it *beyond* the Reynold's critical number. So C_D for the hammer is only about 0.2, compared to that of the shot put, which is nearer 0.5. So, all things being equal, the factors cancel and the range reductions due to drag and wind are similar between the two throwing implements.

7.4.3 Other factors

Weight of the throwing arm

Tabor cited in Huntley and James (1990) states that the only parts of the body movement which impacts on the direction of the shot launch are the arm explosion phase, and to a much lesser extent, the wrist flick. They make the following assumptions:

a   The shot putter's shoulder will occupy the same position at the time of the release, whatever the angle of release.
b   Just before the arm explosion, the shot is travelling in the same direction that it is finally released into, i.e. the arm explosion does not change the direction of the shot.
c   The speed of the shot just before the arm explosion is independent of the angle of release.

They confirmed the appropriateness of these assumptions by studying high speed video footage of shot putters in action.

d   They further assumed that the shot cannot change direction significantly over the duration of the arm explosion.
e   The launch velocity is independent of the angle of launch.

f   A constant force is exerted on the shot throughout the explosive phase.

By balancing the forces over the explosive phase:

$$a = \frac{F}{m} - r_m g \sin v_0 \tag{7.18}$$

where a is the acceleration of the shot during the explosive phase, F is the force applied to the shot by the arm and r_m is a factor (> 1) that accounts for the weight of the arm. If the assumption is made that the center of mass of the arm is $\frac{1}{3}$ out from the shoulder, and the arm weighs about 1.5 times that of the shot, then that corresponds to a value for r_m of 1.5.

Tabor derived an equation for the range of the short put which includes the arm's weight:

$$R = \frac{[U^2 + 2(\frac{F}{m} - r_m g \sin\theta_0)l_2]}{2g} \sin 2\theta_0 \left[1 + \sqrt{1 + \frac{2g(h + l_1 \sin\theta_0)}{[U^2 + 2(\frac{F}{m} - r_m g \sin\theta_0)]}}\right] \tag{7.19}$$
$$+ l_1 \cos\theta_0 - l_3$$

where l_1 is the length of putter's arm, l_2 is the distance of travel of arm explosion, l_3 is the distance of shoulder behind the stop-board, h is the height of shoulder above ground, F/m is the acceleration of the arm, and U is the speed of shot just prior to arm explosion phase.

Equation 7.19 is modelled on the second tab of Model 7.1 – Shot put range reduction through drag. The value of r_m is acknowledged as a 'fudge factor'. It is calculated on the basis of the shot putter's arm weighing 10 per cent of their overall body weight, which is obviously a major overestimate. However, the somewhat larger value of r_m is justified on the basis that the explosive acceleration of the arm will be less than if it were thrown with a horizontal arm, as the pectoral muscles lose efficiency with increased θ_0. This is accommodated by assigning the shot putter with the heavier appendage.

Tabor acknowledges one deficiency in this model. He found that the arm actually extends to an angle which is a few degrees lower than θ_0. He suspects that the arm is pulled down by the weight of the shot. He goes on to suggest that, although the final wrist flip adds little to the increase in v_0, it does raise the angle of release by those few extra degrees towards the ideal release angle, θ_m.

Other environmental factors

Mizera and Horváth (2002) have considered other environmental factors which may, to a greater or lesser extent, affect the range of the shot put and hammer. They studied deviations in the gravitational acceleration caused by the rotating Earth and its resultant centripetal and Coriolis (inertial) forces. These forces, in turn, are dependent on the Earth's latitude at the event field, and also the direction of shot launch (north, south, east or west). Mizera and Horváth also addressed changes in air pressure due to altitude, wind, and temperature.

They base their analysis on the collective equation derived by Landau and Lifschitz (1976) which, in a rotating reference system with the origin set on the Earth's surface, the Cartesian coordinates are defined by: x pointing northwards across the event field surface, y pointing westwards and z pointing vertically upwards ($90°$ to the geoid surface). The motion equation of a thrown shot or hammer is then given by:

$$m\frac{d^2\mathbf{r}}{dt^2} = m\mathbf{g}(\varphi, H) + 2m\frac{d\mathbf{r}}{dt} \times \omega - \frac{C_D A \rho(p, T)}{2}\left(\frac{d\mathbf{r}}{dt} - \mathbf{w}\right)^2 \tag{7.20}$$

where $\times$ represents the vector cross product, $\mathbf{r} = (x, y, z)$ is the the position vector of the shot or hammer, ϕ is the latitude (positive on the northern hemisphere, negative on the southern hemisphere), H is the altitude of the event field, ω is the Earth's angular velocity vector, A is the projected area of the shot/hammer, ρ is the density of the air, p is the air pressure, and T is the air temperature, and $\mathbf{w}$ is the wind velocity vector.

The term $2m\mathbf{d}\mathbf{r}/\mathbf{d}t \times \omega$ accounts for the Coriolis force, while

$$\frac{C_D A \rho(p, T)}{2} \left(\frac{d\mathbf{r}}{dt} - \mathbf{w} \right)^2$$

represents the drag force due to air resistance. Mizera and Horváth (2002) used values of $C_D = 0.47$ for the shot and $C_D = 0.7$ for the hammer.

We see that the gravitational field is dependent on both height and latitude. It does so in the following manner:

$$g(H) = g_0 \left(1 - \frac{2H}{R} \right) \tag{7.21}$$

where $g_0 = 9.81 \text{ m s}^{-2}$, and $R =$ average radius of the Earth's surface $\approx 6.368 \times 10^6$ m. In fact, the Earth is approximately an ellipsoid; wider at the equator and flattened at the poles. The radius at the equator, $R_e = 6.378 \times 10^6$ m, while the radius at the poles, $R_p = 6.357 \times 10^6$ m. The correct value of R may then be calculated from knowledge of the latitude.

We can say that the force of gravity acts only to the center of the Earth, and so:

$$g_x = 0, \quad g_y = 0 \quad \text{and} \quad g_z = -g(\phi)$$

as suggested by Equation 7.20.

It may be shown that (Caputo, 1967):

$$g(\varphi) = 9.78049 \left(1 + 0.005\,288\,4 \, \sin^2 \varphi - 0.000\,005\,9 \, \sin^2 2\varphi \right) \text{ m s}^{-1} \tag{7.22}$$

which accounts for the reduction in gravitational force due to (a) the ellipsoidal shape of the Earth, and (b) the latitudinal variation in the Earth's rotation.

The air density, ρ, at standard temperature (293 K) and pressure (101.325 kPa) may be derived from:

$$\rho(p, T) = \frac{p}{QT} \quad \text{where} \quad Q = 287.05 \text{ J kg}^{-1} \text{ K}^{-1} \tag{7.23}$$

So, Equation 7.20 includes all the relevant environmental (physical and meteorological) factors that govern the trajectory of shot and hammer. It was solved using a fourth-order, Runge–Kutta numerical integration method. A selection of results is delineated below. The baseline distance for the hammer throw is taken as the male world record distance of $R = 86.74$ m, set in 1990. The values for $\phi = 38°$ and other environmental parameters were derived from the location of Athens (Olympic site, 2004). Obviously, any range variations are going to exhibit more in the hammer than the shot put, since a greater launch velocity is achievable with the hammer, and the range is consequently much greater. You may wish to refer to the original paper for further details and results charts.

a   The variation in hammer range with latitude is significant, reducing by 0.45 m for the hammer, and 0.11 m for the shot put, between the equator and the poles. This is due to the Coriolis force which changes g by 0.53 per cent between the extreme latitudes.

b   The variation with changing direction (north, south, east and west) is minimal; it exhibits a change of only 5 cm between east and west for the hammer at the equator and reduces to nothing at the poles or, indeed, as a consequence of swinging the throw sector around into the north/south direction. The change in range of the shot put is negligible in all cases.

c   The range of both shot put and hammer increases with altitude, because gravity is less and the decreased air pressure reduces drag. In going from 0 m to 1000 m altitude, the hammer range increases by 0.55 m, while the shot put range increases by only 0.022 m.

d   These variational factors seem virtually independent of release angle over the scope of normal throwing angles, although there is slightly more sensitivity to change (latitude, direction or altitude) at, or close to, the ballistically optimal angle of 44.2°.

e   An increase in air temperature reduces air pressure and leads to longer throws. A change in temperature from 10°C to 35°C leads to an increase in range of 0.42 m for the hammer and 0.013 m for the shot put.

The main argument behind the importance of these calculations is that there is a case, some believe, to normalize the range values against these physical and environmental parameters, so that throw distances can be recorded for posterity in a more equitable manner. However, it is also worth noting that extreme environmental deviations will also have physiological effects on athletes; these variations will be much harder to model or predict.

7.5 Trajectory diversions: centripetal or centrifugal?

You will have noticed that, in this chapter, I referred to the force which is created when things spin, as a *centripetal* force. I mentioned it when the hammer was swung around, and also the reduction in the effect of the gravitational force due to the Earth's spin.

Many think that, when you spin things with a mass around a fixed point, a force is generated which pushes it outwards; in the case of the hammer, keeping the wire taut and away from the vertical. This is the force that the hammer thrower has to counteract using his own weight on the opposite side of the pivot. Yes?

Actually, *no*. That force, which is oft termed the centrifugal force, is illusory. It does not exist – *no way; no how.*

When a body is moving, it wants to stay travelling in a straight line. For it to move in a circle, it must be acted on by a force *acting inwards* towards the center to keep its trajectory circular. This inward acting force *does* exist and is termed the centripetal force.

The term centrifugal force appears to have come about because of a mistaken perception that there is a force that operates in the opposite direction (a reactive force) to the centripetal force. But that is a misconception. The 'pull' that is felt by the hammer thrower is the force that has to act toward the center, to keep the ball from flying off tangentially, and not radially – until it is launched, that is.

So the centrifugal force is sometimes referred to as a virtual force; but your true 'hardened' physicists even hate that concept, and refer to it as a fictitious force.

The same argument can be made for the Coriolis force. Yes, that's another fictitious force. It comes about, not because there is a force acting on a projectile, but precisely because there *isn't* one. When a body moves in a rotating reference plane, the frame coordinates move while the body is travelling in a linear manner. This gives the appearance that the body is moving in a circle, whereas really, it's only the reference frame which is moving. The linear equivalent of this might be the way in which a train station appears to up and shoot off, as the train accelerates out of the station. Nobody asks where the force that moves the station, the ticket office, the coffee shop and platform (and all as a single entity) comes from!

Summary

In this, our first of the 'specific sports chapters', we look at both the shot put and the hammer throwing events, for which it is clear that, between them, there are many similarities, as well as some significant differences. We commence by looking at the histories and the differing throw techniques for these field events, and follow this with an abridged account of the regulations and formal dimensions of the implements as laid down by the IAAF. The methods of manufacture for both shot put and hammer are briefly described before addressing analysis of the trajectory dynamics.

The analysis starts by considering the simpler dragless case, before moving onto the more complex issues raised by incorporating drag. It is interesting to highlight the differences to both range and optimum angle of *incorporating drag into* the analysis. It is shown that, notwithstanding the fact that such throwing implements possess high weight to drag ratios, the difference in range as a consequence of drag is still significant.

Finally, further factors which affect the trajectory are considered, such as the weight of the throwing arm, as well as environmental parameters such as the Earth's spin and the atmospheric pressure.

Problems and questions

1 A shot (mass $= 7.26$ kg, projection area $= 0.011$ m^2) is released at a height of 1.9 m, with a speed of 14 m s^{-1} and at an angle of $39°$. Assuming no drag on the shot, calculate the maximum height reached, the range and the time of flight.

2 What is the range if the shot is fired with the same launch velocity and angle, but at ground level?

3 Calculate the range for the example of Question 1, but now include the effects of drag into the calculations where the drag coefficient, C_D, is 0.45 and air density $\rho = 1.2$ kg m^{-3}.

4 Repeat the calculation of Question 3 but for a women's shot, where $m = 4.5$ kg, $A = 0.0078$ m and $C_D = 0.42$.

5 In Question 1 above, calculate the optimum release angle for maximum range. Why does the athlete not throw at this angle?

6 'Play' with Model 7.1 – Shot put range reduction through drag – Tab 2, and determine how the range of the shot varies as the weight of the shot putter's arm increases. Choose values of r_m from, say, 1.0 to 2.0.

Of course, you can always 'cheat' to obtain any of these answers by inserting appropriate values into the Excel models provided.

Chapter 8

Discus

8.1 History of the event

Discus throwing dates back several hundred years BCE to the ancient Greek Olympiad Games, although it was also played for other reasons: non-competitively for recreational purposes, to complement funeral ceremonies, and in other festivals. The 'diskos' was one of the original pentathlon events, together with throwing the javelin, running, jumping and wrestling. Pentathlon scoring was a complex affair and varied between Games, but generally, throwing diskoi, together with wrestling, was thought to be one of the 'heavy' events. Greater score-weighting was given to these two events – a win in both would usually guarantee victory. Furthermore, a win in the diskos would enable the athlete to gain precedence in the order of participation in the jump event, which would further advantage the athlete.

Although rules varied between local and national events, usually the discus result was taken from the best of five throws. Athletes competed in the national events for the glory, and prizes would be symbolic in nature such as wreaths or trophies. By contrast, prizes in local competitions were either of a financial nature, or were weapons. Youths competed for meat, though it is thought that there was some symbolism associated with this, as the animal would have been sacrificially killed and would represent the youth's acceptance as a member of 'the citizen body' (or adulthood).

Eight diskoi have been discovered at Olympia with weights varying roughly between two and three kilograms. These were manufactured from either stone or unwrought iron or bronze. However, an 8.6 kg discus has recently been discovered at Nemia, Korinthea. The average diameters of these diskoi seem to be around the 30 cm mark. A statue carved by the sculptor Myron in the fifth century BC entitled 'Discobolus' (now housed in the British Museum) is an emotive depiction of the discus athlete in action. His pose is not so dissimilar from what we would now term the 'power position', although most similar images found on vases or amphoras tended to portray the warm-up swings rather than the actual throws.

The diskoi were released from a small rectangular area known as a 'Balbis' using a whole body twisting action, but, it is thought, without the turns. There is little firm evidence of the ranges achieved in any of the Games, but knowledge of athlete's strength and throwing style, together with information concerning weights and sizes of the diskoi, lead experts to suggest ranges just short of the 100 foot mark would be a reasonable result. This is supported by at least one preserved papyrus, an anonymous epigram, but quoted by Herodotos, who tells of one, Phaÿllos, three times victor at the Pythian Games and commander of a ship from Kroton at the battle of Salamis, who not only achieved a diskos throw of $100 - 5$ feet, but also a jump of $50 + 5$ feet! The 'jump' here would have been a complex affair roughly akin to our current triple-jump. The jump varied between venues but may have involved two hops, two steps and a jump onto hard ground: a sand landing would have been considered cowardly!

Yet a different type of diskos known as a 'Solos', sometimes made from bronze, had a hole bored in it with cord or leather running through it by which the athlete could grip the equipment, swing it around and release (similar to the hammer event, perhaps). Homer described the hum from this device as it was released. Cicero stated that students of teachers preferred to hear the hum from the diskos rather than listen to their professor!

The first modern Olympics in 1896 included two types of discus throw. The first was a standing throw from a sloping platform (known as the 'ancient' style), and the other was a 'free' style from a square area. Gradually the former method was phased out and, by 1912, the discus throw was launched from a 2.5 m diameter circle. Originally, the surface was simply dirt or grass, with the athletes required to wear spiked 'field' shoes. Since the spikes were 2 to 3 cm long, the surface quickly became ruined, and groundsmen spent many hours tending the surfaces in order to make them usable again. In the 1950s the natural surface gave way, first to asphalt, and then to concrete, with the athlete's shoes now spikeless flats.

Today's circles are far from perfect. Frictional coefficients not only vary between different throwing circles, but also on the same throwing circle under different weather conditions. However, the concrete throwing circle has stood the test of time and has remained unchanged in both construction and dimension for almost 60 years.

An American, Robert Garrett, won the first modern Olympics. Women entered the track and field events in 1928, although the discus was the only women's throwing event at this time. Another American, Lillian Copeland, won the silver in that year and followed this with a gold in the following Olympic Games. Al Oerter, again an American, dominated Olympic discus from 1956–68, winning four consecutive gold medals and entering the record books on each occasion. However, since then, the Eastern Europeans have dominated the game in both male and female events, although it is fair to say that no records have been broken since the late 1980s.

As a final thought in this section, it is worth noting that women were deferred consent from taking part in the modern Olympics discus by 32 years (i.e. eight Games). However, there is no evidence of such prudery in the original Greek Olympics. The Augustan love poet Sextus Propertius described both boys and girls training for the diskoi competitions together and, as was the custom of the day, totally naked and oiled. There is, however, no evidence that they competed together. There is a report that one particular training exercise for both sexes involved something called 'bibabis'. In this routine the athlete jumped as high as possible and curled their legs back so that their heels touched their buttocks, before straightening their legs for the landing. An epigram celebrates one young lady who is reported to have carried out one thousand such drills without a break!

8.2 Throw techniques

Discus throwing is a complex affair and is usually taught in stages beginning with lessons in holding the discus (not by gripping, but by resting on the outermost finger joints), followed by the standing throw, the pivot, the step-in (or South African) and finally, the full cross-ring throw.

There are two main styles of throw; the *reversing technique* and the *non-reversing technique*. The term 'reverse' stems from the naturally occurring action of the feet to reverse their positions on delivery. So a right handed thrower drives to the right side while the left side rotates to the rear. In the non-reversing style, this only happens after the discus is released. The East German males and females have used the non-reversing technique to good effect, although it would appear that a greater number of women favour the style than men.

In detail, then, the throw technique for a male, right-handed thrower is as follows.

The discus is supported with the fingers close together and held by only wrapping the ends of the fingers (at the distal joints) around the rim. The hand is slightly cupped so that only the edges and the soft base of the thumb contact the discus. The hand is spread over the rear two-thirds of the discus.

The thrower commences with the athlete's trunk facing directly away from the target 'V', and at the far edge of the circle (let us call this the 12 o'clock position). The knees should be flexed so the discus is between waist and shoulder height, with the feet spaced at shoulder distance. The discus is swung back and forth as low to the ground as possible, relaxing the muscles and setting up a rhythm. The hips are rotated clockwise, and the throwing arm is extended as far clockwise as possible; ideally to the 8 or 9 o'clock position with the hips at 6 o'clock. The weight is transferred to the right foot.

The thrower's body begins the anti-clockwise (from above) rotation with the discus hand trailing behind his body and almost over his left foot. The left arm is relaxed, slightly flexed and opposite the throwing arm, to keep the center of rotation stable. The left leg begins to pivot on the ball of the foot, maintaining the same angle of the knee. The right foot is kept flat on the ground at this point as the discus arm is withdrawn ready to start its forward rotation. The thrower's weight is now shifted over the right foot as the discus is drawn back. Weight is then shifted over, and towards, the left foot which causes the discus to start on its path (not the other way round!) and the thrower begins to rotate the left knee anti-clockwise.

The throwing arm is now 'latched' back as far as possible, as the turns begin. The complete left side of the body starts to turn in unison (i.e. the foot, knee and arm) at this point. Power is now generated by the right leg sweep as the foot is raised when the chest reaches the 10 o'clock position. The thrower now drives towards a point slightly to the right of the center of the circle as the wide right leg pulls him to the left. The knees should be kept wide apart for as long as possible, ensuring that power comes from the body and left thigh.

The right leg should now sweep outside the ring, as the foot sweeps along the back of the circle and the thrower leads with his thigh (rather than his knee). Next, the left leg accelerates anti-clockwise, round and forward, leading with the inside of the thigh, while holding the left arm just above the horizontal. The turn continues until a line drawn between the thrower's right heel and his left toe is aligned with the 12 o'clock–6 o'clock diameter line of the throwing ring. At exactly this point, the thrower hits his power position (ideally!). It is important now for the right hip and knee to be as far in front of the discus as possible, to allow for a long 'pull' on the discus. This is known as 'the separation'.

The body should now be upright but with the right thigh pulled under the body. This increases rotational velocity due to conservation of angular momentum, as the axis of rotation changes from the vertical, to an angle tilting towards the back of the circle and the counterbalancing discus curls towards the front. At this point, with the hips square on to the direction of throw, athletes often focus their eyes on the target; imagined or otherwise. This helps them to drive straighter and more linearly across the ring, and reduces the likelihood of over-rotation of the discus and/or the thrower.

Now the right foot makes contact with the center of the ring with the foot pointing towards the 3 o'clock direction. The left foot should now be pointing 180° away from the right foot.

The thrower has now reached the power position and emphasis shifts to the vertical lift of the discus. On release of the discus, the thrower should continue to rotate the right foot so that, for a non-reverse throw, it turns beyond 6 o'clock as the left foot is still at the

6 o'clock position. The left arm also follows through with speed until it rests in a bent position at the left shoulder. The discus is launched with an extended arm at shoulder height, and knees locked, providing a 'block' against which to throw. At the moment of release, the discus should be tilted slightly downwards giving it a negative angle of attack to the trajectory direction, which provides the all important aerodynamic advantage over the majority of the flight without the possibility of stalling. It is important to pull the discus off the index finger which both increases the contact time (marginally) and increases the spin and hence stability.

As in the case of the shot put and hammer of Chapter 7, the key to a skilled throw is a constant acceleration of the discus from entry to the moment of release.

Most discus launch analyses divide the action into five distinct stages, first defined by Bartlett (1992):

1   *Preparation*: a double support phase starting from the change in discus direction at the end of its backward swing and ending when the right foot breaks contact.
2   *Entry*: a single support phase which finishes with the left foot braking contact.
3   *Airborne*: which finishes with the right foot re-contacting.
4   *Transition*: a single support phase which ends as the left foot lands.
5   *Delivery*: which starts as a double support phase, and ends at the release of the discus.

Considerable video analysis research has focused on the respective durations of each of these phases and their possible correlations with range [e.g. Bartlett (1992), Panoutsakopoulos (2006)]. Figure 8.1 shows the variation in discus speed in each of the five phases. An indication

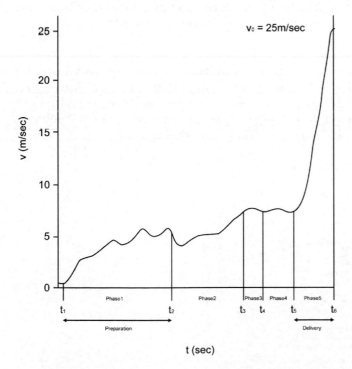

Figure 8.1 Variation of discus speed with throwing phase [adapted from Bartlett (1992)].

of the respective durations of each phase can be seen from the sections marked on the *x*-axis of this graph.

8.3 Notes on discus manufacture

8.3.1 International Association of Athletics Federations Rules Extract from IAAF Rule Book 2008

RULE 189

DISCUS THROW

THE DISCUS

1   *Construction*. The body of the discus may be solid or hollow and shall be made of wood, or other suitable material, with a metal rim, the edge of which shall be circular. The cross section of the edge shall be rounded in a true circle having a radius of approximately 6 mm. There may be circular plates set flush into the center of the sides. Alternatively, the discus may be made without metal plates, provided that the equivalent area is flat and the measurements and total weight of the implement correspond to the specifications. Each side of the discus shall be identical and shall be made without indentations, projections or sharp edges. The sides shall taper in a straight line from the beginning of the curve of the rim to a circle of a radius of 25 mm to 28.5 mm from the center of the discus. The profile of the discus shall be designed as follows. From the beginning of the curve of the rim the thickness of the discus increases regularly up to the maximum thickness D. This maximum value is achieved at a distance of 25 mm to 28.5 mm from the axis of the discus Y. From this point up to the axis Y the thickness of the discus is constant. Upper and lower sides of the discus must be identical, also the discus has to be symmetrical concerning rotation around the axis Y. The discus, including the surface of the rim shall have no roughness and the finish shall be smooth (see Rule 188.4) and uniform throughout.

Figure 8.2 shows the discuss. Note that these rules do not preclude the athlete from treating the upper and lower surfaces of the discus differently. In much the same way that a cricket ball may be polished on one side by a bowler to affect the aerodynamic flow across the ball's surface, and hence the trajectory (Chapter 12), if the upper surface of the discuss is polished, greater lift is created, which can impact positively on the achievable range.

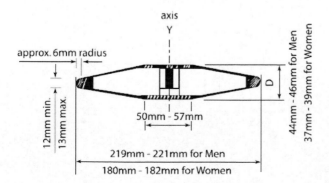

Figure 8.2 The discus.

Table 8.1 IAAF discus regulations

Discus				
Minimum weight for admission to competition and acceptance of a record:				
	1.000 kg	1.500 kg	1.750 kg	2.000 kg
Information for manufacture: Range for supply of implement for competition:				
	1.005 kg	1.505 kg	1.755 kg	2.005 kg
	1.025 kg	1.525 kg	1.775 kg	2.025 kg
Outside diameter of metal rim:				
Minimum	180 mm	200 mm	210 mm	219 mm
Maximum	182 mm	202 mm	212 mm	221 mm
Diameter of metal plate or flat center area:				
Minimum	50 mm	50 mm	50 mm	50 mm
Maximum	57 mm	57 mm	57 mm	57 mm
Thickness of metal plate or flat cenre area:				
Minimum	37 mm	38 mm	41 mm	44 mm
Maximum	39 mm	40 mm	43 mm	46 mm
Thickness of metal rim (6 mm from edge):				
Minimum	12 mm	12 mm	12 mm	12 mm
Maximum	13 mm	13 mm	13 mm	13 mm

8.3.2 Discus manufacture

The discus is constructed from two, or sometimes three, sections; the main body, known as the 'sides', the outer rim and, occasionally, a central hub is used. They are designed with a range of specified rim weights, such as; very high spin, high spin, low spin and center-weighted discuses. Elite throwers want a very high rim weight, (maybe between 80 and 90 per cent of the total weight of the discus) since they spin the discus very fast on release. This high spin rate, together with the greater weight around the circumference of the implement, increases its moment of inertia. This improves the gyroscopic effect and stabilizes the discus to a greater degree, suitable for its, hopefully, longer flight duration. However, if a less competent thrower were to impart too low a spin on a discus designed for high spin, it will be prone to stalling in flight.

Rims have been constructed from bronze alloy and brass alloy, but better quality discuses have steel rims which are cut from tubes of the correct diameter. They are then spun-shaped using a lathe to give the required outer curve of 6 mm radius. A 'shelf' is also carved into it, on to which the side walls are attached.

The sides of very high spin discuses may be made from aluminum, fiber-glass reinforced plastic (GRP), reinforced carbon or acrylonitrile–butadiene–styrene copolymer (ABS), an impact resistant polystyrene. Although the carbon composites have some weight/strength advantages, it is a costly option and the product can be comparatively brittle. This is a perennial problem with the very high spin discuses; to get the weight to the outer rim, the sides have to be very thin, and are therefore, prone to cracking especially when they land on hard ground or in collision with the throwing cage.

High spin discuses, for which the rim contributes 75–80 per cent of the total discus weight, are manufactured in much the same way and with the same materials as the very high spin variant, except that the carbon composite option is exclusive to the high-end equipment. Low spin discuses may use laminated or solid wood, although there may be balance issues with these materials, as their density can fluctuate around the central axis. These, often cheaper devices, have a moulded, rather than a spun rim, and they may incorporate a central hub with aluminum, GRP or wood forming the sides. Some training discuses have replaceable hubs so that the weight can be changed. In this way, the discus can be made intentionally over-weight to add resistance and build the athlete's strength, or alternatively, underweight to improve the athlete's speed.

Finally, there are available, light coloured, vulcanized, rubber discus intended for indoor training use.

8.4 Trajectory dynamics

8.4.1 The basics

The discus, more than any of the major projectiles described in this book, is the projectile that is most influenced by aerodynamic lift. Exceptions are the Frisbee, the flying ring and the boomerang.

Early treatments of discus trajectory models considered the projectile to be a wing – a symmetric aerofoil – with an aspect ratio of 1 : 1. In aircraft design terms, the justification stated that, if the discus flies under the condition of a positive angle of attack, lift is produced as the stagnation point (the point of zero relative velocity) moves from the center line of the discus to the lower surface. This produces a higher pressure on the lower surface than the upper surface, thereby producing the lift.

This portrayal certainly explains the phenomenon of discus stalling, but models based solely on this reasoning have proved inaccurate and offer no elucidation of the enigmas concerning the properties of discuses thrown into winds. It is a fact that, under certain conditions, a discus can travel considerably further when thrown into headwinds of significant force. This is surely most surprising to anyone with a most basic knowledge of physics and the balance of forces on a body. Furthermore, a discus thrown into a tailwind will fly further with a negative angle of attack (i.e. thrown at an angle lower than the launch angle). To fully explain these phenomena, an alternative approach to that of the aerofoil analysis of discus flight must be adopted. The discrepancy between the two analyses occurs because, (a) a discus is not powered by an engine over the duration of the flight, and (b) an airplane does not usually fly over a tight parabolic-like trajectory!

A somewhat inelegant term, aerodynamicity, has been defined in the publications as the ratio of the maximum aerodynamic force in flight to the gravitational force. In these terms, the discus is ten times more aerodynamic than the hammer, but only one-seventh as aerodynamic as the javelin (Hubbard, in Vaughn, 1989). In fact, it is because the drag forces can be compensated by the lift forces that, even on a still day, the range of a discus may equate to the theoretical optimal range achievable in a vacuum.

8.4.2 The trajectory model including drag and lift

Figure 8.3a depicts the resultant velocity, v_{rel}, which is the vector sum of the wind velocity (in this case, taken as a horizontal headwind), v_w, and the projectile's velocity direction vector, v_d. Figure 8.3b indicates that the drag force opposes v_{rel}, while the lift force is perpendicular to v_{rel}. Notice that both v_{rel} and v_d point in a direction above the plane of the discus (i.e. $\alpha < \beta$), indicating a situation soon after launch when the trajectory angle is still large. However, over the duration of the flight, β reduces while α remains roughly constant. Therefore, soon after launch, β will reduce to a value less than α and will continue reducing over the duration of the flight, creating a positive angle of attack over almost the whole of the flight trajectory.

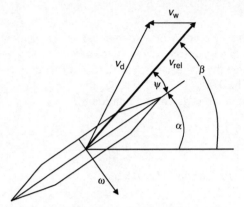

a) Velocity vectors acting on a discus in flight (soon after launch)

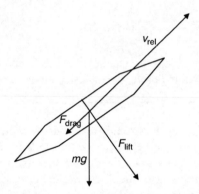

b) Force vectors acting on a discus in flight (soon after launch)

Figure 8.3 (a, b) Velocity and force vectors acting on a discus in flight.

The discus is affected by the gravitational force, mg, the lift force F_{lift} and the drag force, F_{drag}. The drag force, F_{drag} will act along the direction of the relative velocity vector, v_{rel}, given by:

$$\mathbf{v}_{rel} = \mathbf{v}_d - \mathbf{v}_w \qquad (8.1)$$

As we have seen many times, the drag and lift forces are represented by the dimensionless coefficients, C_D and C_L:

$$F_{drag} = \tfrac{1}{2}C_D\rho A v_{rel}^2; \quad F_{lift} = \tfrac{1}{2}C_L\rho A v_{rel}^2 \qquad (8.2)$$

where ρ is the density of air and A is the maximum cross-sectional area of the discus. The acceleration due to the aerodynamic forces is given by:

$$a = \tfrac{1}{2}\rho A v_{rel}^2 \sqrt{C_D^2 + C_L^2} \big/ M \qquad (8.3)$$

where M is the mass of the discus.

Worked Example 8.1

Calculate the instantaneous acceleration due to dynamic forces on a men's competition discus at the top of its flight, if it is subject to a headwind of 5 m s^{-1} and its instantaneous velocity is 14 m s^{-1}. Assume the air density to be 1.15 kg m^{-3}, drag coefficient to be 0.06 and lift coefficient to be 0.9.

The instantaneous acceleration is given by Equation 8.3:

$$a = \tfrac{1}{2}\rho A v_{rel}^2 \sqrt{C_D^2 + C_L^2} \Big/ M$$

For a men's discus $A = 0.038 \text{ m}^2$ and $M = 2$ kg.

$$a = \tfrac{1}{2}\rho A v_{rel}^2 \sqrt{C_D^2 + C_L^2} \Big/ M = \frac{0.5 \times 1.15 \times 0.038 \times (14-5)^2 \times \sqrt{0.06^2 + 0.9^2}}{2}$$

$$= \frac{1.77 \times 0.9}{2} = 0.8 \text{ m s}^2$$

This acceleration will act on the discus in addition to the usual acceleration due to gravity and should be vectorially added to it to derive the total acceleration on the discus.

The lift and drag coefficients, C_L and C_D depend strongly on the attack angle, ψ. However, we will assume that, for the following analysis, largely based on the work of Frohlich (1981), their values remain independent of v_{rel}, ρ and also ω (the discus velocity of rotation which itself is assumed to be perpendicular to the plane of the discus). If the discus is assumed not to be thrown into a crosswind, and so all forces and velocities lie in the vertical plane of the trajectory, the equations of motion will be given by:

$$\left. \begin{aligned} \frac{d^2 x}{dt^2} &= -\tfrac{1}{2}\left(\rho A v_{rel}^2 / M\right)\left(C_D \cos\beta + C_L \sin\beta\right) \\ \frac{d^2 y}{dt^2} &= -g + \tfrac{1}{2}\left(\rho A v_{rel}^2 / M\right)\left(C_L \cos\beta - C_D \sin\beta\right) \end{aligned} \right\} \qquad (8.4)$$

where β is the angle between the horizontal plane and v_{rel}. If α is assumed constant for the duration of the flight (i.e. the orientation of the discus to the horizontal remains constant), then the path of the discus is determined solely by Equation 8.4.

The initial conditions are given by: v_0, the release velocity; v_w, the wind velocity; θ_0, launch angle; α, the discus inclination angle (see Figure 8.3); and y_0, the release height.

At a time T after release with relative velocity v_{rel0} at launch given by:

$$\mathbf{v}_{rel0} = \mathbf{v}_0 - \mathbf{v}_w$$

we can integrate Equation 8.4 with respect to time to yield:

$$\frac{dx}{dt} - v_w = v_{rel_x} = v_{rel_{x0}} - \tfrac{1}{2}(\rho A/M)\int_0^T v_{rel}^2 \left(C_D \cos\beta + C_L \sin\beta\right)dt'$$

$$\frac{dy}{dt} = v_{rel_y} = -gT + v_{rel_{y0}} + \tfrac{1}{2}(\rho A/M)\int_0^T v_{rel}^2 \left(C_L \cos\beta - C_D \sin\beta\right)dt'$$

and integrating a second time yields:

$$
\left.
\begin{aligned}
x &= v_w T + v_{rel_{x0}} T - \tfrac{1}{2}(\rho A/M) \int_0^T \int_0^t v_{rel}^2 (C_D \cos \beta + C_L \sin \beta) dt' dt \\
y &= -\tfrac{1}{2} g T^2 + v_{rel_{y0}} T + \tfrac{1}{2}(\rho A/M) \int_0^T \int_0^t v_{rel}^2 (C_L \cos \beta - C_D \sin \beta) dt' dt
\end{aligned}
\right\} \quad (8.5)
$$

Equation 8.5 clearly illustrates the separate effects of gravity, g, and the wind, v_w (which, in turn, affect v_{rel}), on the aerodynamic forces,. Unfortunately, without the aid of a powerful mathematical computer package these equations do not yield up their solutions easily. However, if instead we choose to integrate Equation 8.4 over small time increments, Δt, and making use of the third-order differentials, $\dddot{x}$ and $\dddot{y}$, on the basis that C_L and C_D are approximately piecewise linear with the angle of attack, ψ, we obtain:

$$
\beta = \tan^{-1}\left(\frac{v_{rel_y}}{v_{rel_x}} \right)
$$

$$
\left.
\begin{aligned}
\dddot{x} &= (\rho A/M)\left\{ 2\left(v_{rel_x}\ddot{x} + v_{rel_y}\ddot{y} \right)\left(-C_D \cos \beta - C_L \sin \beta \right) + \left(v_{rel_x}\ddot{y} - v_{rel_y}\ddot{x} \right) \right. \\
&\quad \left. \times \left[\left(\frac{\partial C_D}{\partial \psi} - C_L \right) \cos \beta + \left(\frac{\partial C_L}{\partial \psi} - C_D \right) \sin \beta \right] \right\} \\
\dddot{y} &= (\rho A/M)\left\{ 2\left(v_{rel_x}\ddot{x} + v_{rel_y}\ddot{y} \right)\left(C_L \cos \beta - C_D \sin \beta \right) + \left(v_{rel_x}\ddot{y} - v_{rel_y}\ddot{x} \right) \right. \\
&\quad \left. \times \left[\left(-\frac{\partial C_L}{\partial \psi} - C_D \right) \cos \beta + \left(\frac{\partial C_D}{\partial \psi} - C_L \right) \sin \beta \right] \right\}
\end{aligned}
\right\} \quad (8.6)
$$

where, over each time increment, Δt:

$$
\left.
\begin{aligned}
\Delta x &= \dot{x}\Delta t + \tfrac{1}{2}\ddot{x}\Delta t^2 + \tfrac{1}{6}\dddot{x}\Delta t^3 \\
\Delta y &= \dot{y}\Delta t + \tfrac{1}{2}\ddot{y}\Delta t^2 + \tfrac{1}{6}\dddot{y}\Delta t^3
\end{aligned}
\right\} \quad (8.7)
$$

Equations 8.6 and 8.7 are modelled in the first worksheet of Model 8.1.

Some notes on discus rotation and pitching moment

As stated, the primary purpose of the discus spin is to maintain a constant plane of orientation of the discus over the duration of the flight. This it does quite successfully. However, there is some discrepancy in the literature with regard to the rate of rotation achieved. It is recognized that, the faster the spin, the further the discus travels; but this is a consequence rather than a cause. Soong (1976) uses a spin rate of 36.9 rev s^{-1} in his calculations, derived from the rate at which a discus rim would roll along the ground if it were released with a typical launch velocity of 25.5 m s^{-1}. This is now known to be a gross over-estimate, with modern values being measured in the range of 5–8 rev s^{-1}. Frohlich (1981), on whose work the analysis in this chapter is based uses the widely accepted value of 7 rev s^{-1} as the discus's typical rate of rotation.

This discus rotation will create a small, but significant torque force which, for a right-handed thrower, will cause the plane of the discus to gently tilt down on the left-hand side, with respect to the right-handed thrower. This pitching moment is especially noticeable towards the end of flight and may be as much as $10°$, causing the discus to pull to the left as it approaches the ground. Hubbard and Cheng (2007) calculated the rolling effect based on pitching moment forces, F_M, represented in much the same way as drag and lift were in the previous section:

$$F_M = \tfrac{1}{2}C_M \rho A d v_{rel}^2$$

where C_M is the dimensionless pitching moment coefficient and d is the diameter of the discus. For the higher-spin discuses, the angular momentum $I\omega$ is high, since both terms are high. Also, since the discus sides are smooth, the surface friction drag will be low, keeping the shear stresses to a minimum. It follows that the in-flight, spin-down torque is negligible. This, together with the high angular momentum, ensures that the axial spin rate remains essentially constant over the whole duration of the flight.

Finally, in this section, a reminder about the effects of a mis-throw in which a 'lazy trailing finger' induces a torque between the rim and the spin axis which causes a 'wobble' in flight, known as the torque-free precession. The precession frequency is defined by the previously derived Equation 6.25:

$$\tau_p = \frac{2\pi}{\omega \, \cos\alpha \left[(I_s/I_\perp) - 1 \right]}$$

where τ_p is the period of precession, ω is the angular velocity of the discus spin, α is the precession angle (the angle the axis of the precessing discus makes with the spin axis), I_s is the moment of inertia along the spin axis, and $I_\perp$ is the moment of inertia perpendicular to the spin axis.

Variation of lift and drag forces with angle of attack

It has been stated that the lift and drag coefficients, C_L and C_D, depend strongly on the attack angle, ψ. A number of studies and experiments have been carried out in order to characterize the variation of C_L and C_D with ψ; most using a wind tunnel, but generally operating on non-spinning discs. Ganslen (1964) considered the discus to be a wing with an aspect ratio (span : chord) of 1. All investigations report a steady increase in drag coefficient with increasing attack angles, from $0°$ where $C_D \approx 0.06$, up to $90°$ where $C_D \approx 1.07$.

The variation of lift coefficient, C_L, with attack angle varies in a more complex way, but is generally characterized by a rapid increase from $C_L = 0$ at $\psi = 0°$, up to $C_L \approx 0.9$ at about $\psi = 30°$. It then drops back, somewhat more slowly, to 0 again at $\psi = 90°$. It would appear that the abrupt decrease in lift coefficient at around $\psi = 30°$ coincides with the formation of a turbulent wake behind the discus. The values of both C_L and C_D reduce by about 30 per cent for an increase in wind (or discus) velocity from 21 to $30 \, \text{m s}^{-1}$.

Figure 8.4 graphs the variations of both C_L and C_D with ψ portraying the means of the major studies in the field.

Effect of wind velocity

We have stated that greater range can be achieved when the discus is released into a headwind. In this section we will endeavour to quantify the phenomenon. The model, which is based on Equations 8.4, 8.6 and 8.7, indicates that the worst case scenario is when throwing into a headwind of $7.5 \, \text{m s}^{-1}$. However, the discus may travel up to 6 m further the into the headwind than with the wind. For winds up to $20 \, \text{m s}^{-1}$, throwing into the wind will always increase the range, albeit minimally for the slower release velocities.

The question may be asked; is there a limit to the strength of headwind, beyond which the range will start to reduce. We may obtain an approximate answer to this by considering an idealized throw in which the angle of release is horizontal ($\theta_0 = 0°$) and the angle of attack,

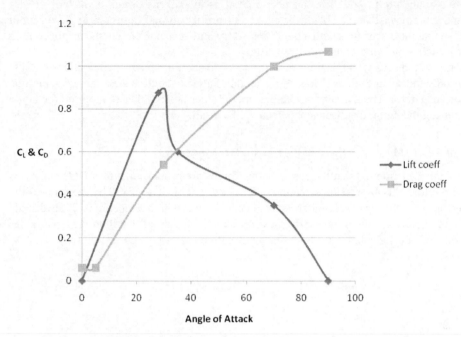

Figure 8.4 Variation of lift coefficient, C_L, and drag coefficient, C_D, with angle of attack, ψ, of discus.

$\psi = 0°$ so that the drag is a minimum. If the wind speed is high enough, the discus will stop its forward motion and begin to head back to the thrower before it hits the ground.

Equation 8.4 will then simplify to:

$$\frac{dv_d}{dt} = -\frac{1}{2}\frac{\rho A C_D}{M}(v_d - v_w)^2$$

where v_d is the velocity of the discus at any time in the flight, t.

This can be integrated with respect to time to yield:

$$v_d = v_w - \frac{1}{(K - Xt)}$$

where K is the constant of integration and equals $-1/(v_0 - v_w)$ with v_0 being the initial horizontal velocity, $X = \frac{1}{2}(\rho A C_D/M)$ and, of course, t is the time. The wind will decelerate the discus until $v_d = 0$ and the maximum distance X_{max} will be reached, where:

$$X_{max} = \frac{1}{X}\left(\frac{-1}{(1 - (v_w/v_0))} + \ln\left|1 - \frac{v_0}{v_w}\right|\right)$$

Remembering that we have assumed minimal drag, an optimal throw with typical values for C_D, v_0, ρ, A and M may produce a range of about 60 m under conditions of a still wind. However, if we were to throw the discus into a headwind which is, say, three times that of the throw velocity, then $v_w/v_0 = 3$ and $X_{max} = 51$ m. This confirms that a headwind which is too high will, in fact, reduce the range. This equation is modelled in Model 8.1 – Discus trajectory, on worksheet 2 – entitled wind effects.

In conclusion then, when throwing into a headwind, the attack angle should be 10–15° less than the release angle. This ensures that, for the majority of the travel, the attack angle is positive enough to provide lift, but not so great as to stall the discus flight. For greater headwinds, the release angle should be reduced from its normal optimum angle of about 35–37° so that the discus flies with a flatter trajectory and does not hang in the air for so long that it to catches the wind at the end of the throw.

In contrast, a discus which is thrown with the wind should be launched at a larger angle and, for tailwinds greater than $20\,\mathrm{m\,s^{-1}}$, the angle of attack should have a larger magnitude (i.e. more negative). Then, assuming there is enough inertia from the spin to hold the angle steady during the flight, the discus may behave like a sail.

Effect of ρ, A, M, g, y_0, C_D and C_L on discus trajectory

We see from Equation 8.3 that the aerodynamic forces are proportional to $\rho A/M$. Only ρ is a true variable in this expression since the other two 'variables' are fixed by the rules of the sport. Density, ρ, can vary by as much as 50 per cent between high temperature, high altitude and low temperature, low altitude extremes. A convenient way of handling these variables within our existing calculations is to define an effective mass, M_{eff} given by:

$$M_{eff} = M \left(\frac{\rho_{STP}}{\rho} \right) \left(\frac{A_{mens}}{A} \right)$$

where $\rho_{STP} = 1.29\,\mathrm{kg\,m^{-3}}$ and $A_{mens} = 0.038\,\mathrm{m^2}$. This expression allows us to conveniently judge the effect of variations in air density and size of discus. In general, increases in range are small for normal decreases of M_{eff}, but is about six times larger when the discus is thrown into a $10\,\mathrm{m\,s^{-1}}$ headwind. Changes in altitude and temperature may only affect the range by 1m on a still day, but several metres under stiff wind conditions. Furthermore, under reduced M_{eff}, it can be shown that the optimum release angle, θ_m, is reduced; and note that M_{eff} will be lower in the case of the women's event, so women should aim to throw at a somewhat lower release angle than men.

Variations in gravity, g, and release height, y_0, result in small changes in range. Over the Earth's surface, gravity varies by approximately 5 per cent (or $0.05\,\mathrm{m\,s^{-2}}$). An increase in g of this amount results in a decrease in range of only $0.34\,\mathrm{m}$, all other variables remaining fixed. It is almost impossible to increase the release height significantly without completely ruining the throw technique. However, an increase in release height of 1 m would increase range by about 2 m (see Equation 3.15) which goes some way to explaining why discus record holders tend to have ectomorphic (long-limbed) frames. See Model 8.1 – Discus Trajectory, Worksheet 3 – Effective mass.

Finally in this section, we can see from Equations 8.3 and 8.4 that varying either C_D or C_L will have exactly the same effect on range as variations in g and M_{eff}. There will, therefore, be small variations under still conditions, but we would expect larger effects when thrown into a wind. An equal increase in both C_D and C_L will result in an increased range when $v_w = 10\,\mathrm{m\,s^{-1}}$. However, increasing C_D while holding C_L constant will, understandably, decrease the range. In fact, if C_D is doubled, the range is reduced by some 15.8 m, while halving C_D will reduce the range by 19.4 m, *irrespective of the angle of attack*.

Summary of release optimization

Bartlett (1992) eloquently summarizes the collective findings from numerous papers on the subject of the most optimum values to maximize range:

- release speed should be as great as possible and this is by far the most important release parameter,
- release height should be as high as is consistent with other optimal release conditions,
- the discus should be thrown to maximize its spin but with zero rolling and pitching angular velocities at release,
- in still air, the release angle should be $35–37°$, with an attitude angle about $10°$ smaller, for elite throwers (i.e. angle of attack equal to $-10°$),
- in headwinds, a lower release angle should be used with an increased negative angle of attack, with an opposite trend for tailwinds.

8.5 Trajectory diversions: Ultimate

The flying disc, often called a Frisbee, a name registered to the Wham-O toy company, is aerodynamically much more efficient than the discus. Although the drag-to-weight ratio is not especially impressive (where it wins on lower drag, it loses on much lower weight), its lift coefficient may be double that of the discus. The flying disc was invented in 1871 in Bridgeport, Connecticut where a certain William Frisbie owned a bakery and pie company – can you see where this is leading!

The students at the nearby Yale University enjoyed tossing the pie tins around and they referred to them as 'Frisbies'. The first 'real' plastic flying discs were manufactured in 1958 by Fred Morrison. The Wham-O toy company bought the patent later that year and the rest, as they say, is history (although one cannot help but wonder how close we came to referring to these projectiles as 'Morrisons', which I guess hasn't quite got the same impact!).

Since that time, apart from the obvious basic catching games, many games have been invented around the flying disc (Wikipedia currently lists eighteen variants, but I haven't checked out just how playable they all are), but by far the most popular and serious sport is the one now called 'Ultimate' (formally 'Ultimate Frisbee').

Ultimate was created in the fall of 1968 by Joel Silver, a student of Columbia High School, New Jersey, adapting it from a game known as Frisbee football. The rules of the game were codified there and were tested by an eclectic bunch of academics and students who were studying politics, drama and creative writing. One, Walter Sabo, is now a major media mogul. Although both the rules and scoring remain essentially unchanged from that time, the regulation pitch size is now $100\,m \times 37\,m$, rather than that defined by the parking lot of the school. Further, the team size is now officially seven per side rather than however many happened to turn up for the game.

The sport is considered by many to be a counterculture activity and differs in a number of ways from other team sports. Fair game play, 'gentlemanliness' and gracefulness are considered important. Unique to such games, there is no referee: the game is self-officiated although 'observers' may be employed. A foul is defined as contact 'sufficient to arouse ire of the player fouled'.

Ultimate is a limited-contact team sport played with a 175 g flying disc, the object being to score points by passing the disc to a player in the opposing end zone, similar to American football or rugby. Players are banned from running with the disc.

The game flows as follows:

1   A coin is tossed to decide which team begins defense and which offense.
2   The teams line up at the end of their respective end zones and the defense begins by launching the discus to the offensive team. This is known as the pull, and is designed to give the defense team time to reconfigure their positions.

3   The thrower cannot move with the disc but must throw it within 10 seconds, hopefully to a team-mate in a strategically strong position.
4   The game progresses, a point being earned when an offensive player catches the disc in the end zone being guarded by the defense.
5   Control of the disc is passed to the defense if a thrower fails to complete a pass, maybe because they have waited more than 10 seconds, dropped the disc, thrown it out of bounds or thrown an interception.
6   The winning team is the one with the highest score at the end of two 24-minute halves. Games may also be played to a set number of points agreed on before the start of play.

Summary

This chapter opens with an account of the long and interesting history of the sport, dating back several hundred years BCE with the original Olympiad competitions, up to the present day. The discus event represents arguably the most complex of the throwing activities and a brief description of the throw is provided, followed by a summary of the IAAF regulation rules pertaining to the projectile itself, and how that influences, and limits, the throwing styles.

A short section on discus materials and manufacture precedes the trajectory dynamic analyses. The most important force in flight is the lift force which arises due to a type of drag known as form drag; a drag which is created because of its shape. This force is analytically quantified, and its effect is seen as so significant that, so long as optimum release parameters are adhered to as defined in this chapter, the range can be increased, even when thrown into a headwind. Spin maintains the necessary gyroscopic stabilizing forces required to keep the orientation of the projectile close to the optimum values necessary for efficient flight. The chapter concludes with a consideration of the effects that other, less significant parameters, such as air density, have on discus flight trajectory.

Problems and questions

1   Which unpowered projectile can travel the greatest distance through the air when thrown?
2   A discus flies at $20\,\mathrm{m\,s^{-1}}$ at 23° to the horizontal into a headwind of $10\,\mathrm{m\,s^{-1}}$. Calculate the relative velocity of the discus to the wind.
3   When a discus is thrown into a steady headwind of up to about $20\,\mathrm{m\,s^{-1}}$ speed, it actually travels further than on a still day (or even in a vacuum), and yet all the headwind can do is create a force acting against the motion of the discus flight. Describe, in non-mathematical terms, how this does not contravene the laws of physics.
4   If the angle of attack of the discus is too great, it is prone to stalling. What is stalling and how does it come about?
5   A poor discus throw results in a wobble of the discus with a precession rate of 0.2 precesses per second. If its spin rate is $5\,\mathrm{rad\,s^{-1}}$, calculate the angle of wobble around the spin axis.
6   Calculate the theoretical maximum distance a discus may be thrown when launched into a headwind which is four times its launch velocity. State the assumptions you make in your calculations.
7   Calculate the effective mass of a women's discus with diameter of $180\,\mathrm{mm}$ and mass of $1\,\mathrm{kg}$. Assume standard temperature and pressure conditions and the diameter of a men's discus of $220\,\mathrm{mm}$.

Chapter 9

Javelin

9.1 History of the event

Spear throwing, in its various forms, must be one of the oldest sporting events in history. The obvious association of javelin throwing with war and hunting is no coincidence. Early forms of the sport included both throwing for accuracy at a target and throwing for optimum range; either unaided as with the modern day event, thrown while riding a charging horse or with the help of a sling-shot chord mechanism attached to the spear, this latter equipment being known as a mesagkylon.

It is surprising the amount of detail that is known about the earliest developments of the event, although there are some significant gaps in our knowledge, and it must be stated that some particulars concerning the sport are closer to intelligent supposition than known fact. Unlike the discus, the javelin is seldom a subject featured on ancient artifact artwork. Where it is seen on vases and such like, the athletes are posed, exercising, polishing their tip or preparing for their throw, rather than an 'action posture', which would have been more instructive for sport historians. Further confusion is created because the Greeks used measuring rods to measure the ranges of jump, discus and javelin throwing events. On ancient artwork, it can sometimes be quite difficult to tell these rods apart from the actual throwing implement. In some cases, however, the illustration is definitely a javelin as the metal tip can clearly be seen. It is also worth noting that the best artists of the period were not employed to decorate vases and amphora, so in some cases, certain details are lacking on the surviving artifacts.

It has been suggested that the first javelin thrower was Hercules, son of Zeus. As early as 600 BC, at the Panathenaea festivals which were staged in honour of Athena's birthday, mounted adolescent teenagers known as ephebes, competed by throwing javelins at targets. However, the measure of importance of javelin throwing can be illustrated by the size of the prizes awarded. Winning contestants of the *stadion* race (an equivalent of our modern running event; completing laps around what we now term a stadium) might win upward of 50 jars of olive oil, winners of the two horse chariot race may be lucky enough to win a prize of 140 jars. However, our lowly winning javelin thrower would typically take only 5 jars home with him. It is known that, at this time, the competition was a target event with the throw taking place from a galloping horse.

One of the lyric poets, Pindar (born 522 BC) quotes:

> As for this bronze-pointed javelin which I am shaking in my hand, I hope I will not, as the expression goes, throw it out of bounds but rather hurl it a long distance, so as to surpass my competitors.

> Pindar, Pythian Odes 1.44–45(470 BC)

Clearly this is representative of our modern event with throw-line constraints and the objective being a competitive distance throw.

The javelin contest was one of the Olympic pentathlon events. It is thought to be the fourth event, following running, discus and jumping, and preceding the wrestling, although even this is not certain. It is also unknown if the event consisted of more than one throw with the best throw counting, as in the modern day event. During this somewhat more modern era, the javelins were in the region of 2 m in length and about 1.5 cm diameter. They were made of wood; dark leaved Elder in one account dating around the fifth century BC, with a metal ferrule forming the tip. It would be much lighter than those designed for war and it is thought, although far from proven, that ranges over 90 m were achieved. Just as in today's event, there was a 'balk-line', striding beyond which would result in a foul throw, and a 'fair area' in which the javelin had to land.

In the third century, another account of the javelin competition on the Island of Ceos, stated that adult victors won either weapons or money, while, as in the case of the diskos competitions, boys would compete for sacrificial meat, a symbol of acceptance to the citizenship.

Women were known to compete with men from around AD 80. Lycurgus, an orator and administrator of Athens, insisted on both male and female competitions, thereby building strong teams in both genders, which, he believed would, in turn, create strong progeny. However, it was actually the Roman historian Plutarch who added discus and javelin to the female repertoire of sporting activities. Propertius, the Augustan love poet, described naked girls working out in the presence of males and even being rubbed down with oil in a similar manner.

Turning now to modern history, by 1780 the javelin was adopted as a sporting event in Scandinavia, and in particular, Finland. They utilized the throwing technique and actions that we would clearly recognize today. The Scandinavian countries dominated the early modern Olympic events. Women were allowed to compete from 1932.

The dimensional and weight specifications for the javelin have essentially remained unaltered since that first modern Olympics of 1896 for men and the 1932 Games for women. The men's javelin is 800 g in weight and 2.6 m long, while the women's javelin weighs 600 g and is 2.2 m long. However, as discussed later in this section, comparatively recently, the specifications for both men's and women's javelins have been changed in terms of the positions of the centers of pressure.

Up until 1953 javelins were made from wood, usually birch. These were weak and unreliable, either breaking or warping very quickly with use. In 1953, the first hollow metal javelin was developed by Franklin and Dick Held in the United States, and this marked the turn of fortune for the Americans on the competitive scene. In 1952, the Americans Cy Young and Bill Miller won gold and silver medals in, rather ironically, the Helsinki Games. The following year Franklin Held himself was the first American to break the world record with an 80 m throw. It should be noted that his hollow metal javelin shaft was much wider than the traditional ones for the required weight, and so could benefit from the corresponding increased aerodynamic lift.

In 1966, an athlete broke the 100 m barrier, but his throwing technique which involved a discus-styled spin at the end of the run-up and before release, was deemed unsafe and outlawed by the International Amateur Athletic Federation (IAAF). Rubberized mats were introduced around this time which improved and regulated foot grip especially in wet conditions. In the 1970s, distances were safely in the 90–100 m region; by 1980, the record lay at 96 m. Then, in 1984, the 100 m barrier was broken for the second time, but on this occasion the East German, Uwe Hohn, used the traditional throwing style to achieve an awesome 104.8 m.

In respect of the women's event, the Soviet bloc dominated immediately after the Second World War. In the 33 years following 1949, only one non-Soviet athlete broke the world record: Kate Schmidt, an American, achieved 69.32 m in 1977. The women's distances also increased rapidly from the early 1950s because they too benefited from the Held-style javelins. The distance records went from 55 m in 1954, to 74 m in 1982. The 1980s saw the demise of the Soviet domination with excellent athletes from Finland, England and Greece challenging the Soviet bloc titles.

In April 1986, the IAAF instigated 'the rule change' which placed two new demands on the male javelin design. Put simply – and the latter sections in this chapter consider the issue in greater aerodynamic and mathematical detail – the center of mass of the javelin was moved forward by 0.04 m, while the rearward tail section was reshaped and made thinner, to reduce the amount the center of pressure moves forward over the duration of the flight. The reasons for the changes were fourfold:

1   With the world record now lying close to 105 m, it was becoming increasingly difficult to safely accommodate the event within the standard stadium running track dimensions. The range had to be reduced if the event was to continue being staged in the vicinity of other field and track events.
2   There was a tendency to instability in the throws as the yaw (sideway horizontal movement) would go out of control, not only missing the landing sector but possibly endangering other contestants, officials and even spectators.
3   Although clearly the older javelin was capable of greater range, foul throws were prevalent as the javelin would tend to land either horizontally or tail-first. A near-horizontal landing would often require a difficult judgement call from the officiators since the old-rules javelins tended to pitch, nose-down only when in close proximity to the ground (see Figure 9.3)
4   An indirect consequence is that the new-rules javelin is less susceptible to environmental variations. Although clearly this is an advantageous side-effect, as will be explained later, many feel the sport has lost something as a consequence.

The women's javelin was subject to a similar rule change in April 1991. It is, therefore, important when studying any aspect of the javelin event, be it range-tables and records or trajectory modelling analyses, that one is aware of which era, pre or post rule-change, one is actually examining. As an indication, the male 1986 rule change typically took 10 m off the range and in that interim period before the female rule-change was instigated, male and female ranges were similar.

Petra Felke of East Germany was acknowledged to be the last great women's thrower of this era, setting four records from 1985 to 1988 when she threw 80 m. From 1988 to 2000, consistent gains were witnessed particularly by the male athletes as they mastered new throwing techniques more suited to the new-rules javelin. Three athletes emerged as the dominant throwers of the period; Seppo Raty from Finland, Steve Backley from England and Jan Zelezny of Finland. The latter, with 75 throws recorded at above 90 m as well as three Olympic and two world championship titles, is now acknowledged as the greatest javelin thrower in history. His current record of 98.48 m obtained at Jena, Central Germany in 1996 remains unbeaten by the substantial margin of 5.4 m.

It has to be said that the women's throws do not seem to have made quite the same headway as the men's since the advent of their new-rules javelin although there is evidence of a steady improvement to the present time. The best throwers over the last ten years have been European, and mainly Eastern European, with the German, Tanja Damaske, achieving nearly

67 m in 1999. The current world record holder is Barbora Špotáková, a Czechoslovakian, who threw 72.28 m in September 2008 at Stuttgart.

9.2 Throw technique

Although throw variable sensitivities are discussed in fuller detail later, it is clear that the three key throw parameters are going to be:

1   velocity of throw,
2   angle of throw, and
3   height of throw.

And, in fact, they are listed above in order of importance (i.e. variable sensitivity) with throw velocity by far the most important. As will be shown, although range is sensitive to throw height, it is not a variable which, in practice, the athlete has much control over. Good throw technique is about maximizing items 1 and 3 above, while maintaining a value close to the optimal range of item 2. The athlete focuses their application of physical ability in terms of speed, power and rhythm into each of these areas. To this end, the javelin is not thrown with the arm alone, but 'pulled' through the combined effort of the entire body.

As with many sports throws and hits, the kinematic action begins at ground level where the body segment inertia is largest. It then works up through the segments which generally get smaller and lighter as the movement progresses; the speed increasing until, the last segment, the javelin, is set into motion. The analogy to the cracking whip action is again illustrative of the dynamics involved here, with the whip's start of motion, near the wrist, possessing the characteristics of high power, low speed, and then, travelling along the tapering length of the whip, culminating in the sonic 'crack' due to the low power, high speed action at the whip end. In summary, and with cognizance of impulse requirements, the athlete needs to haul the javelin over as long a distance in as short a time as possible. The throw technique for the right-handed athlete is outlined below:

1   The run-up must generate as much momentum as is contingent with stability and control requirements. After about nine strides, the arm is laid back. There will now be about three or four strides to the throwing arc, with the hand at about shoulder height and the body turned sideways. If the javelin point is on a level with the forehead, the angle of attack will be about correct.
2   When the right foot is planted (before the release), the right knee is bent to soften the impact, allowing the body weight and run momentum to be maintained in the throw direction. The right arm should be held back and straight as the left arm is raised, also in a straight and pointing motion towards the perceived 'target'. It is now the left arm which maintains the center of gravity of the body through the trunk. The left leg is extended forward, waiting for ground contact.
3   The body weight swiftly passes over the right foot and ground contact is made with the left foot. At this point, the whole of the left side is anchored by the bracing action of the left foot; it acts as a fulcrum to allow the right side to continue its forward motion into the release. The right foot should have been planted with the heel out, while the left elbow is positioned over the left foot. The body should be balanced and stable in its movement (if that makes sense!).
4   Furthermore, the posture allows for a series of stretch reflexes to progress, starting with the right hip which rotates forwards, and then stops. This causes the ribs and chest to pull forward to their full allowable extent, which then stretches the shoulder as the javelin

still remains behind the athlete. As the chest stabilizes, the shoulder is whipped up over the body and finally the arm follows on from that movement.

5   As the arm follows through, the elbow is bent in much the same action as a tennis serve. The arm and hand continue to accelerate through the release of the javelin to the extent that the throwing hand may finish with a slap on the thigh. Although the recovery phase is principally on the left leg, slower throwers succeed in staying upright by transferring weight to their right leg. However, faster throwers such as Zelezny often finish on the floor with their face to the ground.

In summary, the overall action is one of an explosive horizontal movement in which:

- the momentum is maintained from the run-up into the throw,
- the body leans back as the legs 'run away' from the upper body,
- the throw is initiated from the legs,
- the hip and shoulder axes rotate with respect to each other,
- the left side of the body is braced, and
- the arm strike is delayed, thereby increasing the feeling of a whipping of the whole body.

In order to carry out video scrutiny of throw technique, Bartlett (2007, p.81) initially divided the actions of the complete javelin throw into four categories:

1   the run-up: creating a controllable maximum speed,
2   crossover steps: 'withdrawal' of the javelin to extend the acceleration path: a transfer from forwards to sideways action,
3   the delivery stride, known as the action phase, similar to the shot, the hammer and, in fact, the discus, and
4   the recovery: to avoid crossing the foul line.

In a further analysis, in a paper by Best et al. (1993), the motion is further sub-divided in a manner immediately associated with the throwing action as follows:

1   time T_1: instant of left foot contact to begin the cross-over stride,
2   time T_2: instant the left foot leaves the ground during the cross-over stride,
3   time T_3: instant of right foot contact with the ground to begin the delivery stride,
4   time T_4: instant the right foot leaves the ground (or starts to drag along the runway),
5   time T_5: instant of final (left) foot plant preceding the release, and
6   time T_6: instant of javelin release.

As a consequence, the phase periods are then given by:

1    $T_1 - T_2$: the contact time of the left foot in the cross-over stride,
2    $T_2 - T_3$: the non-contact (flight) time in the cross-over stride,
3    $T_3 - T_4$: the contact time of the right foot in the delivery stride,
4    $T_4 - T_5$: the flight (and/or right foot drag) time preceding final left foot plant and
5    $T_5 - T_6$: the time between the final foot plant and release of the javelin.

In throw analyses, these phase intervals are used to correlate relationships between the durations, or more often, video frame numbers, against different athlete's throws, with particular reference to ranges achieved.

9.3 Notes on javelin manufacture

9.3.1 International Association of Athletics Federations Rules
Extract from IAAF Rule Book 2008

RULE 193

JAVELIN THROW

THE JAVELIN

3   *Construction.* The javelin shall consist of three main parts: a head, a shaft and a cord grip. The shaft may be solid or hollow and shall be constructed of metal or other suitable material so as to constitute a fixed and integrated whole. The shaft shall have fixed to it a metal head terminating in a sharp point.

 The surface of the shaft shall have no dimples or pimples, grooves or ridges, holes or roughness, and the finish shall be smooth (see Rule 188.4) and uniform throughout.

 The head shall be constructed completely of metal. It may contain a reinforced tip of other metal alloy welded on to the front end of the head provided that the completed head is smooth (see Rule 188.4) and uniform along the whole of its surface.

4   The grip, which shall cover the center of gravity, shall not exceed the diameter of the shaft by more than 8 mm. It may have a regular non-slip pattern surface but without thongs, notches or indentations of any kind. The grip shall be of uniform thickness.

5   The cross-section shall be regularly circular throughout [see Note (i)]. The maximum diameter of the shaft shall be immediately in front of the grip. The central portion of the shaft, including the part under the grip, may be cylindrical or slightly tapered towards the rear but in no case may the reduction in diameter, from immediately in front of the grip to immediately behind, exceed 0.25 mm. From the grip, the javelin shall taper regularly to the tip at the front and the tail at the rear. The longitudinal profile from the grip to the front tip and to the tail shall be straight or slightly convex [see Note (ii)], and there shall be no abrupt alteration in the overall diameter, except immediately behind the head and at the front and rear of the grip, throughout the length of the javelin. At the rear of the head, the reduction in the diameter may not exceed 2.5 mm and this departure from the longitudinal profile requirement may not extend more than 300 mm behind the head.

Note (i): Whilst the cross section should be circular, a maximum difference between the largest and the smallest diameter of 2 per cent is permitted. The mean value of these two diameters shall correspond to the specifications of a circular javelin.

Note (ii): The shape of the longitudinal profile may be quickly and easily checked using a metal straight edge at least 500 mm long and two feeler gauges 0.20 mm and 1.25 mm thick. For slightly convex sections of the profile, the straight edge will rock while being in firm contact with a short section of the javelin. For straight sections of the profile, with the straight edge held firmly against it, it must be impossible to insert the 0.20 mm gauge between the javelin and the straight edge anywhere over the length of contact. This shall not apply immediately behind the joint between the head and the shaft. At this point it must be impossible to insert the 1.25 mm gauge.

6   See Figure 9.1 and Table 9.1.

7   The javelin shall have no mobile parts or other apparatus, which during the throw could change its center of gravity or throwing characteristics.

8   The tapering of the javelin to the tip of the metal head shall be such that the angle of the point shall be not more than $40°$. The diameter, at a point 150 mm from the tip, shall not exceed 80 per cent of the maximum diameter of the shaft. At the midpoint between the center of gravity and the tip of the metal head, the diameter shall not exceed 90 per cent of the maximum diameter of the shaft.

9   The tapering of the shaft to the tail at the rear shall be such that the diameter, at the midpoint between the center of gravity and the tail, shall not be less than 90 per cent of the maximum diameter of the shaft. At a point 150 mm from the tail, the diameter shall be not less than 40 per cent of the maximum diameter of the shaft. The diameter of the shaft at the end of the tail shall not be less than 3.5 mm.

9.3.2 Javelin manufacture

Today's javelins are constructed from either steel or high grade and anodized aluminum. The regulations, as specified above, allow little variation on the dimensions and make-up of the equipment, and so manufacturing methods are fairly standardized.

The javelins are cold-drawn from tubular ingots which are the diameter of the javelin in the grip area. The tube is then drawn out in both directions towards the point and the tail, such that the correct regulation taper is maintained. Some slight variation in taper is allowed, dependent on the distance rating. Javelins designed for greater distances can be more aerodynamic. However, a short throw with such a javelin may result in a trajectory stall and a tail-first, foul landing. A further design variant is the javelin design for headwind or tailwind throws. However, the jury is still out as to whether the correct choice of these alternatives actually makes any difference to the range.

It should be noted that the throw may create a longitudinal axis spin up to 25 rev s^{-1} which helps to stabilize the javelin in flight. It will also be subject to a vibration along its length as the final release necessarily creates cross-forces. Elite throwers want these vibrations reduced to a minimum, as it is detrimental to the throw and gives them a better impression of the standard of their attempt. By contrast, beginner throwers prefer a more flexible, forgiving implement.

The javelin point is turned from steel and is either a heat-shrink fit or glued into place with epoxy resin. Rubber-tipped javelins are often the equipment of choice for schools and colleges for obvious reasons. Soft, non-marking rubber tips can also be used for indoor games.

The javelin is coated with a polymeric paint such as coloured polyurethane, either by spraying or using a powder coat and heat treatment. The final stage in the construction is the application of the cotton grip which is hand-wound onto the implement.

Very high specification javelins are now available made from glass/carbon composite. Although in the hands of an elite thrower, greater ranges may be achieved, they are unforgiving in the hands of lesser mortals. For such athletes, flexibility in the javelin is important to prevent shoulder or hip injury, since 'something has to give' in the throwing action, and usually, the javelin being set into vibration reduces impact injury on the athlete. As well as being exceedingly stiff, these javelins are relatively fragile and expensive and, as such, are normally reserved for the elite (and usually, sponsored) thrower.

Training javelins are created from a variety of cheaper materials and are designed to exercise differing aspect of the event. They may be made from plastics such as polyethylene with soft or hard elastomer tips. One style uses cone-shaped 'wiffleball-like' materials located in several positions along the length of the implement. These flight resistors increase the drag so that the throw improves the athlete's strength and style – a form of resistance training.

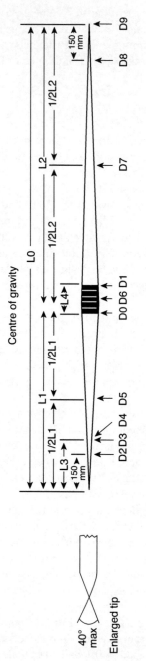

Figure 25 - International Javelin

40° max

Enlarged tip

Lengths (all dimensions mm)

Serial	Detail	Men Max	Men Min	Women Max	Women Min
L0	Overall	2700	2600	2300	2200
L1	Tip to C of G	1060	900	920	800
1/2L1	Half L1	530	450	460	400
L2	Tail to C of G	1800	1540	1500	1280
1/2L2	Half L2	900	770	750	640
L3	Head	330	250	330	250
L4	Grip	160	150	150	140

Diameters (all dimensions mm)

Serial	Detail	Men Max	Men Min	Women Max	Women Min
D0	In front of grip	30	25	25	20
D1	At rear of grip	–	D0–0.25	–	D0–0.25
D2	150 mm from tip	0.8 D0	–	0.8 D0	–
D3	At rear of head	–	–	–	–
D4	Immediately behind head	–	D3–2.5	–	D3–2.5
D5	Half way tip to C of G	0.9 D0	–	0.9 D0	–
D6	Over grip	D0 + 8	–	D0 + 8	–
D7	Half way tail to C of G	–	0.9 D0	–	0.9 D0
D8	150 mm from tail	–	0.4 D0	–	0.4 D0
D9	At tail	–	3.5	–	3.5

Figure 9.1 The javelin.

Table 9.1 IAAF javelin regulations

Javelin			
Minimum weight for admission to competition and acceptance of a record (inclusive of the cord grip)	600 g	700 g	800 g
Information for manufacture: Range for supply of implement for competition:			
	605 g	705 g	805 g
	625 g	725 g	825 g
Overall length:			
Minimum	2.20 m	2.30 m	2.60 m
Maximum	2.30 m	2.40 m	2.70 m
Length of metal head:			
Minimum	250 mm	250 mm	250 mm
Maximum	330 mm	330 mm	330 mm
Distance from tip of metal head to center of gravity:			
Minimum	0.80 mm	0.86 mm	0.90 mm
Maximum	0.92 mm	1.00 mm	1.06 mm
Diameter of shaft at thickest point:			
Minimum	20 mm	23 mm	25 mm
Maximum	25 mm	28 mm	30 mm
Width of cord grip:			
Minimum	140 mm	150 mm	150 mm
Maximum	150 mm	160 mm	160 mm

9.4 Trajectory dynamics

9.4.1 The basics

If the javelin is considered as a simple un-aerodynamic point source mass of 0.8 kg (the men's regulation javelin mass), then its range will simply be given by Equation 3.11:

$$s_m = \frac{v_0^2}{a} \sin 2\theta_0$$

With a throw velocity of $30\,\mathrm{m\,s^{-1}}$ at the optimum angle of $45°$, this equates to a range of about 90 m.

It has been shown (Chapter 7) that this equation provides range values which are not too dissimilar from that of the most accurate shot-put models. However, the javelin, the most aerodynamic of the sports throwing implements, portrays many flight differences to devices such as the shot, where the mass is consolidated closely around the center of gravity or which have a lower drag-to-weight ratio (see Table 9.2).

To commence our analysis, we must make some basic assumptions; some of these will be refined during later, more in-depth analyses:

1   Although the javelin is allowed to pitch in flight (rotate in the vertical plane), it is assumed that there is no yaw rotation, i.e. the javelin stays in the plane of the trajectory (see Figure 9.2a).
2   The javelin is assumed not to spin about its own axis in flight. In fact, spin is imparted and it does alter the aerodynamics of the flight but the effects are second order and will be considered later.

Table 9.2 Physical characteristics of the shot, hammer, discus and javelin

Properties	Shot	Hammer	Discus	Javelin
Mass (kg)	7.260	7.260	2.0	0.80
Volume (l)	0.70	0.70	0.90	1.25
Density (kg/l)	10.37	10.37	2.22	0.64
Typical velocity (m s^{-1})	15	30	25	30
Max vacuum distance (m)	22	94	66	94
Typical throw velocity (m s^{-1})	15	25	25	30
Projected area (m^2)	0.0095	0.0138	0.039	0.063
Inverse mass (k g^{-1})	0.138	0.138	0.50	1.25
Reynold's no.	1.1×10^5	2.0×10^5	3.7×10^5	0.6×10^5
Drag coefficient	0.47	0.7	1.0	1.2
Drag-to-weight ratio	0.01	0.02	0.15	0.85
Aerodynamicity (F_{aero}/F_{drag})	0.0086	0.075	0.764	5.33

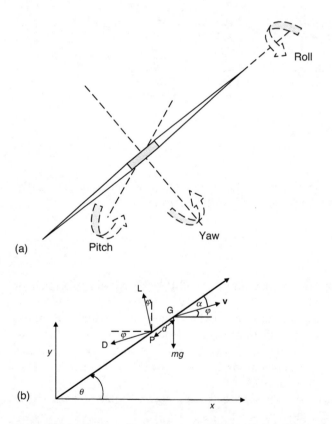

Figure 9.2 (a) Javelin toll, pitch and yaw (b) Javelin coordinate system.

3   For now, we will assume the influence of the wind is negligible.
4   The coefficients of lift and drag depend only on the angle of attack (the angle between the javelin orientation and the relative wind vector). (*Note*: in general aerodynamic analysis this is often called the 'yaw angle'. However, somewhat confusingly, in throwing trajectory dynamics, the yaw is the swing off to left or right.)

5   The lift and drag forces are assumed to be proportional to the square of the speed at each angle of attack. Applying this relationship to the lift and drag force Equations 5.1 and 5.2, we see that the lift and drag coefficients must be velocity independent.

Bartlett and Best (1988) have empirically shown that the variation of drag and lift may be modelled as:

$$\left.\begin{array}{l} K_D = 0.00024e^{0.09\alpha} \\ K_L = 0.0000561\alpha^{1.34} \end{array}\right\} \tag{9.1}$$

where $K_D = \frac{1}{2}\rho C_D A$, $K_L = \frac{1}{2}\rho C_L A$, and α is the angle of attack. See Model 9.1 – Javelin K_L and K_D variation with attack angle. We may now calculate the drag and lift forces once the angle of attack and velocity is known.

6   Even though lift and drag are effective over the whole of the javelin area, it is mathematically expedient to consider that both appear to act from a single point on the javelin known as the *center of pressure*. The precise point of the center of pressure may be found from wind tunnel experiments. It does vary with angle of attack, and so moves along the length of the javelin over the duration of the trajectory flight. It is this movement in relation to the javelin's center of gravity which accounts for the aspect and movement of javelin while in flight. For this analysis, however, the center of pressure is considered to be static over the flight duration, lying a fixed distance behind the center of gravity.

The horizontal and vertical forces resolve respectively as follows:

$$\left.\begin{array}{l} -F_D \cos \varphi - F_L \sin \varphi = m\ddot{x} \\ -mg - F_D \sin \varphi + F_L \cos \varphi = m\ddot{y} \end{array}\right\} \tag{9.2}$$

where φ is the angle the velocity vector makes with the horizontal. As before, the drag opposes the direction vector, $\mathbf{v}$, while the lift is perpendicular to the drag.

For the pitching moment, the rotational forces equate thus:

$$\left(-F_D \sin \alpha - F_L \cos \alpha\right) d = I\ddot{\theta} \tag{9.3}$$

where I is the moment of inertia, d is the distance between the center of pressure and the center of gravity, and θ is the pitching angle (angle of javelin to the horizontal) at any moment in the flight. So, $\theta = \alpha + \varphi$ and $\varphi = \tan^{-1}(\dot{y}/\dot{x})$. See Figure 9.2b.

Equations 9.2 and 9.3 are three second-order linked ordinary differential equations. In order to simplify the solution we convert them to six first-order equations and adopt a numeric approach to the solution.

$$\left.\begin{array}{l} m\dfrac{dv_x}{dt} = -F_D \cos \varphi - F_L \sin \varphi \\[2ex] m\dfrac{dv_y}{dt} = -mg - F_D \sin \varphi + F_L \cos \varphi \\[2ex] I\dfrac{d\omega}{dt} = \left(-F_D \sin \alpha - F_L \cos \alpha\right)d \\[2ex] \dfrac{dx}{dt} = v_x \\[2ex] \dfrac{dy}{dt} = v_y \\[2ex] \dfrac{d\theta}{dt} = \omega \end{array}\right\} \tag{9.4}$$

where ω is the angular velocity of the pitching rotation.

This equation system can be solved to obtain unique solutions from a given set of initial conditions at suitable discrete time intervals (e.g. $\Delta t = 0.05$ s). The standard Runge–Kutta method may be tuned to find optimum variable values to maximize range. The model will also ascertain if $\theta > 0°$ when $y = 0$; which would correspond to a foul throw due to the tail hitting the ground before the tip.

In this model, typical initial values for variables may be: $m = 0.8$ kg, $d = 0.255$ m, $I = 0.42$ kg m^{-2}, $y = 2$ m, $x = 0$ m, $\theta = 30°$, $\phi = 30°$, $V = 30$ m s^{-1}, and $\omega = 0°$ s^{-1}. For these values, the model calculates a range of 82.7 m and confirms that the throw is legal.

9.4.2 Overview of more advanced analyses

In all probability, one of the most accurate optimizing javelin models is delineated by Hubbard and Rust (1984), the details of which are beyond the scope of this text. However, using their model, parameter variables are either provided or derived as follows: mass $= 809.6$ g (a 'Held-90' javelin), transverse moment of inertia $= 0.4094$ kg m^{-2}, distance from center of mass to tip $= 1.08$ m, the wind velocity is assumed to be zero, height to center of mass $= 2$ m, release velocity $= 31.1$ m s^{-1}, release angle $= 30.11°$, angle of attack $= +5.85°$ and the javelin's rotation in the x–y plane $= -9.2°$ s^{-1}. These values are considered to be typical for a good throw. The model not only shows the trajectory of the javelin to a good accuracy, but also the angle of attack of the javelin at each point in its trajectory.

As has been stated, there is an optimum theoretical angle of release ignoring biomechanical limitations, and an optimum realistic angle of release which takes into account the fact that the release velocity is dependent on release angle. This factor has to be incorporated into the model to generate the chosen stated values of v_0 and θ_0 in our model. The expression first verified by Red and Zogaib (1977) and used by many models since is:

$$v_0 = v_{35°} + 0.127\left(\theta_0 - 35°\right)$$

They considered maximum release velocity to occur at $35°$ so the equation is normalized to that angle, $v_{35°}$ being that top release velocity. It would seem that 0.127 times the difference in angle from $35°$ would result in a tiny variation in v_0, but the fact that it does make a significant difference in the model only serves to highlight the extreme sensitivity of v_0 to the range.

A different approach utilizes the impulse-momentum equation on the release phase:

$$F\Delta t = m\Delta v$$

From this we see that, for a fixed mass, m, the change in velocity (from zero to v_0) is equal to the product of the applied force and *the time over which that force is applied*. On the old-rules javelin, many athletes found it advantageous to throw at a lower than theoretically optimum angle of release. This allowed them to hang onto the javelin for the longest possible time, increasing Δt and hence the velocity of release. However, this advantage was largely negated with the arrival of the new-rules javelin.

Figure 9.3 shows the output from the Hubbard and Rust model (1984) applied to both the old- and new-rules javelin. Several interesting aspects from these plots can be considered.

1 Each javelin point marked on the graph represents an equal time increment of 0.524 s. It can be seen that, for both types of javelin, the instrument slows down towards the end of the flight; however, the new rules javelin commences the deceleration earlier in the flight.

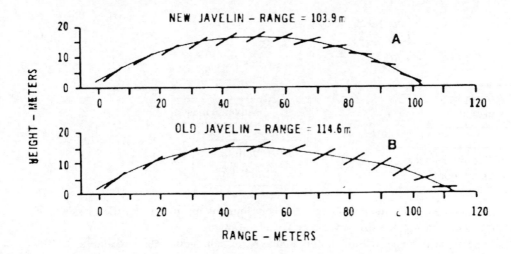

Figure 9.3 Comparison of old- and new-rules javelin trajectory.

2   The angle of attack is maintained, and in fact increases, much longer in the case of the old-rules javelin.

3   The added lift of the old-rules javelin as a consequence of the larger angle of attack is clearly visible over the latter half of the trajectory as a distortion in the basic parabolic trajectory.

4   It is unclear in the case of the old-rules javelin, for this run of the model, if the javelin hits point first or is a foul throw.

5   The difference in range between old and new javelin is of the order of 10 per cent.

The aerodynamicity of a projectile is defined as the ratio of maximum possible aerodynamic forces which might occur during flight to the gravitational force.

So:

$$\frac{F_{aero}}{F_{grav}} = \frac{(\rho v^2/2) A C_D}{mg} \tag{9.5}$$

For comparative purposes we see (Table 9.2) that, for the shot put, aerodynamicity equals 0.0072, for the discus it is 0.68, while for the javelin it is a much higher value of 4.49. The javelin is about five times more aerodynamic than the discus which, in turn, is nearly 100 times more aerodynamic than the shot put. The implications of this are that, for the shot put, drag is negligible, maybe reducing the range by about 1 per cent over the dragless case. Although drag on the discus is far from what may be considered negligible, the additional lift factor all but compensates for the drag, and so the range in air and in vacuum are broadly similar. Now, in the case of the javelin, the lift is still considerable, but the weight and drag are much reduced compared with the other two throwing implements. The consequence is that the range is calculated to be 18 per cent greater than the range of a throw in a vacuum.

Equations 9.5 may be decomposed to its factors:

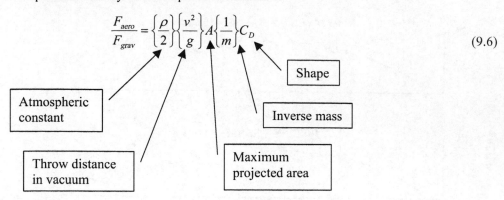

$$\frac{F_{aero}}{F_{grav}} = \left\{\frac{\rho}{2}\right\}\left\{\frac{v^2}{g}\right\}A\left\{\frac{1}{m}\right\}C_D \qquad (9.6)$$

Shape

Atmospheric constant

Inverse mass

Throw distance in vacuum

Maximum projected area

Equations 9.6 indicates respectively the contributions of air density, the mass specific initial kinetic energy, the size, inverse mass and the shape.

One can perform a second-order polynomial regression on the Hubbard and Rust (1984) model, allowing us to simply derive an expression for range from optimum practical (including the angular pitch velocity variation) release velocity:

$$R = 1.248 + 0.0397v_0 + 0.1088v_0^2 \qquad (9.7)$$

This can be compared with the standard equation for the vacuum, zero-drag conditions modified for release height, h, and distance of the center of mass to the tip, d:

$$R = \frac{v_0^2}{g} + d + h = 0.1041v_0^2 + 1.16 + 2.00 \qquad (9.8)$$

Clearly there is a similarity in these two expressions. However, the difference in the multiplier to the quadratic term (0.1088 compared to 0.1041) and the linear term ($0.0397v_0$) are purely measures of the contribution of the aerodynamic process.

Worked Example 9.1

Calculate the difference in range for a javelin released with a velocity of $30\,\mathrm{m\,s^{-1}}$ under otherwise optimum conditions, first including the drag/lift effects, and second excluding it.

For the case including drag, from Equation 9.7:

$$R = 1.248 + 0.0397v_0 + 0.1088v_0^2 = 1.248 + 0.0397 \times 30$$
$$+ 0.1088 \times 30^2 = 100.36\,\mathrm{m}$$

For the dragless case, from Equation 9.8:

$$R = \frac{v_0^2}{g} + d + h = 0.1041 \times 30^2 + 1.16 + 2.00 = 96.85\,\mathrm{m}$$

A change in range of 3.51 m.

Hubbard and Alaways (1987) noted a highly non-linear step function in several parameters for a release velocity in the region of 21–23 m s^{-1}. First, the angle of attack suddenly jumps from $-11°$ up to around $0°$ at about $22.5°$. Second, the angular velocity of the pitch drops from around $18°$ s^{-1} down to $0°$ s^{-1} at 23 m s^{-1}. It then creeps up in a linear fashion to $3°$ s^{-1} at a release velocity of 35 m s^{-1}. Finally the release angle rises from $26°$ to $30°$ over a release velocity range of 20–24 m s^{-1}. Thereafter it levels out over the rest of the increasing release velocity range.

One assumption made in order to simplify the basic analysis of Section 9.4.1 is that d, the distance between the center of pressure and the center of gravity (Figure 9.2) remains constant throughout the flight. This is not the case. De Mestre (1990) described movement of the center of pressure as being; first behind the center of gravity, then moving forward (and for a short time, it actually moves in front of the center of gravity) before moving backwards again towards the end of the flight, so that the javelin undergoes a final pitch downwards as it nears the ground, thereby ensuring a legal point-first landing. The more complex models incorporate this dynamic center of pressure movement which, although not a sensitive variable in the model, is nevertheless an appropriate modelling step-wise refinement.

Arguably, the most in depth studies of aerodynamic fluid analysis have been carried out in the field of aircraft design. In fact, a number of modelling insights can be gleaned when the javelin is modelled as a simple aircraft fuselage (Bartlett and Best, 1988). The javelin is considered as an elongated axisymmetric fuselage with a grip. For such bodies, the drag coefficient is approximately proportional to the square of the angle of attack, the pitching moment is linear with respect to the angle of attack and the aerodynamic lift coefficient is exponentially related to the angle of attack.

Using such aviation-derived models, javelin performance under conditions of headwind and tailwind may be analysed. It is found that, generally, throwing under either conditions show benefits in terms of the range attained. The effect of a headwind, on release, is to increase the speed of the javelin relative to the air, and cause the angle of attack to be greater than the difference between the attitude angle and the release angle. The reverse applies to a tailwind throwing situation. A thrower experiencing a headwind will release the javelin with a slightly negative angle of attack, but the headwind will result in a slightly positive angle of attack and hence positive lift.

Best *et al.* (1995) performed a full three-dimensional optimization model from which the contour map shown in Figure 9.4 is produced. It is interesting that the model suggests *two* optimization peaks for maximum range; one with a release angle of just over $32°$ and an angle of attack of something in the region of $-4.2°$ (the upper left maximum contour), and the other with a release angle nearer to $33°$ and a corresponding angle of attack of something in the region of $-5.5°$ (the lower right maximum contour). Both alternatives provide a calculated range in the region of 93 m.

By now it should be clear that for all these attempts to fine-tune the trajectory model, it is the release velocity, v_0, which is by far the most sensitive variable. Hubbard (1984a) took the approach that the advantage gained by throwing an 'aerodynamic' javelin over the 'vacuum' thrown one, to a first approximation, creates a constant range improvement of approximately 20 per cent for typical throwing velocities. In a vacuum, the range is given by:

$$R = \frac{v_0^2}{g} + d + h \quad \text{so} \quad \frac{\partial R}{\partial v_0} = \frac{v_0}{g}$$

and for an aerodynamically thrown javelin, therefore:

$$\frac{\partial R}{\partial v_0} = (1+\varepsilon)\frac{v_0}{g}$$

tail-landing foul throw. It is, therefore, better for the athlete to throw well, with an appropriate style and within his/her known capabilities, while ensuring the sensitive parameters are optimized. However, in training, it may be a good strategy to pay less attention to those less sensitive parameters in order to favour the more sensitive ones.

9.5 Trajectory diversions: including the kitchen sink

In June 2009, a throwing contest with a difference occurred between two great javelin throwers at either end of the competitive age spectrum: Roald Bradstock, aged 47, a British double Olympian, though now a United States resident, and Matti Mortimore, aged 15, one of Britain's brightest young throwers who has recently achieved a personal best of 69.63 m.

In the contest, not a javelin was seen, although ten rather unusual projectiles (from a sporting perspective) were competitively dispatched. Perhaps the most familiar projectile to them would have been the snooker cue. Bradstock managed 60.04 m, some 6.5 m further than the youngster. Similar in shape, though rather a different size, was the next projectile: the toothpick. In this round, Mortimore trounced the more experienced athlete with a remarkable 15.1 m. Mortimore was also clearly the more confident in the welly-wanging challenge (*Note*: competitively throwing a boot is a common competition in many parts of the United Kingdom, particularly Yorkshire where it goes by the term welly-wanging). Mortimore attained a creditable 40.6 m against Bradstock's feeble 25 m.

One can only assume that Bradstock has had some discus throwing practice since he managed to dispatch the vinyl LP record a remarkable 112.1 m against Mortimore's trifling 13 m. Could this be explained by the fact that the teenage Mortimore likes to download his music, while Bradstock perhaps prefers his melodies in the more traditional analogue record form? In support of this conjecture it is worth mentioning that, on a previous occasion, Bradstock has been recorded throwing an iPod 150 yards.

Table 9.3 shows the complete results table and Bradstock winning the competition 8–3, with particular competences in the delivery of the flatter objects such as the mullet, the *Guinness Book of Records*, and the bar of chocolate. On the other hand, Mortimer clearly excelled with a golf ball and the toothpick. We will never know what he might have achieved in throwing the water balloon; while Bradstock only managed 3.8 m, sadly with Mortimore's attempt, the balloon exploded on launch!

Table 9.3 The throws decathlon

Item	Bradstock	Mortimore
Snooker cue	60.04	53.65
Welly	25.00	40.60
Water balloon	3.80	Exploded!
Mullet	33.10	27.4
Golf ball	96.60	108.90
Guinness Book of Records (1976)	50.55	39.75
Toothpick	13.30	15.10
Paper aeroplane	33.65	2.80
Chocolate bar	89.00	39.23
LP record	112.10	13.00
Kitchen sink	14.40	12.40

Summary

The chapter opens with a discussion of the long and fascinating history of the javelin event, from Ancient Greece to the present day. It is interesting that, in 1986, rule changes regarding the positioning of the javelin's center of mass were instigated, which were intended to reduce the range of the contemporary javelins, primarily for safety reasons. It should be noted that this would not have been successful without a detailed qualitative and quantitative understanding of the javelin's flight.

The javelin throw technique is described with particular reference to the athlete's performance and their adherence to optimized release parameters. Included in this section are brief notes of the IAAF regulations pertaining to the javelin and how these rules inform the manufacturing processes.

Trajectory analysis of the javelin follows, with emphasis placed on the role of both drag and lift in the flight. Although the derived equations of motion are complex, polynomial regression methods reduce the equations to an easily usable form. Analysis under headwind and tailwind conditions follows, based on aircraft flight calculations and the results of a full 3-D optimization model are presented in the form of a contour map. Aspects such as rate of spin, grip methods and variations in tail shape are addressed in qualitative terms and their influence on flight is revealed.

Problems and questions

1 In 1953, the first hollow metal javelin was developed by Franklin and Dick Held in the United States. However, we have the technology to manufacture javelins with exactly the same physical properties (dimensions, weight, weight distribution, etc.) but solid, rather than hollow. How would you think such a javelin would fly compared with a regulation javelin?

2 Plot a graph of lift and drag coefficients against angle of attack ranging from $0°$ to $20°$ (from Equation 9.1). What can you say about these coefficients for small and large angles of attack?

3 Calculate the difference in range for a javelin released with a velocity of $25\,\mathrm{m\,s^{-1}}$ under otherwise optimum conditions, first including the drag/lift effects, and second excluding it.

4 Calculate the aerodynamicity for an Amazonian blowdart with a mass of 150 g, a projected area of $2\,\mathrm{cm^2}$, and a drag coefficient of 1.0 when it leaves the blowpipe at a velocity of $40\,\mathrm{m\,s^{-1}}$. Take air density to be $1.2\,\mathrm{kg\,m^{-3}}$.

How does the aerodynamicity compare with that of the shot put, the hammer, the discus and the javelin?

5 An athlete can throw a javelin 83 m. If she manages to maintain all the same throw parameters but raise her release height by 0.04 m, by much will her range increase?

6 Hubbard and Alaways (1987) investigates and questions the 1986 IAAF javelin rule change. He states:

As already noted, the main effects of the rule change have been (a) to decrease the achievable range by about 10 per cent, by reducing the potential to utilize aerodynamic lift to increase range and (b) to make the flight much less sensitive to initial condition.

Thus by the rule modification, the javelin throw has been changed from an event in which finesse and skill were important (perhaps too much so) to one for which strength and power are once again preeminent.

Discuss.

Chapter 10

Golf

10.1 Introduction

10.1.1 The game

Of all the sports discussed in this book, golf is, in all probability, one of the earliest to be devised. The Romans had a game called Paganica, which involved hitting a stone with a stick. The French had a similar game called chole, while the English had cambuca, which used a ball made of wood. Possibly the strongest claim to golf comes from the Dutch, who were known to play a game called kolf as early as 1296. In its original form, kolf was played on any available terrain, including churchyards, highways, and frozen lakes. The object was to hit a succession of targets by striking the ball with a long-handled wooden club. To allow a clear shot, the ball was slightly elevated on a pile of sand called a tuitje, from which we get the modern term, tee.

An alternative theory suggests that the Scots were actually the originators and earliest exponents of the game we know today. However, it is definitely known that golf, in some form similar to the present day game, has existed for at least five hundred and fifty years. James II of Scotland, in an Act of Parliament dated March 6th, 1457, had the game banned on the grounds that it was interfering with his subjects' essential archery practice (Chapter 15 discusses the trajectories of arrows!).

The first surviving written reference to the game of golf at St Andrews is contained in Archbishop Hamilton's Charter of 1552. This reserved the right of the people of St Andrews to use the 'linksland' (the name given to a stretch of land near the coast which was characterized by undulating terrain, often associated with dunes, infertile sandy soil and grasses) 'for the purposes of golff, futball, schuteing and all gamis'. Indeed, as early as 1691, the town had become known as the 'metropolis of golfing'.

In fact, the word golf may be of Scottish origin, derived from the words glove, gowl, or gouf, possibly borrowed from the medieval Dutch word colf, meaning club ('spel metten colven' translates as 'game played with club'). It has been suggested that shepherds near St Andrews became adept at hitting round stones into rabbit holes with their wooden crooks. However, it was an enthusiastic Scottish Baron, James VI, who brought the game to England when he succeeded to the English throne in 1603.

The Honourable Company of Edinburgh Golfers was formed in 1744. At this time, the first rules of golf (numbering thirteen in all) were drawn up for an annual competition open to sportsmen from any part of Great Britain and Ireland. A few years later, the Society of St Andrews Golfers was formed, and in 1834, when King William IV became the Society's patron, the title was changed to the Royal and Ancient Golf Club of St Andrews.

The Royal Blackheath Golf Club of England was founded in 1766, followed by the Old Manchester Golf Club on Kersal Moor in 1818. Surprisingly, golf established its roots in

Canada prior to the United States, with the formation of the Royal Montreal Club in 1873, the Quebec Golf Club in 1875, and a golf club at Toronto in 1876. So, it was not until 1888 that golf arrived in the United States. A Scotsman, John Reid, first built a three-hole course in Yonkers, New York near his home, and then later that same year, he formed the St Andrews Club of Yonkers on a nearby 30 acre site.

Today the Royal & Ancient, St Andrews for the United Kingdom, and the United States Golf Association (USGA) for the USA, are the two major governing bodies committed to promoting and developing the game of golf worldwide.

10.1.2 Golf clubs

The earliest club makers were thought to be the skilled artisans who produced bows and arrows, and other implements of war. The earliest reference to a set of golf clubs was again in Scotland. A set would consist of; 'play clubs' (long-noses) for driving, 'fairway clubs' (grassed drivers) for medium range shots, 'spoons' for short range shots, 'niblicks' that are today's sand wedges, and finally, a 'putting cleek'. The first authenticated record of a club maker was in 1603, when William Mayne was appointed to the Court of James I of England, to make golf clubs for the king and his courtiers. Two well-known Scottish club makers of the time were Andrew Dickson of Leith and Henry Mill of St Andrews.

The club heads were made from tough woods such as beech, holly, dogwood, pear or apple. By contrast, the shafts were made of the more flexible, ash or hazel wood, which provided the necessary whip to the drive. Once the parts had been carved into shape, the club-maker would attach the club head to the shaft with a splint and then bind them together using leather straps. Later improvements included hollowing out the club-head and back filling it with lead-shot, and also applying a harder front face to the club using bone. The clubs thus produced were not cheap and many survived only one game. So already, the game was becoming associated with the upper classes.

The modern day club consists of the grip, the shaft and the head. Grips are made from rubber or leather with different combinations of small holes, grooves or ridges, and they come in different sizes to suite each golfer. These materials are designed to make it easier for the golfer to hold on to the club when striking the ball, without the necessity of gripping it too hard. Any tendency of the hand to slip during the club's impact with the ball will result in unwanted extra muscle tension during the swing phase. Furthermore, if the hand is allowed to slip on the shaft, the torque produced on impact may cause the club head to twist around the shaft, producing an undesired slice or hook shot.

Modern shafts may be manufactured from chrome-plated steel, stainless steel, aluminum, carbon or graphite fiber-reinforced epoxy, boron-fiber reinforced epoxy or titanium. Shafts created from carbon-fiber have the advantage of being lighter than steel, so can create a greater angular acceleration. This results in a larger velocity and impact with the ball, which in turn, carries the ball further on a drive. Most manufacturers rate their shafts in one of six degrees of stiffness. If a golfer has good technique and strength, then a very stiff shaft would be preferred, so that all the energy generated in the swing is delivered to the ball. The stiffer shafts can also be more accurate in their delivery of shot.

However, some advantage can be gained from the more flexible shafts. Although they will never provide as accurate a shot as the stiffer ones, a well-timed whip motion can store energy at the top of the swing. This energy will then be released on impact with the ball, resulting in an efficient energy transfer into the launch. This can be used to the advantage of the less muscular, yet skilled exponents of the game.

There are three main types of golf club head; woods, irons and putters. Additionally, hybrid clubs have recently found favour amongst the golfing professionals. Hybrids are a fusion of the wood and the iron, and purport to offer the control of the iron with the forgiveness of the club. Golfing regulations allow for a maximum of fourteen clubs to be carried around the course. Each head will have different parameters of weight distribution and sweet spot (center of percussion) profile.

Club heads for drivers and other woods may be made from stainless steel, titanium, or graphite fiber-reinforced epoxy. Face inserts may be made from zirconia ceramic or a titanium metal matrix ceramic composite. Oversize metal woods are either filled with synthetic polymer foam, or left hollow. Traditionalists can still buy woods that are actually made of real wood. Persimmon, laminated maple, and a host of exotic woods may be used in the head. Wooden club heads are usually soaked in preserving oil or coated with a synthetic finish such as polyurethane to protect them from moisture.

Club heads for irons and wedges may be made from chrome-plated steel, stainless steel, titanium, tungsten, beryllium nickel, beryllium copper, or combinations of these metals. Heads for putters may be made from all of the same materials as irons, but additionally softer materials, such as aluminum or bronze may be used, since the energy of the ball impact is much less when putting.

10.2 The golf ball: development and production

10.2.1 History

Ignoring the undoubted attempts at using stones, pebbles and such-like as the projectile in the very early days of the game, the first golf balls were carved from wood. Information is scant, but these inefficient balls were most likely made of hardwoods such as beech or boxroot. Wooden clubs were often used in conjunction with these balls, which would have made the game of golf a somewhat jarring experience! It is known that these balls were used from at least the mid-fifteenth century all the way up until the seventeenth century, when the 'featherie' ball was invented.

In 1618, this new type of golf ball was created by handcrafting a cowhide sphere stuffed with goose feathers. The balls were manufactured while the leather and feathers were wet. As the leather dried, it shrunk and the feathers expanded to create a hardened, compact ball. Once coated with paint, these balls were sold, often for more than the price of a club. The time-consuming processes involved in creating a featherie ball ensured that the price was certainly out of reach of the masses. James I of England commissioned James Melvill and an associate to make featherie balls for the court. It was an exclusive grant for 21 years with the balls stamped by Melvill. Any other ball found in the Kingdom not bearing his trademark would be confiscated. Though expensive, this type of ball had great flight characteristics, capable of being driven up to 175 yards (160 m) and made the wooden ball obsolete almost immediately. Sadly, the balls were rendered useless once wet. It is interesting to note that much effort went into attempting to make the surface of the balls as smooth as possible, using many thick coats of paint, in the mistaken belief that this would allow for a greater distance. In fact, it was discovered much later by accident, that the opposite was the case.

For over three centuries, the featherie was the standard ball in use, only to be replaced with the advent of the 'gutta percha' ball. The gutta percha ball, made from the sap of the Sapodilla tree of the same name, was discovered in 1848 by the Reverend Adam Paterson (some references would have you believe that a Reverend Doctor Robert Adams was the true inventor: and in the same year). However, the story goes that Rev. Paterson received a gift from India, which was packed and surrounded by gutta percha for protection. He discovered that, when it was heated, the material softened and could then be shaped into a ball. When it cooled, it hardened and it was suitable to be used in the game he loved. It soon became known as 'the gutty'. Even though the ball was easier to produce than the featherie, as it could be formed in a mould (a favourite Victorian pastime), the ball was not an instant success. Although good driving distances could be achieved, it rather tended to duck in flight. Almost by accident, it was soon discovered that used, scratched balls, or ones that were improperly smoothed, often had a truer flight than their smoother counterpart could achieve. Thus the hand hammered gutty ball was formed. Balls were hammered with a consistent pattern throughout with a sharp-edged saddler's hammer. A ball, well constructed in this manner could achieve distances of 225 m, and, furthermore, it could be used in the wet British weather. After a few years, handmade gutty balls gave way to metal pressed gutties, which in turn, made golf affordable for the golfer who had to manage on a lower income. Golf truly became the sport for the masses. 'The Bramble' design, with its minute bulges resembling a blackberry fruit, became the most popular design of the gutta percha era. This pattern was carried over with a few brands of rubber balls.

The advent of the rubber ball changed the face of golf as we know it. A new design was invented by an American dentist, Dr Coburn Haskell, in association with the BF Goodrich Company in 1898. Featuring a solid rubber core with high-tension rubber thread wrapped around it and then covered in gutta percha, this design was universally adopted in 1901, after it proved so effective in the British and US Opens. The first automatic winding machine was patented in 1900 by John Gammeter, allowing the rubber core balls to be economically mass-produced. However, due to lack of standards, size and weight varied widely. Although the balls looked like the gutty balls, they did give the golfer a clear extra 20 m. In 1905, a breakthrough came when William Taylor discovered that, by applying a dimple pattern to the Haskell ball, it would travel an even greater distance, as drag was reduced.

Prior to this, the balls featured a multitude of outer designs for better airflow. The mesh, reverse mesh and bramble designs eventually gave way to the dimple pattern, first used in competition in 1908. Dimples refer to the depressions on the surface of the ball. Once the ball is hit with a backspin, dimples cause currents of air moving above the ball to move faster, thereby reducing air pressure and promoting lift (see Section 5.6). Dimples also facilitate the movement of air to the back of the ball, reducing air pressure at the front face, thereby further reducing drag (see Section 5.2).

The reason dimples are a better option than, for example, a sand roughened surface, comes down to the critical Reynold's number, R_{ecr}. For a sand roughened ball, the reduction in drag at the R_{ecr} is greater than that of the dimpled golf ball. However, as the Reynold's number continues to increase, the drag increases. The dimpled ball, on the other hand, has a lower R_{ecr}, and the drag is almost constant for Reynold's numbers greater than R_{ecr}. Therefore, the dimples cause R_{ecr} to decrease, which implies that the flow becomes turbulent at a lower velocity than on a smooth sphere. This, in turn, causes the flow to remain attached longer on a dimpled golf ball implying a reduction in drag. As the speed of the dimpled golf ball is increased, the drag changes little. This partly explains why an old ball will not have the range of a new one; as the surface wears with use, the dimples become shallower and less capable of moving the air, to reduce drag and provide lift. Although round dimples were accepted as the standard, a variety of other shapes have been tested, including squares, rectangles, and hexagons. The hexagons actually result in a lower drag than the round dimples and is currently manifest in the 'Callaway HX' ball.

Many other ball designs were created and tested during the early twentieth century, including mercury, cork and metal cores. Only in 1972 did Spalding introduce the first two-piece ball, which was a significant improvement on the Haskell.

Size and weight standards were established in 1930 by the British Golf Association, prompting the United States Golf Association to create their own standard in 1932. Both organizations specifications differed until 1990, when a united standard was set.

10.2.2 The modern designs

Below are the official standards as set by the USGA and in use today.

Weight

The weight of the ball shall not be greater than 1.620 ounces avoirdupois (45.93 gm).

Size

The diameter of the ball shall not be less than 1.680 inches (42.67 mm). This specification will be satisfied if, under its own weight, a ball falls through a 1.680 inches diameter ring gauge in fewer than 25 out of 100 randomly selected positions, the test being carried out at a temperature of $23 \pm 1°C$.

Spherical symmetry

The ball must not be designed, manufactured or intentionally modified to have properties which differ from those of a spherically symmetrical ball.

Initial velocity

The initial velocity of the ball shall not exceed the limit specified (test on file) when measured on apparatus approved by the United States Golf Association.

Overall distance standard

The combined carry and roll of the ball, when tested on apparatus approved by the United States Golf Association, shall not exceed the distance specified under the conditions set forth in the Overall Distance Standard for golf balls on file with the United States Golf Association.

The modern ball structure can be divided into two categories; wound balls which are a direct descendent of the Haskell ball, and solid balls. The most recent wound balls, modelled on the ball created by Zieglar Natta in the late 1950s, were constructed from rubber thread wound around one of two types of core; either a liquid core, or a solid core made from synthetic rubber. The ball was covered in either balata, a type of latex sap derived from the Manilkara tree, or a high restitution synthetic rubber called Surlyn developed by Du Pont. The wound balls excel in spin performance, but fall short in distance and durability. These balls are seldom used today on the professional circuit.

The one-piece ball, invented in 1967 by James Bartsch, is moulded completely from Surlyn, with dimples impressed directly from the mould. Because these balls deform to such a large degree when struck, they lose a considerable amount of energy, which in turn reduces the driving distance. Compounds may be added to tailor performance, but payoffs are encountered. For example, when a compound is added to increase the coefficient of restitution to counteract the energy loss on impact, there is a reduction in durability. On the other hand, when a compound is added to improve durability, the coefficient of restitution reduces, leading to even shorter driving distances. Today, the one-piece golf ball is the most basic ball and is primarily for beginners, although they are occasionally used as driving range balls. This type of ball is seldom used as a playing ball today. It is, however, inexpensive and very durable and, on impact with the clubface, it unmistakably has a softer feel.

The two-piece golf ball, invented by Bob Moliter in 1968, is used by most ordinary everyday golfers because it combines durability with the maximum distance. The balls are made with a single solid sphere (core), usually a hard plastic, enclosed in the ball's cover. The solid core is, typically, a high-energy acrylate or resin, and is covered by a tough, cut-proof blended cover, often Surlyn, that gives the two-piece ball more distance than any other ball. The configuration allows the energy on impact to be transferred efficiently into ball flight. Although the firmer feel of the golf ball does produce more distance to a player's game, it is not so easily controlled as a softer ball. The two-piece is virtually indestructible and, with its high roll distance, it is by far the most popular golf ball among ordinary golfers.

Multi-layer balls were patented in 1984. In these, the core material is wrapped in several layers. As a result of the latest advances in technology, manufacturers are able to flexibly combine materials, degrees of hardness, specific gravity, and so on, in ways that enhance the ball in a variety of performance related features.

Three-piece golf balls have either a solid rubber or liquid center (the core), a layer of enhanced rubber, and over that is moulded a cover of durable Surlyn, Surlyn-like material, or balata. Sometimes tungsten weights are used in the middle of the synthetic core to optimize weight distribution. These balls are softer and take more spin, allowing a skilful golfer more control over the ball's flight when hit. They typically have a higher spin rate than a two-piece ball and are more controllable by good players. The layered construction combined with a soft synthetic cover produces the very high spin rate, providing maximum control and feel.

A recent addition to ball construction is the four-piece golf ball. Although not yet in common use, it could be the way ball design is heading in the future. Each layer or component of the golf ball has a specific and different purpose. All the layers work together to offer the longest hitting, softest feeling golf ball. The solid rubber center is designed primarily to provide the explosive distance. This is surrounded by a layer that transfers energy to the 'hot' core. An extra layer compared to that of the three-piece ball is then added. This layer provides the increased distance, whilst also producing the mid-iron spin and feel around the green. However, the feel of the golf ball comes from the outer cover. Usually containing between 300–500 dimples, it is the thinnest layer. In the case of the four-piece ball, it is usually made from urethane since it must be both durable and soft. In summary, multi-layer balls tend to have much better control and feel around the green, they provide more spin, but do not travel as far and are less forgiving on hooks and slices as the traditional two-piece ball.

Balls are tested using a combination of wind tunnels and human assessors. Typical of the type of wind tunnel experiments carried out, are attempts to find the ultimate dimple arrangement. Prototypes are tested for lift and drag characteristics over a range of dimple numbers, diameters, depths and patterns. Measurements are performed over a selection of wind velocities and spin-rates to allow simulation of the variety of driver shots possible. Key constants in the equations of motion for the ball can then be derived so that accurate mathematical models of ball trajectories can be constructed. These computer models will then allow precision predictions of shot trajectory for given launch speed, spin and angle.

The latest advance in the golf ball's development was patented by the 'Precept' company in 1997 and is called muscle-fiber core technology. The core material is made from a unique polymer rubber compound, which bonds tightly with the outer mantle – but only at the time of club impact. The core then 'eases back' during the flight. Another original feature is the fully seamless polyurethane cover. As a consequence, the soft core deforms more when struck by a driver, providing an efficient transfer of energy and the desirable explosive response parameter. As the core loosens against the mantle in flight, the spin is reduced to improve target accuracy. Finally, the responsive outer cover provides the desired 'greenside' feel.

It is generally thought that golf ball performance may now be nearing the limit of its design potential, with only minor improvements and tweaks possible in the future. In fact, improvements in both core and outer layers are still possible, but not easily without overstepping the regulations as laid down by the governing bodies on ball specification. Understandably, the major ball manufacturing companies are always trying to plead the case for rule changes to allow them the freedom to expand into new areas of development.

10.3 Flight parameters and launch technique

(*Note*: Where relevant, all directions stated assume a right-handed golfer.)

10.3.1 Impact dynamics

The impact dynamics of the game of golf are simpler to consider than many other games since the ball is always static when struck. The basic mechanics of impact are considered in Section 4.6 and the velocity imparted to the ball is taken from Equation 4.22 with some symbols changed for convenience.

$$\text{Ball velocity } v = \frac{MV(1+e)}{M+m} \tag{10.1}$$

where M is the mass of the club head, m is the mass of the ball ($= 46\,\text{g}$), V is the velocity with which the club strikes the ball, and e is the coefficient of restitution between the ball and club head.

Obviously, we are looking to maximize v, and to this end, it is worth noting that the ball's initial velocity will be directly proportional to V. The harder you hit it, the further it goes! However, the influence of the masses is a little less obvious as the initial ball velocity is proportional to the fraction $M/(M+m)$. Now, if we take a typical club head mass of 200 g, and with the ball's mass fixed by regulation at 46g, then the fraction $M/(M+m)$ will be equal to approximately 4/5. If we consider, for a moment, a club head of infinite mass, then m would be considered negligible and the $M/(M+m)$ ratio will approach unity. So, although doubling the club head's velocity will significantly increase the driving distance, altering the club head's weight can only vary the velocity by an absolute maximum of 20 per cent. In computer simulation terms, M would be termed an insensitive variable and V a sensitive variable in the model. Furthermore, since V will be greatly reduced by increasing M (because the angular acceleration of the swing will be reduced), in purely practical terms, maximum distance will be obtained with the lightest available club head.

In consideration of the coefficient of restitution (CoR), e, high-speed photography has shown that, on impact, the club head shows negligible deformation; nearly all energy loss on impact is due to the ball. Ball manufacturers make good use of this fact in their marketing promotions. Although the CoR does, indeed, vary with ball quality, it is usually found to be in the region of 0.7. Furthermore, the regulations effectively place a maximum limit on the value of e that manufacturers must not exceed, by specifying a maximum speed the ball must leave a standard machine that provides a fixed impact force to the test balls.

Taking $M = 200$ g, $m = 46$ g and $e = 0.7$, Equation 10.1 gives a value of v/V of 1.4. This indicates that a golf ball leaves the club following impact about 40 per cent faster than the club actually hits the ball. Only the use of high compression (large CoR) balls will improve on this figure, and that will be accompanied by the unsatisfactory loss of 'feel' discussed earlier. Incidentally, an increase in ambient temperature will also increase the CoR of golf balls. Although very much dependent on the type of ball, typically, e can range from 0.64 to 0.75 over the golf-playing annual temperature range of, say, $0°$C to $27°$C. As a consequence, a drive of 220 m on a hot summer's day may only achieve 200 m on a frosty winter's morning.

Over the last 30 years or so, it would seem that golf courses have, seemingly, been reduced in size, as a consequence of equipment improvements and the ease with which long distances can be achieved. Many tournament courses have had to be lengthened in order to maintain their competitive challenge. It is no wonder, therefore, that the governing bodies attempt to limit equipment improvements by regulating equipment design parameters.

The use of a lofted club (club with a face angled upwards) has two effects. First, if we assume the club head is travelling horizontally with the ground at the point of ball contact, then it follows that the contact will be oblique and a vertical component will be generated which raises the ball at an angle to the ground. This provides the height required to achieve a useful distance. Section 4.6.1 discusses this impact in some detail. For the perfect lossless impact ($e = 1$) the launch angle is found to be equal to the pitch of the club. For the lossy case, where on impact, the ball begins by sliding up the face, and then gripping it and rolling up until it leaves the face, Equation 4.30 defines the launch angle.

The second and more complex factor concerns the bottom-spin (or back-spin) which will be imparted to the ball hit with a lofted club. As a consequence of the spin thus applied, there will be an upward force or lift, known as the Magnus force, which will act at right-angles to its direction of flight (Figure 10.1).

For a golf ball driven from the tee with a linear speed of, say, 70 m s^{-1} and an initial rotational backspin of, say, 300 rad s^{-1}, it is found that the amount of lift generated is approximately equal to the force of gravity. Therefore, in the first stages of flight, and until the drag reduces the angular velocity of the spin, the ball may fly in an approximately straight line. A 5-iron can produce up to 630 rad s^{-1} (or 100 rev s^{-1}) causing the ball to initially

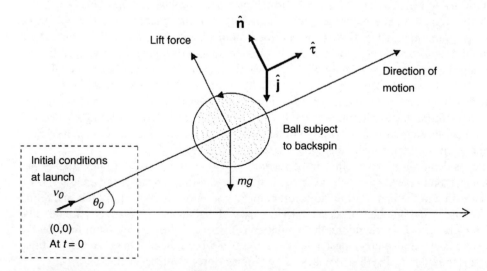

Figure 10.1 Lift on an undercut golf ball.

curve upwards against the force of gravity. After 5 seconds of flight, the ball may have lost 80 per cent of its initial spin and the flight tends back to the parabolic trajectory.

10.3.2 Flight dynamics

Bottom-spin is essential to achieve shot distance. Without spin or drag, the optimum launch angle would be $45°$ (given by Equation 3.11). However, in the case of a golf ball being driven, the biomechanical and physical limitations prevent launching with that amount of lift, while still maintaining the forward component velocity required to achieve good distance. If we now factor the Magnus effect into the equations of motion, we find that the typical optimal launch angle is reduced considerable to around $20°$.

Using the values stated above for linear speed and rotational backspin for a golf ball being driven from the tee, we find the Reynold's number to be 2×10^5, the dimples keeping the drag to a low value at these relatively high speeds. As the flight path proceeds, both linear and angular speeds reduce. The lift and drag coefficients (C_L and C_D) are not typical of the simpler, non-spinning projectiles, but change considerably with the spin parameter $a\omega/v$, where a is the radius of the ball, ω is the angular backspin (rad s^{-1}) and v is the speed of the ball (see Figures 10.2 and 10.3).

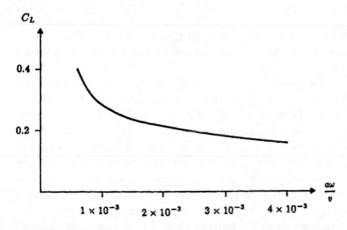

Figure 10.2 Lift coefficient versus spin parameter for a golf ball (from Bearman and Harvey, 1976).

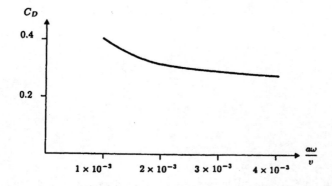

Figure 10.3 Drag coefficient versus spin parameter for a golf ball (from Bearman and Harvey, 1976).

Combining the three possible forces on a golf ball; those of gravity (Chapter 3), backspin and sideways spin (Chapter 6), leads to the summative equation:

$$m\frac{dv}{dt} = m\mathbf{g} - \tfrac{1}{2}\rho A C_D v^2 \hat{\boldsymbol{\tau}} + \tfrac{1}{2}\rho A C_L v^2 \hat{\mathbf{n}} + \tfrac{1}{2}\rho A C_S v^2 \left(\hat{\boldsymbol{\tau}} \times \hat{\mathbf{n}}\right)$$

where $\hat{\mathbf{g}} = -g\hat{\mathbf{j}}$, $\hat{\boldsymbol{\tau}}$ is the unit vector in the ball's direction of travel, and $\hat{\mathbf{n}}$ is the unit vector perpendicular to $\hat{\boldsymbol{\tau}}$. C_S is, of course, the sideways coefficient. (see Figure 10.1).

Because of the variation in values of C_L and C_D, it is best to analyse the golf ball motion using numerical techniques. Assuming no hook or slice shot, the equation above reduces to the following:

$$\left.\begin{aligned}
\frac{d}{dt}(v \cos \theta) &= -\frac{\rho A}{2m} v^2 \left(C_D \cos \theta + C_L \sin \theta\right) \\
\frac{d}{dt}(v \sin \theta) &= -\frac{\rho A}{2m} v^2 \left(C_D \sin \theta + C_L \cos \theta\right) - g \\
\frac{dx}{dt} &= v \cos \theta \\
\frac{dy}{dt} &= v \sin \theta
\end{aligned}\right\} \tag{10.2}$$

Initial conditions state that $x = y = 0$, $v = v_0$ and $\theta = \theta_0$ on impact when $t = 0$ (Figure 10.1). A Runge–Kutta step-by-step method may be applied to these equations, inserting appropriate values of C_L and C_D at each step.

However, Tait (1893) suggested simplifications which certainly seem to produce an accurate model over the substantial portion of the flight. First, he stated that θ is so small over the duration of the flight that we can safely let $\sin \theta = \theta$ and $\cos \theta = 1$. Further, he considered that, as the flight proceeds, both ω and v decrease in such a way as to keep the term $C_L v$ approximately constant. The final simplification, which is credited to Bearman and Harvey (1976), stated that, although ω and v decrease over the duration of the flight, C_D may be taken as approximately constant (i.e. the approximately linear portion of Figure 10.3). Accepting these conditions means that the lift force has been linearized with respect to translational speed, while the drag force remains non-linear, but, at least, in a simpler, quadratic form. The conditions imply that the range of the golf ball is actually approximately proportional to v_0 rather than v_0^2, as would be the case in a vacuum.

Equation set 10.2 now reduces to:

$$\left.\begin{aligned}
\frac{dv}{dt} &= -\frac{\rho A C_D}{2m} v^2 \\
v\frac{d\theta}{dt} &= \frac{\rho A C_L^*}{2m} v - g \\
\frac{dx}{dt} &= v \\
\frac{dy}{dt} &= v\theta
\end{aligned}\right\} \tag{10.3}$$

C_L^* is actually a constant based on $C_L v^2$, justified because C_L decreases with v^2. Setting $K_D = 2m/(\rho A C_D)$ and $K_L^* = \rho A C_L^*/(2m)$ the solution of Equation 10.3 obtained with initial conditions (v_0, θ_0) at $t = 0$ is:

$$\left. \begin{aligned} x &= K_D \ln\left(1 + \frac{v_0 t}{K_D}\right) \\ y &= \left[\theta_0 K_D - \frac{K_D^2}{v_0^2}(v_0 K_L^* - g) - \frac{K_D^2 g}{2v_0^2}\right]\ln\left(1 + \frac{v_0 t}{K_D}\right) + \frac{K_D(2v_0 K_L^* - g)}{2v_0}t - \frac{g}{4}t^2 \end{aligned} \right\} \quad (10.4)$$

This model (see Model 10.1 – Golf trajectory model) based on Tait's, and Bearman and Harvey's approximations, is a very accurate representation of the golf ball trajectory up until just before it lands. The inaccuracies manifest in a slightly exaggerated range and height. However, by adjusting K_D by trial and error, an almost perfect representation can be obtained.

Key features of the resulting trajectory path are: a roughly straight-line initial flight path, and a peak reached at about two-thirds the way to the ball's first contact with the ground. One other limitation in the use of the model is that Tait's first assumption that $\sin\theta = \theta$, and $\cos\theta = 1$, breaks down when the launch angle is greater than $20°$.

Figure 10.4 shows an output from Model 10.1 – Golf trajectory model, i.e. Equation 10.4 for a ball launched at $15°$ with a velocity of $70\ \mathrm{m\ s^{-1}}$ and a rotational backspin of $300\ \mathrm{rad\ s^{-1}}$.

As a contrast to this complex model, D. G. Christopherson cited by Daish (1972) has developed a much simpler empirical set of equations which are rather elegant only if imperial units are used. My attempts at converting the equations to SI units seemed to spoil the graceful simplicity somewhat!

$$\begin{aligned} &\text{Carry in yards} = \tfrac{3}{2}v_0 - 105 \\ &\text{Carry plus run in yards} = \tfrac{5}{4}v_0 - 27 \end{aligned} \quad (10.5)$$

For these equations only, v_0 is then measured in feet per second ($1\ \mathrm{m} = 3.28\ \mathrm{ft}$).

One area that has not, thus far, been considered is the amount of run the ball attains once it has made first contact with the fairway, following its main flight. This will be governed by three factors.

First, the CoR between the ball and the ground must be considered, as the ball carries out its multiple bounces before it settles into its final rolling phase. In respect of the CoR, the golfer can do little about the properties of the course, such as surface composition, dampness, roughness of the surface, and temperature. Of course, balls do have widely varying CoRs, so the player can exert at least a little control over how much bounce is obtained.

The second aspect is the forward momentum remaining after the ball lands following its main flight and first bounce. This is dependent on the mass of the ball (fixed by regulation)

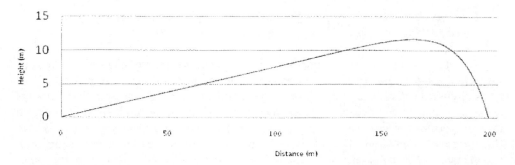

Figure 10.4 The trajectory of a golf drive based on Equation 10.4.

and the angle of impact with the ground. The impact angle, in turn, is imposed by the launch angle (θ_0) and the amount of backspin imparted. As stated, the greater the back-spin, the larger the angle of approach to the ground.

The third factor is the moment of inertia of the ball; the way the mass is distributed from the center core out to the surface. A ball with more weight distributed around the outer surface (a larger moment of inertia) will tend to continue to move forward, (either along the ground, or through the air on its bouncing trajectories) than one which has its weight concentrated in the core. The disadvantage of a ball with a high moment of inertia is that it will not pick up as much backspin from the drive as a ball with a low moment of inertia. It will not, therefore, benefit from the lift that contributes to the greater range. Generally, a ball landing with a low angle of approach (clubs with a loft angle less than that of, say, a 5-iron) will have its backspin cancelled on its first bounce and its future flights resulting from the remaining bounces will be in top-spin mode. The steeper approaches normally seen by the use of, say, a 9-iron, will cancel backspin on its second bounce and will 'top spin' into its third bounce. Wedges usually only experience two bounces before the run-on and the ball is often initially in backspin mode as it slides into its run-on roll. Actual 'spin-back' of the ball under such circumstances on the green is not unknown.

Horizontal ball deviations

As with many of the projectiles discussed here, two primary factors contribute to side-ways deviations of the ball. First, there is the effect of crosswind to consider; and second, sidespin. Sidespin can be, either accidental in nature (the amateur!), or premeditated by the more experienced player, and employed to accommodate course doglegs, other obstacles or obstructions.

The terminologies of the varied shots involving sideways drifts are:

Draw; the ball begins heading to the right of target but heads back to the target.
Fade; the ball begins heading to the left of target but heads back to the target.
Slice; ball begins heading to the left of target and finishes too far to the right of the target.
Hook; ball begins heading to the right of target and finishes too far to the left of the target.
Pull hook; ball begins heading to the left of target and finishes too far to the left of the target.
Push slice; ball begins heading to the right of target and finishes too far to the right of the target.

And for the sake of completeness:

Block push; ball travels in a straight line but stays right of the target.
Block pull; ball travels in a straight line but stays left of the target.

It is quite usual for amateurs to impose rather too much spin and it may suggest the need for them to choose a harder 'distance' ball, which is less prone to spin.

As discussed in Chapters 2 and 5, assessing the effect of windage on any ball trajectory is always a complex issue. We do, however, know the wind effects on golf balls, in comparison to other sporting projectiles, by looking at the value of the (diameter)^2/mass ratio see Table 5.1 (Reciprocal of 4th column). The golf ball falls midway between the cricket ball (negligible), and the table tennis ball (very large, if it were ever subject to wind!). In fact, the only other common sporting projectile with a similar (diameter)^2/mass ratio is the American baseball.

We also know from Chapter 5 that, for the 'well-behaved' wind (i.e. constant wind vector over the duration of the flight), to a first approximation, the horizontal component of the path will resemble a parabola with a constant sideways force, of the form $F_W = \frac{1}{2}\rho A v^2 C_S (\hat{\boldsymbol{\tau}} \times \hat{\mathbf{n}})$ where C_S is a sideways coefficient (c.f. Equations 5.1 and 5.2).

However, because the tee-driven ball travels high, far and fast, it can be subject to widely and rapidly varying wind magnitudes, with the possibility of wind vectors acting in all directions over the duration of the flight. In the case of a golf ball, the wind vector may even act upward or downward. Although, to be fair, this is only likely to occur where the ball travels close to large obstacles, which can produce localized eddying. This is most unlikely on a typical golf course, but spectator stands placed on a course for championships may produce such oddities.

Calculating the sidespin Magnus force is similar to the topspin model previously discussed, but without the added complexity of having to continually vector-sum the force with gravity. The equation is, again, of the form $F_M = \frac{1}{2}\rho A v^2 C_S$. C_S will, of course, reduce over the duration of the flight, as the spin reduces. The Magnus force is always at right angles to the direction of motion. So, if the value of C_S were constant, the motion would be the arc of a circle. In the more realistic case of the spin of the ball reducing over the duration of the flight, however, the arc will degenerate to that of a section of a spiral.

One most interesting phenomenon arises when a ball is influenced by both crosswind and sidespin simultaneously. For a ball subject to a true crosswind (at $90°$ across the line of flight), and also subject to sidespin, the vector sum of the crosswind and airflow acting against the ball's direction results in an airflow vector with a component acting at an angle across the line of flight, and pointing back to the launch point (Figure 10.5). Now, if the direction of spin is such that the direction of rotation with respect to the front face of the ball is the same as the direction of the crosswind, the Magnus force will have a component in a direction *towards the target*. As a consequence, under such circumstances, the ball will actually travel further. Distance is gained by creating a forward directing force vector out of the combination of the crosswind and the Magnus force, neither of which, alone, actually possess a forward-directed component.

The convex face of the club

If one lays a straight edge along the face of a wooden club in a direction parallel with the ground, it will be noted that the club face has been cast or moulded with a slightly concave curve in that direction. This seems rather counter-intuitive in respect of achieving an accurate shot on target. Surely, the player would expect to impact the ball with a club face as normal to the intended ball direction as possible? For the convex face, this would only be

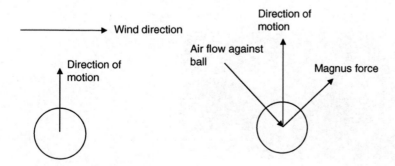

Figure 10.5 The Magnus effect with a crosswind.

true if contact were made with the exact center of the club face. One would imagine that a ball hit slightly off-center on the right side of the face would send the ball in that direction, while a ball hit slightly to the left of center would do the opposite.

The problem arises because of the torque forces generated on the club by the miss-hit, and the solution is to utilize the sidespin Magnus force. A ball hit off-center is going to cause the club to twist in the player's hands. A ball hit on the shaft side of the center of the clubface will tend to twist the club in an anti-clockwise direction (looking from above). A ball hit off-center towards the toe of the club will cause it to twist clockwise. So, even if the club face were perfectly plane, an off-center impact would still cause the ball to push or pull in a direction off target, as the club head twists in the player's hands.

A slightly convex head counteracts this by setting the ball into a sidespin. As the club head twists because of a miss-hit in, say, the clockwise direction, a reaction force is imparted on the ball, sending it spinning in the opposite, anti-clockwise, direction. The reverse would be true for a ball hit on the other side of the center line. Now we have set up a spin on the ball, it will curve through the air. The direction of that swing, in either of the miss-hit cases (i.e. left or right of center line), is such that the ball will be brought back on to target. In summary, a ball hit on the shaft side of the clubface will cause the ball to pull to the left of target but gradually spin back to the right, as the trajectory progresses to the target. A ball hit on the toe side of the center line will initially head right, but swing back left to the target.

10.4 Trajectory diversions: GolfCross

GolfCross is a new game; best described as a cross between golf and rugby. The game is played with standard clubs but the target resembles a rugby goal. The aim is to hit the ball up into a net, strung between vertical and horizontal bars.

However, for the purposes of this treatise, we are interested in the ball. And, yes, the ball is a cross between a golf ball and a rugby ball. It is approximately the same size and weight as a championship golf ball; it is made of similar materials. It even possesses those all too recognizable dimples. The obvious difference is that it is severely egg-shaped!

Now the question is: how well does this ball fly? Does it spin? It does indeed spin, and in a variety of complex ways; tumbling and twisting on its multiple asymmetric axes (see Figure 10.6).

Moreover, reports state that, in fact, it flies truer than any traditional ball, although not so far. There are many modes of flight, dictated by the positioning of the ball on the tee, or 'tee cup' as it is called. The tee cup is made of rubber and actually fits over a traditional golfing tee.

If maximum distance is required, the ball is placed upright but set a little back on the tee cup. There will be minimum backspin, an efficient trajectory and a good run on. To promote

Figure 10.6 Spin orientation modes of a GolfCross ball.

backspin, the ball is placed perfectly upright on the tee cup. The club can now 'get under' the ball and it will tumble away rapidly with a high pitched hum – the higher the pitch, the greater the backspin!

A torpedo shot is produced by pointing the long axis of the ball to the target and impacting on the sharp end. The ball will follow a long, low, yet accurate, flight.

Hitting fades or draws and controlled slices and hooks are simply a matter of correctly setting the spin axis of the ball. The further the ball is angled to one side, the more the ball will curve in that direction. The more it is angled, the flatter the trajectory will be, resulting in a more oblique angle of descent, with greater run on.

The GolfCross ball is reported to be more aerodynamically stable than the traditional round golf ball. Enthusiasts of the game claim even new recruits can:

* hit the ball straight every time;
* perform controlled slices and hooks, when they want to, with ease;
* adjust the degree of fade or draw as required;
* generate backspin – even with a wood, or out of the rough;
* apply top-spin to achieve long low running shots, ...

... and if they really want to show off, they can perform double-curves and even play tunes with the ball!

Summary

The chapter commences with a discussion on the intriguing history of the game of golf, from which one may deduce an insight into the way the ordinary folk lived in centuries long gone, and their simple desire for sport. The way the modern golf club is manufactured is discussed in the light of optimizing ball launch parameters through judicious club design. A detailed history of the golf ball is presented in terms of its development through successive improvement through the ages.

Flight parameters are addressed, starting first with a mathematical treatment of the club–ball impact. Once the ball leaves the club, it flies subject, as usual, to the forces of gravity, lift and drag. The effect of the dimples on the surface of the ball significantly alters the flight characteristics, and this is examined. The results show that, under certain launch conditions, over a large section of the flight immediately following impact, the force of gravity is nearly matched by the lift force. This will result in a ball that rises in a near straight line trajectory, before curving and falling to earth quickly towards the end of its flight. Further trajectory analysis shows that a ball, hit into a crosswind with the aid of spin, may gain range over one hit under still air conditions.

The dynamics of golf ball bounce on the green is considered in terms of its coefficient of restitution and residual momentum. Conditions for multiple bounces are presented and issues of ball 'carry' on the green addressed. Finally, it is shown that, contrary to expectations, a club that presents a slightly convex face to the ball will, in fact, reduce the possibilities of mis-hit errors.

Chapter 11

Tennis and squash

11.1 Introduction: the game

11.1.1 Tennis

> It is well known to tennis players that a rapidly rotating ball in moving through the air will often deviate considerably from the vertical plane. There is no difficulty in so projecting a ball against a vertical wall, that after rebounding obliquely it shall come back in the air and strike the same wall again.

So begins the paper entitled, 'On the irregular flight of a tennis ball' by Lord Rayleigh, dated 1877; arguably the first rigorous mathematical treatise on spin ball flight which was essentially 'correct'. One could, in fact, say that all the future analyses of trajectory modelling involving lift and drag are merely developments and refinements on this paper.

Although tennis can be traced back to the Greek game of sphairistike (meaning 'playing with a ball'), the game is generally thought to have originated in the twelfth century, first created by French monks to keep themselves entertained during religious ceremonies! History does not recall how the early balls were constructed, but it may be assumed that they were quite soft, since the game was played by hitting the ball with the bare hand. Originally called 'jes de paume' (the game of the palm), the game evolved into 'real tennis' over the next three centuries, incorporating concepts from games such as 'Palla', 'Fives', 'Pelota' and 'Handball'. Over that period, the bare hand later became protected by a glove and then replaced by a racquet. It is thought that the early court dimensions were defined by monastery cloisters. The French players would begin the game by shouting the word 'tenez!' which meant 'Play!'

In the fifteenth century a series of drama plays called *The Wakefield Cycle* were written by an anonymous playright known as 'The Wakefield Master'. In his 'Second Shepherd' play he contrived that the newborn baby Jesus was presented with a tennis ball by one of the shepherds. A poem, 'The Turke and Sir Gawain' appears in the Percy Folio Manuscript in which the gallant Sir Gawain plays tennis against a group of giants. Although the Folio is dated 1650, the language and spelling would seem to place the origins of the poem at around 1500 in the North West of England.

Rackets of various forms were seen from the sixteenth century onwards, but it was not until 1750 that the present configuration of the lopsided head, thick gut and a longer handle emerged. Real tennis was played on a stone-floored court with four high walls and covered by a sloping roof. The new rackets allowed the ball to be scooped from the corners of the court and could produce powerful shots, even involving spin. The earliest real tennis balls were made from leather stuffed with wool and hair which, when hit with full vigor, could seriously injure a player and was even known to cause the occasional death. Later balls were less dangerous as the leather was discarded and the outer was constructed from cotton and wool tied tightly around the sphere. At this time, only royalty or the very wealthy could play the game: the word 'real' being a contraction of 'regal'. However, today there are about thirty real tennis courts in Great Britain with vibrant club memberships. The oldest and unquestionably, most famous court is the one within Hampton Court Palace, constructed by Henry VIII in 1625, and it is still in regular use to this day.

In 1874, a Welsh army officer, Major Walter Clopton Wingfield acquired the patent rights for the equipments and rules for a game which bore a resemblance to the modern game of tennis. Around that time the first tennis courts emerged in the United States. The game soon spread to different parts of the world including Russia, Canada, China and India. The smooth croquet courts were considered to be ready-made convenient tennis courts during those times. The original court devised by Wingfield was hourglass shaped with a taper in the region of the net, and was shorter than the modern court. Wingfield's version of tennis courts and the rules of games evolved over a fairly short period of time into the game we know today.

Although the first modern tennis club was opened in Leamington Spa, the first championships at Wimbledon in London were played in 1877. On 21 May 1881, the United States National Lawn Tennis Association (now the United States Tennis Association) was created to standardize the rules and to organize competitions. The US National Men's Singles Championship, now called the US Open, was first held in 1881 at Newport, Rhode Island. The enthusiasm for the game in France did not wane with the years and the French Open held its first tournament in 1891. Thus, Wimbledon, the US Open, the French Open, and the Australian Open (dating to 1905) became, and have remained, the most prestigious events in tennis. Together these four events are called the Grand Slam (a term borrowed from bridge).

The global game rules are now presided over by the International Tennis Federation (ITF), and they have changed remarkably little over the last 80 years or so, with the exception of the introduction of the tie-break rules by Jimmy Van Allen, created to limit the time of play in very evenly balanced matches and to prevent 'a lingering death'.

The general consensus is that the greatest male player of the first half of the twentieth century, if not all time, was Bill Tilden. From 1920–1930, Tilden won singles titles at Wimbledon three times, and the US Championships seven times. In 1938, however, Donald Budge became the first person to win all four Grand Slam singles titles during the same calendar year and, in fact, went on to win six consecutive Grand Slam singles titles in 1937 and 1938. Budge himself stated his opinion that Pancho Gonzalez was the greatest player, with Lew Hoad at least in the running for the title.

Rod Laver, Bjorn Borg and Pete Sampras were each regarded by their professional peers to warrant the title of best male player. John McEnroe has said that either Laver or Sampras is the greatest player ever. Roger Federer is now considered to be the greatest all-rounder with great potential still ahead of him. However, on a straight statistical analysis performed by the tennis historian, Raymond Lee, the top-ten league table of greats, in top-down order, is: Laver, then Borg and Tilder (tie), Federer, Gonzales, Rosewall, Budge, Lendl, Connors and Sampras.

No such 'formal' league tables of top women players exist, but a crude web search of top female players revealed the following trends. The top slot goes to Martina Navratilova winning 18 Grand Slam singles titles and 40 doubles. She also played during a period of very stiff competition. Steffi Graf won 22 Grand slam singles and a Golden Slam (a Grand Slam + Olympic Gold in the same year). 'Fraulein Forehand', as she was called, unquestionably possessed the most powerful female forehand shot on the circuit. Chris Evert had an illustrious career winning 18 Grand Slam singles and 3 doubles. Although she dominated the baseline, sadly she never really developed the 'power serve' which her contemporaries had begun to use to great effect. Billie Jean King is arguable the most important women's player of all time.

She won 12 Grand Slam single titles, 14 doubles and 11 mixed doubles. Furthermore, she worked tirelessly to promote the game for women and, in fact, women's sports in general. Finally, in my list is Margaret Court who won no less than 24 singles Grand Slams, 19 doubles and 19 mixed doubles, making a total of 62 Grand Slams. She was ranked No. 1 in the world seven times in the 1960s and 1970s yet still remains acknowledged as one of the best female tennis players of all time.

11.1.2 Squash

Squash was invented in Harrow school around 1830, when the pupils discovered that a punctured rackets ball, which 'squashed' on impact with the wall, produced a different and interesting game. It not only provided a greater range of possible shots and strategies, but it also required greater effort on the part of the players; they had to approach and 'attack' the ball rather than simply wait for the ball to return to them, as in the case of rackets. The game at Harrow proved so popular that, in 1864, the first four squash courts were constructed at the school and squash was officially founded as a sport in its own right.

Although in those early days there was no standardization of rules, two common court dimensions emerged: the English courts were 21 feet wide and players used a 'soft' ball, while in the United States, courts were only $18\frac{1}{2}$ feet wide and a harder ball was utilized. Both courts were 32 feet long.

The first recorded reference to squash other than in Harrow school, appeared in 1890 in the English book *The Badminton Library of Sports and Pastimes* written by the Duke of Beaufort. Eleven years later, Eustace Miles, a world champion at both tennis and rackets, published the first book on squash. In it he stated that squash was enjoyed by thousands of players the world over. By that time, there were courts in schools and universities around England, and some even in private houses. The first professional Squash Championship was held in 1920 in England. H. A. L. Rudd, writing in *Baily's Magazine* in 1923, suggested that, with the arrival of the first English Amateur Championships, rackets might lose many players to squash. He was concerned at this prospect as he considered rackets to be the better of the two games. Apparently he thought that, although squash 'afforded a good sweat', it did not demand the same skill as rackets. Rudd's forecast proved to be only too correct as squash grew rapidly and soon left its parent sport far behind in popularity.

As the game of squash developed and became more popular, the administrative structures were established. The first national associations to be created were the United States Squash Racquets Association (1907) and the Canadian Squash Racquets Association (1911). However, in England, the game was regulated by a squash subcommittee of the Tennis and Rackets Association from 1908, until it achieved full status as the 'Squash Rackets Association' in 1928. A standardized court dimension was ratified in 1923, based on a court built at the Bath Club in London. It was 32 feet by 21 feet (9.75 m × 6.4 m). The point-a-rally scoring system to 15 was used universally in squash until 1926, when the current hand-in, hand-out system to 9 points was adopted. However, an American Hardball form still exists which utilizes the older scoring system, on the belief that it ensures the game's faster progress.

The earliest championship players were Egyptian. In 1933 the great player F. D. Amr Bey, won the first of his five British Open Championships which were considered the World Championships of their day. He was followed in his achievement by M. A. Karim who won the title four times from 1947 to 1950. They were succeeded by the 'Khan Dynasty' from Pakistan; Hashim (1951–1958), Roshan (1957), Azam (1959–1962), Mohibullah (1963), Jahangir (1982–1992) and Jansher (1993–1994) who totally dominated the sport over 40 years. The only players to break into their successes were Jonah Barrington (Great Britain and Ireland) who won the English Open six times between 1967 and 1973, and Geoff Hunt (Australia) who was ranked number one in the world between 1975 and 1980. Arguably, no players influenced the game more than these two, capturing the imagination of sportsmen and women everywhere, and starting a boom in the sport which raised the number of courts to 50,000 worldwide and the number of players to something in excess of 15 million.

The Women's British Open commenced even earlier than the Men's, with Joyce Cave winning the title in 1922, and Nancy Cave in 1924, 1929 and 1930. In fact, until 1960 the title belonged solely to English players, with Janet Morgan (later Shardlow) winning 10 championships between 1950 and 1958. She was followed by a string of antipodean champions to the present day. The most famous woman squash player ever, the Australian Heather McKay, dominated the sport from 1966 to 1977 and remained undefeated throughout her 18 year playing career. She is acknowledged as the greatest women's squash player ever. She was succeeded by a New Zealander, Susan Devoy, who won the title eight times between 1984 and 1992, then by Michelle Martin (three World Opens and six British Opens) and finally, Sarah Fitz-Gerald (World Titles in 1996, 1997, 2001 and 2002) is considered the last of the great consistent winners.

Squash is now played in 130 countries under the auspices of the World Squash Federation which maintains responsibility for the game rules, court and equipment specifications, refereeing and coaching. It schedules a world calendar of events and is currently striving for the inclusion of squash on the Olympic events roster. By comparison, tennis has been an Olympic event since the first modern Olympiad of 1894.

11.2 The rackets

If we are to define the tennis racket as a sport striking implement designed around a looped frame which supports strings, then the first such devices appeared in the sixteenth century. Initially, these wooden rackets were made from one or more branches of ash, bent to the required shape and glued with animal glues. Mechanization in the 1940s allowed a larger number of thinner 'veneers', bonded with urea-formaldehyde adhesives, to be used. This multiplicity of layers could be more easily bent to shape, and so the natural variability in wood was averaged out. In addition to ash, maple, sycamore and hornbeam have been used in the main frame member to provide the required strength and stiffness properties, hickory for wear resistance in the outer layer, beech and mahogany for the throat and handle, and obeche as a lightweight filler in the shaft.

These wooden rackets were formed by: first, a combination of hand crafting to achieve section shapes; second, steam and pressure to soften and bend the material to form the racket profile; and finally, glue to hold the final shape. This construction led to a racket of limited strength which had a tendency to warp. However, by 1936, methods of laminating improved, leading to better quality rackets, and so this form of construction prevailed until the 1970s.

Tubular steel rackets were actually available as early as 1920; these were often combined with steel strings, but the rackets failed to become established on the circuit due to their excessive weight. However, in the 1970s steel rackets did briefly become very popular as new methods of tubular extruding became available. Metal rackets had the advantage that they were durable, even though they did not have the dynamic characteristics of wood. In 1976, Howard Head introduced the oversized aluminum Prince Classic with a typical playing area of $710\,cm^2$, as opposed to the traditional $450\,cm^2$. Although head sizes are now $840\,cm^2$, the benefit of increased head size is to increase the polar moment of inertia and hence the racket's 'sweet spot' (see Section 11.4.1).

Although the original wooden rackets were cheap, relatively durable and easy to make, their inability to absorb impact shock left players prone to the severe complaint of tennis elbow, though, to be fair, even these early wooden specimens were less damaging than their metal variants. Furthermore, the performance of the wooden rackets altered with the prevailing weather conditions. To ameliorate these deficiencies, the new composite material racket was introduced in 1982, designed by Dunlop for John McInroe. This next generation of racket was lighter, it being constructed from graphite injected with foam, which allowed for greater manoeuvrability. Modern rackets are broadly similar in construction, but are refined with the addition of a range of modern materials including Kevlar, boron, ceramics or titanium, giving added levels of lightness, frame flexibility, safety (unlike wood, the composite does not splinter on destructive impact) while remaining cost effective.

Squash racket advances have followed, albeit somewhat lagging behind, the tennis racket developments. Today's squash rackets are roughly half the size and weight of tennis rackets, allowing for greater manoeuvrability in the confined space of the squash court.

11.3 The balls

11.3.1 Tennis balls

Lawn tennis as we know it was instigated in the 1870s, when India rubber balls were utilized, which were made from a vulcanization process invented by Charles Goodyear in the 1850s. The design quickly developed into a hollow core which was pressurized with gas. Originally, core manufacture was based on the 'clover-leaf' principle, whereby uncured rubber sheet was stamped into a shape resembling a three-leafed clover, and this was then bent into the spherical shape, and held there under the combined opposing forces of internal pressure and surface elasticity.

Modern tennis balls must comply with the ITF regulations covering size, bounce, deformation and colour. The ball's performance characteristics are based on a range of dynamic and aerodynamic properties. Tennis balls are classified as: Type I (fast speed), which are used for slow paced court surfaces such as clay, Type II (medium speed) which are used on hard core surfaces, Type III (slow speed) which are used for fast court surfaces such as grass, and finally, high altitude balls intended solely for use above 1219 m. Balls are usually either fluorescent yellow or white, which helps both the players and the spectators to see the ball. However, the primary purpose of introducing the yellow balls was that television cameras are particularly sensitive to that colour and viewers could follow the game more closely from the comfort of their television lounges.

The balls are either of a traditional diameter (6.63 cm for Types I and II), or a larger diameter (67.6 cm for Types I and II) which is 6 per cent bigger. These larger balls were introduced to slow the game down and increase the likelihood of prolonged rallies.

A further variant in the construction is that the balls may be either pressurized or pressureless. The pressurized ball consists of a hollow rubber-compound core with a slightly pressurized gas within, and surrounded by a felt fabric cover. The cover not only speeds up the onset of chaotic flow, and thereby, reduces drag, but it also affects the coefficient of restitution. These balls are replaced every six games in tournaments since, not only does the fabric wear, but the gas leaches through the rubber membrane leading to a loss of bounce characteristics. Cubitt and Bramley (2006) have studied the degradation of pressured balls with storage and with play. They also described a method of re-inflation of under-pressurized balls by holding them in a pressurized container of at least 70 kPa for up to eight days. However, when the balls were subjected to pressures above 250 kPa, they were found to actually reduce the ball's coefficient of restitution.

Pressureless balls have thicker cores or may be filled with microcellular material and are less prone to lose their bounce with play. They usually have a stiffer, woodier feel than pressurized balls. In fact, they get bouncier as they get lighter due to loss of surface fuzz.

The balder they get, the more their flight, bounce, and spin response changes from the expected reaction.

11.3.2 Squash balls

The first squash balls were nothing more than child's rubber play balls as the original racket balls were thought to be too hard and fast for play in a fully enclosed court. For competitive events, the balls used were different for the differing dimensions of courts. When the courts became standardized, so did the balls. Under the auspices of the Tennis and Rackets Association (TRA), Colonel R. E. Compton from the Royal Automobile Club in London was given the responsibility of weighing, measuring and creating the means of comparing the bounce of the various balls in circulation.

The Avon India Rubber Company produced balls of differing sizes from 3.65 cm to 4.3 cm in diameter, and with different surface finishes; matt or varnish-dipped – or even the 'Bath Club Holer', which had a hole in it to create the challenge of an unpredictable bounce! The Silvertown Company produced a popular enamelled black ball measuring 3.9 cm in diameter. The Gradidge Company, which later became Slazenger, marketed balls in a choice of colours; white, red or black. However, in 1923, the TRA adopted the RAC standard ball as the official ball for amateur championships. However, very quickly, this was found to be too fast for safe play, and reduced speed balls were created. Slazenger introduced the first synthetic balls in 1960 made from butyl, but although performance was more stable than the natural rubber variant, the results were disappointing in terms of resilience.

Dunlop continued using natural rubber products, importing raw rubber from Malaysia. They first masticated it to soften it, and then mixed it with a variety of synthetic chemicals, before finally curing it. This produced a ball which played with consistency, was resilient to damage and was of a suitable matt surface and colour: the modern squash ball. The fifteen ingredients they state they add to the compound, are measured in the correct proportions to create the fast ball (blue dot), the medium ball (red dot) and the slower ball (yellow dot).

The resulting compounds are then warmed and placed in an extruder allowing the compounds to be forced though a die. A rotating knife cuts the compound into pellets which are then cooled. The pellets are pressed into the hemispherical shape using a hydraulic press at around 500 kg pressure and at a temperature of about 150°C for 12 minutes. The edges of the two halves are buffed and they are glued together using a rubber solution. The assembled balls are cured again and pressed at around 450 kg to set the rubberized adhesive. Finally, the whole ball is buffed to smooth the adhesive joint, and to give the surface its characteristic matt finish.

Modern advances in material technology are leading manufacturers to search for materials which maintain performance over the usable range of temperatures, thereby reducing the need to warm the ball prior to play.

11.4 Impact considerations

11.4.1 Racket design

Dynamic properties of the racket

The requirements of tennis and squash rackets are:

*   to return or serve the ball with the greatest possible speed;
*   to return or serve the ball with the greatest possible accuracy;

- to minimize the occurrences of player injury (primarily tennis elbow) due to repetitive impact of racket with ball.

Bullet points, rather than a numbered list, have been used above to indicate that the order of importance of these requirements is contentious. One could argue that each one of them is the most important (or, indeed, the least important) from a number of angles. Furthermore, some players might argue that other factors are equally important, such as lightness, flexibility, and that most elusive of player demands; the 'feel' of the racket. However, with regard to the bullet point list above, it is clear that the first two items relate to efficiency of energy transfer from racket to ball, while the last item relates to racket vibration following impact with the ball, which, of course, also has a bearing on the efficient energy transfer on impact.

So, faced with these variable requirements, how do racket manufacturers quantify design aims for their products? The object of racket designers is to improve the dynamic characteristics so as to make ball control more precise and to dampen vibrations caused by the racket/ball impact. Although parameters such as moment of inertia and center of mass play a part in the mechanical construct, a key aspect is that which is associated with the concept of the 'sweet spot' or 'Center of Percussion' (CoP).

Strictly, the sweet spot is a rather unscientific and immeasurable term which is the hypothetical area of the racket where it feels that the ball has been hit efficiently. It leaves the player with the rather satisfying feeling of a 'well hit ball'. By contrast, the CoP is the area on the racket's face where the least vibration results from the impact with the ball. It lies in the area where a combination of nodes in the major vibration modes of the racket coincides.

For a clearer model, imagine the racket to be suspended vertically by a horizontal wire passing through a hole in the racket's handle at the position of the racket's axis of rotation (i.e. where the hand would normally grip the racket). If the ball now hits the racket normal to the plane of the strings *above* the CoP, there will be a principal tendency for the racket to react by moving in a translational manner, and it will duly respond by sliding along the wire in the *same direction* as the incoming ball. By contrast, if the racket is hit by the ball *below* the CoP, the racket will tend to twist around some point near to the center of gravity, such that the bottom of the racket will move away from the impact, while the handle end will slide along the wire *back in the direction the ball came from*. It follows that the CoP corresponds to that strike point where the linear translational motion in one direction perfectly matches the torque induced rotating motion in the other direction. As a consequence, and with regard to the racket hanging from the wire, a ball impacting with a racket at the CoP point will result in the racket initially not moving in either direction along the wire; it will simply be left swinging back and forth in a pendulum-like manner. If the friction between the wire and the racket is low enough, the racket will slowly build up an oscillating motion on the wire; the handle sliding back and forth as a consequence of its pendulum motion. It should, however, be stressed that this slow building of motion occurs *after* the impulse, and so has no bearing on the 'feel' of the stroke of the racket in play. A ball struck at the CoP of the racket will result in the handle apparently generating no reactive force that the player will be aware of.

One further parameter which needs to be defined is the 'radius of gyration'. This is a measure of how the mass of the racket is distributed along the length of the racket with respect to the axis of rotation. The rotational inertia, which is the racket's resistance to angular acceleration, is given by:

$$I = mk^2 \tag{11.1}$$

where I is the rotational inertia, m is the mass of the racket, and k is the radius of gyration.

Now that these terms have been defined, we can state the formula which relates these measures:

$$q = \sqrt{\frac{k^2}{cg}}$$
(11.2)

where q is the distance of the center of percussion from the axis of rotation, and cg is the distance of the center of gravity from the axis of rotation.

Figure 11.1 shows the relative positions of these quantities on a racket. The precise location of the CoP of a modern racket is slightly further away from the handle than the dead center of the racket face.

Brody (1996) maintains there are actually three separate sweet spots.

The first, the CoP may be measured as $q = 0.248 \times T^2$, where T is the period of the racket swinging as a pendulum around the axis of rotation in seconds, and q is the distance in meters from the axis of rotation to the CoP .

The second is the vibration node. The lowest frequency oscillation (known as the fundamental in musical terms) is the most uncomfortable for the player and creates the greatest vibrations. This occurs somewhere between 125–200 Hz in a modern racket. Not only does such vibration result in energy loss, but that frequency causes the fine ligaments in the elbow joint to damage or even tear, leading to the dreaded lateral epicondylitis; otherwise known as tennis elbow. However, if the racket can be designed so that one of the nodes of vibration is positioned in the vicinity of the center of the racket face (i.e. the sweet spot) and the other at the axis of rotation, then, not only is the vibration minimized, but what vibration there is will not be transferred into the player's hand, wrist and arm, and up to the elbow.

The final sweet spot, according to Brody (1996) is the power spot. This is the point on the racket face where the maximum rebound velocity of the ball occurs. In measuring this, the equation:

$$V_{hb} = eV_{ib} + (1 + e)V_r$$
(11.3)

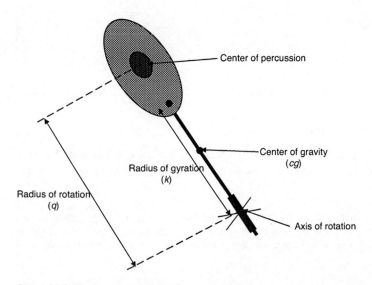

Figure 11.1 Physical properties of a tennis racket.

is maximized, where V_{hb} is the velocity of the hit ball, V_{ib} is the velocity of incident ball, e is the ratio of ball rebound speed to incident speed (not the coefficient of restitution – why?[1]), and V_r is the velocity of the racket at the impact point.

Ideally, in racket design methodology, the aim is to achieve a racket where all three sweet spots perfectly coincide.

Worked Example 11.1

Calculate the distance from the center of rotation of a tennis racket to its center of percussion if its mass is 0.3 kg, its moment of inertia is 0.03 kg m^2, and the distance from the center of rotation to the center of gravity is 0.35 m. If the tennis racket was held in a pivot at the axis of rotation and allowed to swing freely around that point, according to Brody (1996) what would you expect its period of swing to be?

From Equation 11.1 we (have):

$$I = mk^2 \quad \therefore \quad 0.03 = 0.3k^2$$

So, k, the radius of gyration $= 0.316$

Then from Equation 11.2:

$$q = \sqrt{\frac{k^2}{cg}} = \sqrt{\frac{0.316^2}{0.35}} = 0.535\,\text{m}$$

According to Brody (1996), the period of the tennis racket as a pendulum is given by:

$$q = 0.248 \times T^2$$

So:

$$T^2 = \frac{q}{0.248} = \frac{0.535}{0.248} = 2.16$$

Therefore:

$$T = 1.5\,\text{s}.$$

Dynamic properties of the strings

The purpose of the strings is to return the ball with speed and accuracy, to allow ball spin to be maximized in slice and topspin shots and to reduce the injury-inducing impact vibrations.

---

1 The coefficient of restitution quantifies the degree of rebound of an object with a *static* surface. In the case of a ball/racket impact, the surface (in this case the racket) is moving up to meet the ball on impact. For this reason, the effective value of *e* would appear to be much greater than 1 which would, of course, contravene the energy conservation law.

The stiffness in the string plane is dependent on the string tension, the string length and the material choice of the strings. Much has been written in the literature on the effects of string tension variations, but roughly speaking, the optimum tension is in the region of $20\,\mathrm{kN\,m^{-1}}$, which is about the same as the ball stiffness.

There are many varieties of string available, the most common being constructed from nylon, either with a solid core, or woven and sheathed groups of fine filaments. Some strings are made from more than one material; these are known as composite strings. Kevlar has been used in the past, but has recently fallen out of favour as it has an inability to maintain the string tension, even over the short duration of a single game.

Polyester strings are another option, and some players prefer these due to the fact that they have a shorter impact duration which can affect the impact angle of the ball and leads to tighter control. However, again are poor at holding the tension and, furthermore, many players find them just too stiff. An early material used was natural gut. It is a highly elastic string and maintains tension better than many of the more modern equivalents. They are still in use by a select group of elite players as they feel that these strings 'cup' the ball on impact more effectively than their synthetic counterparts.

It used to be thought that high string tension produced high ball speeds. However, the more recent reports suggest that this is not necessarily so. The tennis ball loses a great deal of energy when it compresses and reshapes; mainly by heat dissipation. What is generally acknowledged is that a highly tensioned set of strings, especially on a firm racket will offer the best control of the ball. A lower tension allows the 'trampoline effect', as it is called, to dominate, which creates greater ball rebound velocity. The trampoline effect involves greater string deflection and less ball deformation, which minimizes energy loss and allows for a more efficient transfer of energy.

Let us now turn our attention to the ball spin-action on racket impact. Many studies have been conducted; some results are reliable, others are still open to interpretation. One thing is clear however; lower tensioned strings allow the strings to 'wrap around' the ball, ensuring greater contact area, and certainly, the studies confirm that greater spin is generated from such rackets. Choice of string material also plays a small part.

However, in 1977, a patent was submitted for what was known as a 'spaghetti' stringing system. This design consisted of normal 'cross strings' (those that go across the face of the racket), but they had two layers of main strings (the longitudinal ones). These were not woven between the cross strings as normal, but lay, aligned, above and below the cross strings. Sleeves, known as rollers, surrounded the main strings where they touched and intersected the cross strings. The resultant string system improved frictional contact with the ball without any substantial energy transfer reduction. Goodwill and Haake (2001) showed that the ball spin imparted from such a racket could increase from $200\,\mathrm{rad\,s^{-1}}$ up to as much as $350\,\mathrm{rad\,s^{-1}}$ for equivalent strokes. The spaghetti system of stringing was outlawed by the ITF in 1978 following controversy in a match between Guillermo Vilas and Illie Nastase, when Vilas defaulted the game in which Nastase was using the spaghetti racket stringing system.

Over the years several papers have studied the impact of racket with ball, including; Cross (2000), Haake et al. (2003), and Goodwill and Haake (2004). Here we will look at an early, yet interesting endeavour performed by Groppel et al., (1983).

First, Table 11.1 lists the symbols used in the formulas.

We begin by resolving the impulse during impact into the normal and tangential components with reference to the racket face (Figure 11.2). The expressions for the changes in linear and angular momentum for both the racket and ball are given by:

Table 11.1 Symbols used in tennis racket/ball impact calculations

Symbol	Parameter
p_n	Change in moment in a direction normal to the racket surface
p_t	Change in moment in a direction tangential to the racket surface
Ω	Angular velocity of the ball before impact
ω	Angular velocity of the ball after impact
Ω'	Angular velocity of the racket before impact
ω'	Angular velocity of the racket after impact
u_n	Speed of ball normal to the racket surface before impact
v_n	Speed of ball normal to the racket surface after impact
u_t	Speed of ball tangential to the racket surface before impact
u	Speed of ball before impact
v	Speed of ball after impact
v_t	Speed of ball tangential to the racket surface after impact
U_n	Speed of racket normal to the racket surface before impact
V_n	Speed of racket normal to the racket surface after impact
U_t	Speed of racket tangential to the racket surface before impact
U	Speed of racket before impact
V	Speed of racket after impact
V_t	Speed of racket tangential to the racket surface after impact
r	Radius of tennis ball
m	Mass of tennis ball
M	Mass of racket
β	Angle of racket to horizontal at the point of impact
φ	Angle of linear racket movement
I	Moment of inertia of the ball
I'	Moment of inertia of the racket
e	Coefficient of restitution of racket/ball impact
S	Relative sliding velocity between ball and racket
C	Compression velocity

For the ball   For the racket

$$m\left(v_n - u_n\right) = p_n \qquad M\left(V_n - U_n\right) = -p_n \tag{11.1}$$

$$m\left(v_t - u_t\right) = -p_t \qquad M\left(V_t - U_t\right) = p_t \tag{11.2}$$

$$I\left(\omega - \Omega\right) = -rp_t \qquad I'\left(\omega' - \Omega'\right) = 0 \tag{11.3}$$

If we model the ball as a thin spherical shell, then $I = 2mr^2/3$ (see Table 2.6).

The relative sliding velocity, S, at the point of impact may be defined as the difference in the tangential racket and ball velocities, including any sliding caused by the spin which will occur immediately on impact:

$$S = V_t - \left(v_t + r\omega\right) \tag{11.4}$$

The compression velocity, C, at any instant during the impact process will simply be the difference in velocity in the normal direction between the racket and the ball:

$$C = V_n - v_n \tag{11.5}$$

Substituting Equations 11.1 to 11.3 into Equations 11.4 and 11.5 yields:

$$S = -S_0 + \alpha p_t \tag{11.6}$$

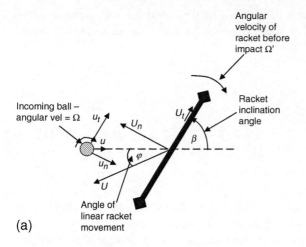

(a)

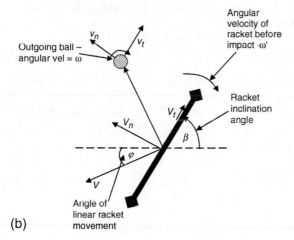

(b)

Figure 11.2 (a, b) Coordinate axes for a racket/ball impact.

where:

$$S_0 = u_t + r\Omega - U_t \quad \text{and} \quad \alpha = \left(\frac{1}{m} + \frac{1}{M} + \frac{r^2}{I}\right)$$

and:

$$C = -C_0 + \gamma p_n \tag{11.7}$$

where:

$$C_0 = U_n - u_n \quad \text{and} \quad \gamma = \left(\frac{1}{m} + \frac{1}{M}\right)$$

The normal component of the *total* impulse, p_{nT}, may be found by graphical means to be:

$$p_{nT} = (1+e)p_{n0} \tag{11.8}$$

where e is the coefficient of restitution and p_{n0} is the normal impulse at the point of zero compression velocity (i.e. the mid-point of the impact when ball compression is a maximum).

Now Equations 11.1 to 11.3 and Equations 11.6 and 11.7 may be solved on the condition that there is no slip between the racket and the ball *at the time when the two separate*, at the end of the impact process, implying that the ball has fully reached the rolling phase before separation occurs (see Section 4.6). High speed video analyses of tennis slice shots show this to be a correct assumption.

The results are found to be:

$$v_n = u_n + \frac{1+e}{m\gamma}C_0 \tag{11.9}$$

$$v_t = u_t - \frac{1}{(m\alpha)S_0} \tag{11.10}$$

$$\omega = \Omega - S_0\frac{r}{\alpha I} \tag{11.11}$$

Although these calculations are derived on the basis of a ball travelling horizontally towards the racket (see Figure 11.2a), balls arriving from any angle can be treated in exactly the same manner by simply redefining the angle β, so the racket impacts at the same relative direction (see Worked Example 11.2). The initial velocities of the center of the ball and racket expressed in terms of normal and tangential components are:

$$u_t = u\cos\beta \qquad U_t = -U\cos(\beta - \varphi) \tag{11.12}$$

$$u_n = -u\sin\beta \qquad U_n = U\sin(\beta - \varphi) \tag{11.13}$$

Finally, substituting the inertia relationship for a hollow ball, $I = 2mr^2/3$, together with Equations 11.12 and 11.13, into Equations 11.9 to 11.11, and assuming $m/M \ll 1$ (i.e. the ball is much lighter than the racket), we find:

$$v_n = (1+e)(U\sin(\beta - \varphi) + u\sin\beta) - u\sin\beta \tag{11.14}$$

$$v_t = -\frac{2}{5(u\cos\beta + U\cos(\beta - \varphi) + r\Omega)} + u\cos\beta \tag{11.15}$$

$$\omega = -\frac{3}{5r(u\cos\beta + U\cos(\beta - \varphi) + r\Omega)} + \Omega \tag{11.16}$$

Model 11.1 – Tennis racket impact calculator, solves these equations.

Worked Example 11.2

Calculate the linear and angular velocities of a tennis ball following an impact with a racket if the ball arrives at $+15°$ to the horizontal and at a speed of 25 m s^{-1}. The coefficient of restitution, ball-to-racket is 0.7, the linear velocity of the racket impacts with the ball at 15 m s^{-1}, the angle of the racket is 80° to the horizontal (with racket face pointing upwards), while the racket face slices downwards on impact at 40° to the horizontal. The ball spin before impact is 150 rad s^{-1} and the tennis ball radius is 0.0325 m.

If we assume the ball arrives horizontally and adjust the angle of the face of the racket and the slicing angle accordingly, we find the following values apply:

$$e = 0.7, \quad U = 15\,\text{m s}^{-1}, \quad \beta = (80+15) = 95°, \quad \varphi = (40+15) = 55°,$$

$$u = 25\,\text{m s}^{-1}, \quad r = 0.0325\,\text{m} \quad \text{and} \quad \Omega = 150\,\text{rad s}^{-1}$$

So, from Equation 11.14:

$$v_n = (1+e)(U \sin(\beta - \varphi) + u \sin \beta) - u \sin \beta$$
$$= (1+0.7)(15 \sin(95 - 55) + 25 \sin 95) - 25 \sin 95$$
$$= 33.825 \, \text{m s}^{-1}$$

From Equation 11.15:

$$v_t = -\frac{2}{5(u \cos \beta + U \cos(\beta - \varphi) + r\Omega)} + u \cos \beta$$
$$= -\frac{2}{5(25 \cos 95 + 15 \cos(95 - 55) + 0.0325 \times 150)} + 25 \cos 95$$
$$= -2.208 \, \text{m s}^{-1}$$

$$v = \sqrt{v_n^2 + v_t^2} = \sqrt{33.825^2 + (-2.208)^2} = 33.9 \, \text{m s}^{-1}$$

$$\text{Angle of rebound given by arctan}\left(\frac{-2.208}{33.825}\right) = -3.73°$$

The actual rebound angle with respect to the horizontal will be $15 - 3.73 = 11.27°$ since the racket has been mapped onto the horizontal line to accommodate the angle of the incoming ball at $15°$ to the horizontal.

Finally, from Equation 11.16:

$$\omega = -\frac{3}{5r(u \cos \beta + U \cos(\beta - \varphi) + r\Omega)} + \Omega$$
$$= -\frac{3}{5 \times 0.0325(25 \cos 95 + 15 \cos(95 - 55) + 0.0325 \times 150)} + 150$$
$$= -148.7 \, \text{rad s}^{-1}$$

This is only a slight reduction in spin. However, the racket is held at an angle which is nearly normal to the incoming ball and it would take either quite a severe angle, or a strong slicing movement, to alter this high degree of incoming spin.

11.4.2 Ball bounce from the court

The bounce of a tennis ball from a court is a complicated dynamic process which is largely influenced by the court surface. Variations in ball deformation upon ground impact act to influence such factors as ball speed, spin and bounce height, thereby effecting the nature and style of play.

Any normal ball impacting with a surface can be modelled using a parallel spring and damper equivalent configuration. The equation of motion of a ball of mass, m, with a spring constant, k, and a damper constant, c, may be represented by:

$$m\ddot{x} + c\dot{x} + kx = 0 \tag{11.17}$$

where x is the position of the center of mass of the ball, $\dot{x}$ is the first time differential of x, and $\ddot{x}$ is the double differential of x with respect to time.

If the center of mass of the ball is at $x = 0$ when $t = 0$, then Equation 11.17 has a solution of the form

$$x = ae^{-bt} \sin \omega t \qquad (11.18)$$

where a and b are constants and ω represents the natural frequency of the system.

This expression has been found to be accurate for low-pressure hollow shell balls such as tennis balls, but not, for instance, solid Superballs.

The spring and damping constant, k and c respectively, are then given by:

$$k = m\frac{\pi^2}{T_c^2} \qquad (11.19)$$

and

$$c = -\frac{2m}{T_c} \ln(e) \qquad (11.20)$$

where T_c is the total contact time of the ball with the surface and e is the coefficient of restitution.

Figure 11.3 shows the deformation of a tennis ball in 1 ms intervals when it collides with a concrete slab at 45 m s^{-1} (100 mph). The ball first makes contact at $t = 0$ s and breaks the surface contact at time $t = \sim 4$ ms.

The ITF incorporates a 'Court Pace Rating' (CPR) value, previously known as a Surface Pace Rating (SPR) to categorize the playing performance of the surface types. The CPR defines both the velocity and the angular behaviour of a ball rebounding from a court surface. Clay courts are classified as Category 1 (slow pace), hard courts as Category 2 (medium/medium-fast pace) and natural grass and synthetic turf are classified as Category 3 (fast pace) surfaces. It should, however, be appreciated that a court's CPR is influenced greatly by its anatomy, which is modifiable. The synthetic turf courts utilized at the Australian Open incorporate a form of infill and a layer of crumbed rubber which lowers the courts CPR to a value similar to, although a little higher than, clay (see Figure 11.4).

The coefficient of friction affects the horizontal component of the tennis ball's velocity. A surface with a low coefficient of friction causes the ball to bounce faster. A surface with a high coefficient of restitution causes the tennis ball to bounce higher than a surface with a lower value. The clay courts of the French Open have both a high coefficient of friction and a high coefficient of restitution, resulting in slow, high bounces. The hard courts incorporated by the Australian and US Opens have moderate coefficient of restitutions, resulting in medium bounce heights. However, the low frictional coefficient of concrete courts results in brisk bounces at the US, whilst the relatively soft synthetic turf of the Australian Open has a high

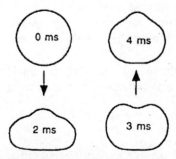

Figure 11.3 Cross-section of a tennis ball during a 100 mph collision with a concrete slab.

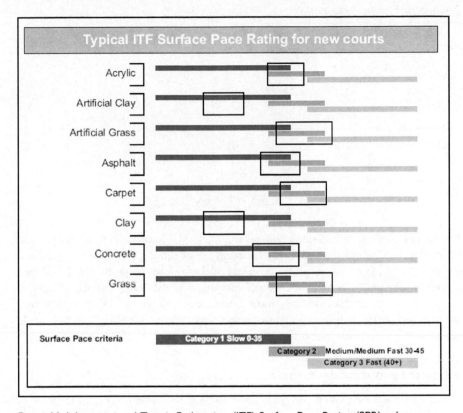

Figure 11.4 International Tennis Federation (ITF) Surface Pace Rating (SPR) values.

frictional coefficient, resulting in reduced bounce speeds. The grass courts of Wimbledon have both a low coefficient of friction and low coefficient of restitution which results in very fast, low bounces.

CPR values are calculated from the court's coefficient of friction and coefficient of restitution properties; both influence the interaction between the tennis ball and the surface of the court:

v_{ix} = horizontal inbound velocity (m/s)

v_{iy} = vertical inbound velocity (m/s)

v_{fx} = horizontal outbound velocity (m/s)

v_{fy} = vertical outbound velocity (m/s)

e = coefficient of restitution (CoR)

μ = coefficient of friction (CoF)

a = pace perception constant ($= 150$)

b = *mean* coefficient of restitution for *all* surface types ($= 0.81$)

CPR = court pace rating (categorized in Table 11.2)

Table 11.2 Court pace rating categories

Category	CPR
Slow	< 29
Medium-slow	30–34
Medium	35–39
Medium-fast	40–44
Fast	> 45

where

$$e = \frac{v_{fy}}{v_{iy}} \qquad \mu = \frac{v_{ix} - v_{fx}}{v_{iy}(1+e)} \qquad CPR = 100(1 - \mu) + a(b - e)$$

In addition to the variations regarding bounce angle and height, each surface has its own distinctive attributes. Sure footing and consistent bounces are among the advantages of hard courts; the bounce is often unpredictable on grass and clay due to surface induced ball degradations. Players find it beneficial to slide into shots on the loose surface of clay; a technique which does not work on hard courts. Furthermore, 'kick serves' (serves with a high arcing topspin) are more difficult to return on the harder courts compared to grass, while other types of serve are actually easier to return.

11.5 Trajectory dynamic considerations

We have studied trajectory motion including drag and lift in general terms in Chapter 5, and so in this section, we will consider only those aspects which are particular to the tennis ball motion through the air. Incidentally, little has been written in the academic press with regard to squash ball trajectories (in fact, despite an extensive literature search, nothing has been found), but again, we only need to apply the ball's physical parameters to the equations previously stated to obtain trajectory results which would certainly be accurate, given (1) the squash ball's low value of drag-to-weight ratio, (2) the short trajectory distances involved within the enclosed court, and (3) the lack of unpredictable short term atmospheric changes such as wind gusts. By contrast, the cotton outer of the tennis ball as well as the nature of the game of tennis make this an interesting ball to study and it is these facets which will be addressed in this section.

We may begin with our standard 'simple' drag and lift force equations with the drag and lift forces calculated using the velocity-squared function:

$$F_D = C_D \rho A \frac{v^2}{2} \tag{11.21}$$

$$F_L = C_L \rho A \frac{v^2}{2} \tag{11.22}$$

and work through to a trajectory profile using a Runge–Kutta analysis on the equations exactly as in previous chapters. In the case of the tennis ball, the interesting aspect is the values and dependencies of the coefficients of drag and lift, C_D and C_L.

Although over 20 years old now, one of the best studies is an experiment carried out by Štěpánek (1987) in which he spun tennis balls at varying speeds in a motorized frame and then released them into an open-ended vertical wind tunnel and, by so doing, measured the deviations over a range of wind velocities. By this means, values of C_D and C_L were obtained over a ball angular velocities range of 800 up to 3250 rpm, and a wind speed range of 13.6 up to 28 m s^{-1}. He deduced that a distinct variation with air velocity, and therefore Reynold's number, could not be obtained.

He then employed a non-linear, least-squares regression to obtain expressions for C_D and C_L. The equations are stated as:

$$C_D = 0.508 + \frac{1}{\left[22.503 + \frac{4.196}{(\omega/v)^{5/2}}\right]^{2/5}} \tag{11.23}$$

$$C_L = \frac{1}{2.022 + \frac{0.981}{(\omega/v)}} \tag{11.24}$$

These equations are used to the present day when values for lift and drag coefficient of tennis balls are required.

Figure 11.5 shows the variations of the lift and drag coefficients with the ratio (ω/v).

Mehta and Pallis (2001) also found that the Reynold's number plays no part in the value of C_D, with the flow operating in what they call the 'transcritical regime', where the air separation location does not move significantly with R_e. It is thought that one reason for this is that the drag is totally of the pressure drag variety over the rough fabric surface of the ball. This surface is very effective at causing a very early transition of the laminar boundary layer and a rapid thickening of the turbulent boundary layer. Because of this, the separation

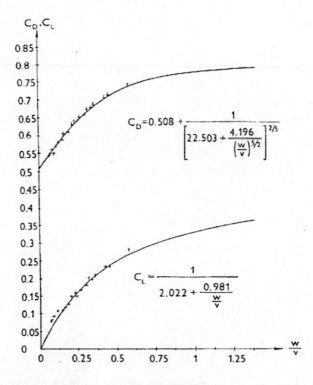

Figure 11.5 Values of drag and lift coefficients as a function of the (ω/v) ratio [adapted from Štěpánek (1987)].

moves up to the apex region similar to that of balls operating in the subcritical region, and the outcome is a value of C_D which is close to 0.5; slightly higher for a new ball. The higher value for new balls is thought to be due to the individual pressure drag contributions from each of the fibers of the 'fuzzy' surface. This is sometimes termed the 'fuzz drag' and is thought to contribute between 20 per cent–40 per cent of the total ball drag. The value of C_D is found not to differ significantly between the standard and the new larger balls; the slowing of the game being caused simply by the added drag of the greater cross-sectional area.

The variation in C_D with wear, seems to show a gradual increase as the fabric 'fluffs-up' with wear. It then steadily reduces as the cover wears to a smooth shell, from whence the usual variation of C_D with R_e is then exhibited, with a critical value of R_e being around 1×10^5 as expected from any normal smooth ball.

Figure 11.6 shows the final trajectory profile output from the model discussed, presented as a range of angles and distances from the net which would result in a first ground impact at the same point on the court.

Tennis serving acceptance window

Brody (2002) has studied the vertical angular window through which a tennis server must aim in order to both clear the net and to land within the legal service area. He considered the acceptance window as a function of both the ball speed and the impact height.

Figure 11.7 shows the variation of acceptance window with ball speeds ranging from $31 \, \mathrm{m \, s^{-1}}$ (70 mph) to $58.1 \, \mathrm{m \, s^{-1}}$ (130 mph). It is no surprise that the acceptance window is inversely proportional to the ball speed. It is found that, by reducing the ball speed from $44.7 \, \mathrm{m \, s^{-1}}$ (100 mph) to $40.23 \, \mathrm{m \, s^{-1}}$ (90 mph), the acceptance window increases by 30 per cent. This corresponds to a successful serve probability increase of about 5 per cent.

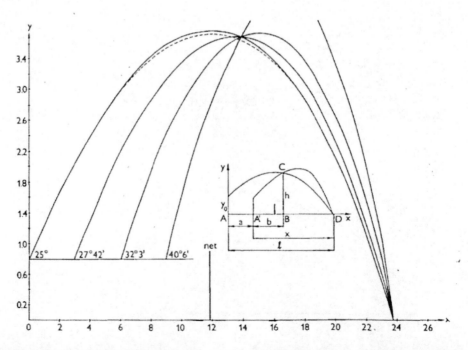

Figure 11.6 Calculated ball trajectories for a variation in launch angle and shot-to-net distances [adapted from Štěpánek (1987)].

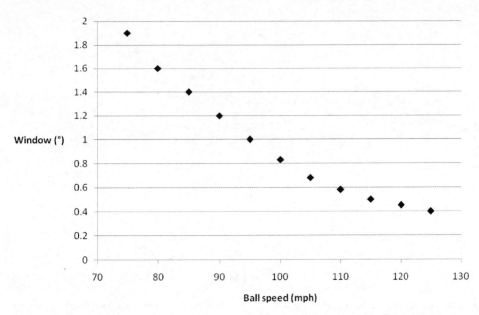

Figure 11.7 Vertical angular acceptance for a serve versus ball speed [adapted from Brody (2002)].

However, this 10 per cent drop in service velocity will considerably increase the probability of a successful return, decreasing the probability of an 'Ace' serve. It should be noted, however, that Brody's detailed analysis assumed a non-spinning service ball which rather detracts from the confidence of the analysis since most strong serves are performed with some topspin; the consequence being to add to the effects of gravity and, thereby, widen the acceptance window even for high speed serves. He does, however, later in his paper, acknowledge the effect of top spin on the acceptance window. He states without proof, that a topspin of 25 rev s^{-1} will effectively double the acceptance window.

In the same paper, Brody proves that (again for a non-spinning ball) the service window of acceptance increases with impact height in an approximately linear fashion. This, after all is why players serve overhand. Figure 11.8 shows the results of his calculations and the disadvantage that short players have when it comes to executing powerful serves can clearly be seen. Indeed, to increase the acceptance window by the same 30 per cent of the previous study, the player must impact the ball 15 cm higher. The biomechanics of the stroke rather precludes this option, although impacting slightly high on the racket string area and 'reaching for the shot' can certainly create some improvement.

Finally, Brody summarizes that optimizing both the reach and the topspin in serving are best accomplished by throwing the ball high in the serving action. Not only does this allow the player to time the impact for the optimum contact height, but also topspin is created by hitting the ball while it still possesses a considerable downward velocity component. Both factors result in an opening up of the acceptance window.

11.6 Trajectory diversions: Miniten

In these trajectory diversion sections I have endeavored to seek out interesting departures from the central theme without being accused of 'pushing the envelope' too far 'outside the box'!

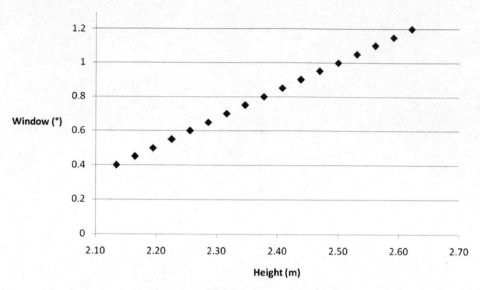

Figure 11.8 Vertical angular acceptance for a 44.7 m s^{-1} (100 mph) ball serve versus height at which ball is struck [adapted from Brody (2002)].

With that in mind, I am sure that you will agree that Miniten was a lucky find; a tennis based game which I had never heard of. Before I discovered this game it was my intention in this section to convey the essentials of the game of 'Mini Tennis' to you (previously termed 'Short Tennis'). Miniten is in many ways similar to mini tennis in that they have both been designed around the need to play a seriously competitive game on a court with small 'real estate'. There are a number of differences however; the main one being that in Miniten *competitors play in the nude*! The game was devised in the 1930s by Mr R. Douglas Ogden, a naturist. Today it is played all over the world, but solely within the naturist communities (at least to my knowledge).

So, both mini tennis and Miniten take the game of tennis and modify it to allow play on a reduced-area court. To accomplish this, mini tennis utilizes a foam ball (of variable density dependent on player age/ability), together with a short-handled racket. The ball has the required high drag-to-weight ratio, and the shortened racket has the reduced hitting capacity to slow the game down and reduce shot lengths. The main disadvantage of mini tennis is that the game must be played indoors, since any wind level would unpredictably disrupt the trajectory of the light foam ball.

Naturists obviously do not want the limitation of playing indoors, so their game uses a standard tennis ball (which is relatively unaffected by atmospheric conditions), but takes the racket design back to the days of the inception of the game (c. f. 'jes de paume' in Section 11.1.1). The hitting device, known as a 'thug', consists of a pair of plywood discs (maximum size 0.27 m) which cover the front and back of the hand; the player's hand grips a sculpted handle which lies between the discs, which are separated by the width of the player's clenched fist around the handle.

Otherwise, apart from the reduced court area, the game rules broadly follow those of standard tennis.

My first thoughts on seeing photographs of the thug (having never played the game!) was that this was some way of protecting any sensitive and unprotected areas of body from the full force of a competition tennis ball. However, this is not the case. The purpose of the thug design is simply to restrict the force of the shots, and hence the trajectory distances. However, apparently balls can still be spun to full effect, and forehand and backhand shots perform the same function as at Wimbledon.

Obviously one disadvantage of the Miniten version is that there is no opportunity for player-sponsored revenue to be derived from company clothing logos!

Summary

This chapter commences with a description of the history of both tennis and squash. Lord Rayleigh, in 1877, carried out what is widely acknowledged as the first treatise on ball flight, with his work on the observably odd flight of spinning tennis balls. Although some conclusions of his work were misdirected, his paper forms the basis on which all future sport flight trajectory analysis has subsequently developed. The technology behind the design and manufacture of the tennis and squash striking implements is explained, followed by a similar treatment for the balls.

The trajectory analysis begins at the point where the racket impacts with the ball to launch it on its trajectory path. Energy and momentum transfer and conservation form the basis of the study, with particular emphasis on the importance of the center of percussion as the most energy efficient point to impact the ball. Equations are derived from which ball velocity can be derived from knowledge of the racket's velocity at impact and its weight. The part played by the racket strings is explained and the effect of differing string tensions on the launch is described both qualitatively and in analytic detail.

Ball bounce from the court is described with reference to bounce efficiency and spin creation in relation to the types of courts commonly encountered. In-flight trajectory analysis follows the general methods as performed in previous chapters, and is seen to throw up few surprises. However, aspects such as the wear on the ball's fabric surface (tennis) and the serving window calculations which are unique to the tennis serve are expounded.

Problems and questions

1   Using the simple Brody (1996) Equation 11.3 (pg 207) for rebound velocity from a normally oriented racket, calculate the speed of the incoming ball if the rebound ball velocity is 25 m s^{-1} and the racket strikes the ball at 35 m s^{-1}. Consider the coefficient of restitution to be 0.7.

2   Calculate the moment of inertia of a squash racket which has a period of swing around its axis of rotation of 2 s, a center of gravity of 0.3 m from the axis of rotation and a mass of 0.15 kg.

3   The Groppel *et al.* (1983) paper which analyses the impact of an angled racket with the ball compares the ball spin derived from the theoretical equations with actual ball spin created from real tennis strokes. Accuracy results are variable. Suggest some reasons why that might be so.

4   Calculate the equivalent spring and damper constants for a tennis ball with a mass of 0.057 kg, a total impact time of 5 ms and a coefficient of restitution of 0.7.

5   State the CPR category for a court when the inbound horizontal ball velocity is 75.44 m s^{-1}, the outbound horizontal velocity is 20 m s^{-1}, the inbound vertical velocity is 40 m s^{-1} and the outbound vertical velocity is 26 m s^{-1}.

Chapter 12

Cricket and baseball

12.1 Introduction: the game

12.1.1 Cricket

The origins of the game of cricket are lost in the annals of time although a number of legends abound. In the twelfth century there is mention of the game of 'Creag'. This may have formed the earliest foundations of the game. A 'Cric' was a shepherd's staff and may have been a readily available and suitable striking implement. The first written record of 'Creckett' or 'Crickett' appears in 1598. In 1611, two players were arrested for playing this 'illegal and immoral' game (one can only assume that there were no fielders or that batting and bowling were considered more serious crimes than fielding!). In the ensuing years, many more arrests followed.

Cricket in those days was a game for the common working man, and the only time they had the opportunity to play the game was on the Sabbath. At the end of the English Civil War in 1648, the new government banned all Sunday matches and, for a time, interest in the sport diminished, regaining popularity once the Puritan Government was superseded and betting on the game's outcome became a popular pursuit. There was a period of poor player behaviour; often resulting in rioting, until, in 1788, the Marylebone Cricket Club compiled the first laws of the game (note that cricket is one of the very few games to have 'laws' rather than 'rules'). These laws have essentially remained unchanged since then, except for the initiation of the over-arm method of bowling in 1864.

Cricket became the 'game of gentlemen' as it gained respectability, and the laws defining the rules of the game and the specifications of pitch and equipment were developed. Before the nineteenth century, the game was played only in England and the Eastern United States. At the turn of the nineteenth century the cricket fields underwent a transformation, as they altered from roughly ploughed furrows to level and manicured 'squares'. South Africa and Australia took up the game around this time. The first international game was played in 1844 in New York, between the United States of America and Canada. England took on Australia in Melbourne in 1877 in what was the first Test Match; Australia won the match by 45 runs.

However, it was when Australia beat England again in 1882 (this time by only 8 runs) that prompted a writer to publish his 'obituary' of English cricket in which he suggested that the game was 'dead' and its body should be cremated, the ashes being spread throughout Australia. The following summer, England played another series against Australia and the press referred to the challenge as 'reclaiming the ashes'. A small trophy was made and filled with ashes, which was given to the English captain. Ever since, all games played between England and Australia are said to be played 'for the ashes'.

The game of cricket has made only one appearance in the Olympic Games, in 1900. England played France (although in actual fact, most of the French team were from the British Embassy). England won the match and the gold medal.

In 1909 the Imperial Cricket Conference (ICC) was formed to govern the laws of the game with founding members from England, South Africa and Australia. Today the organization is known as The International Cricket Council. The laws had to be tightened up quite considerably as a consequence of the infamous 'bodyline tour' of 1932–33. The bodyline was a tactic employed by the English team primarily to counter the brilliant and extravagant batting of the Australian, Donald Bradman. By aiming the ball low and pitching short, following the bounce, the ball would rise from the ground and, although ostensibly aimed at the leg stump, it would invariably fly unpredictably and dangerously towards some part of the batsman's body.

Accompanying this bowling line would be a cordon of close fielders set on the leg side. The result would be that the batsman had to choose, either to take evasive action from the ball, or attempt to fend it away with the bat, possibly giving catching chances to the close-set leg-side field. This tactic created a great deal of bad feeling between the two teams which went beyond the confines of the game and developed into a diplomatic incident between the two countries. The Australian captain famously stated 'There are two teams out there. One is playing cricket. The other is making no attempt to do so'. The English team maintained that its tactic was not dangerous, and no serious injuries had been reported as a direct consequence of the scheme. However, the rules were changed in 1933 to disallow this tactic. Notwithstanding, ill-feeling between the two sides still exist to this day.

Test matches may last up to five days. However, a shortened one-day version of the game was played by some clubs from around 1960, although it was not initially widely accepted. The first one-day International match was played in Melbourne in 1971 and now the Cricket World Cup is held every four years which is a series of one-day matches between 16 countries. The last change to the rules of the game came as a consequence of a one-day match between New Zealand and Australia in 1981. New Zealand required six runs off the last ball to draw the match. The Australian captain, Greg Chappell, ordered his brother, Trevor, to bowl an underarm ball which rolled along the ground effectively preventing the batsman from hitting the required six. This underhand tactic resulted in a banning of underarm bowling from the game.

The concept of the 'third umpire' was initiated in 1992. He or she adjudicates from off the field, when asked to by the two on-field umpires, but the third umpire does have the benefit of television replays. Sachin Tandulkar was the first player to be called out by means of this technology. In 2001 Channel 4 Television in Great Britain was the first to broadcast use of the 'Hawk-Eye' ball tracking system, invented by Dr Paul Hawkins and David Sherry. It is now utilized in a variety of sports, and many believe it is only a matter of time before the equipment is used in the actual adjudication process.

Finally, in 2007, the inaugural ICC Twenty20 World Cup tournament was held in South Africa. Although, immensely popular, it is essentially an evening entertainment event, and like the limited-overs versions of the game, it is considered frivolous and inconsequential by the true cricket devotees.

12.1.2 Baseball

Compared with cricket, baseball has a short history. Its origins are derived from the English game of rounders in the early nineteenth century. It grew in popularity; variously called townball, base and finally baseball. Popular legend suggests that Abner Doubleday invented the game, although Alexander Cartwright formalized the rules in 1845 and is acknowledged as the true father of the game.

In 1846 Cartwright's Knickerbocker Baseball Club of New York City lost to the New York City Baseball Club in New Jersey; the first major contest. Although, at this time the teams most decidedly had amateur status, a convention of such clubs was formed in 1857 creating the first league and forum for rule discussions, and occasionally the teams charged spectators for the privilege of watching the games. And so the National Association of Baseball Players was created.

The American Civil War caused disruption in so many aspects and, over this period, the game lost much of its popularity. Union soldiers did however succeed in disseminating the game around the United States as they travelled on their manoeuvres. So, at cessation of the conflict, the players returned to their game with renewed enthusiasm and the league's annual convention in 1868 attracted over 100 clubs. Charging for games now became the norm and teams would seek sponsorship and donations in order to travel to games. Winning now became more important than ever.

Although players were not meant to be paid, it is naïve to think that they were not; some were given 'jobs' by sponsors, while others were secretly paid. The first fully professional team was the Cincinnati Red Stockings. The team simply bought in the best players and beat every other team, winning 65 games and losing none. Their competitors had no choice but to follow suit; the first professional baseball league, the National Association, was formed in 1871. As in cricket, there followed a period when gambling became prevalent, and that, together with the promotion of alcohol sales among the spectators, actually led to a reduction in crowds. The National Association, which was essentially a consortium controlled by players and the more dedicated fans, was replaced by the National League which was controlled and run by businessmen. They set the organization on a firm footing with standards for ticket prices, schedules and player contracts.

Restrictions on these contracts lead to dissatisfaction among players and some attempted to form their own league; the Union Association. This was short lived and there followed a series of competitive leagues including the American League and the Federal League. This collection of leagues was locked in legal battles relating to monopolies and restrictive contracts, and, in fact, it wasn't until the 1960s that baseball really prospered under, among others, the Continental League.

The game grew and thrived during the roaring twenties and, arguably, the first sports super-celebrity, Babe Ruth, emerged. Ruth revolutionized the game with, first his success as a pitcher, and then his prowess as a hard hitter and homerun hitter. Ruth ushered in an era of economic prosperity for the game. The Second World War understandably created a difficult time for the game but, following 1945, the game again exploded in popularity. Immediately following the war, unofficial racial segregation was prevalent, and although Jackie Robinson was the first player to cross the 'colour barrier' in 1947, it was not until the mid 1960s that all leagues could boast fully integrated African-American players into their teams. It is worth mentioning that two media developments have played a major contribution to the popularity of the game; radio between the wars, and then television mainly during the 1960s.

The Major League Baseball Players Association is the union of professional major-league baseball players and has championed players' rights since 1953, ensuring players get the best deals from teams, leagues, sponsorship and the media. It has not been without difficulty. Players have entered into three strikes since its instigation, the worst battle being in 1992 when even President Clinton was brought in to mediate over the cancellation of the World Series. The conflict continued until 1996 when a deal was finally reached. However, during those ensuing years, the game lost much of its fan base. It remains to be seen if its popularity returns to its former prominence.

12.2 The bats: design, construction and performance

12.2.1 Cricket bats

Cricket bats are invariable constructed from a form of willow, *Salix alba* 'Caerulea', otherwise known as the cricket bat willow. It is cultivated in large plantations, primarily in the counties of Essex, Norfolk and Suffolk in Great Britain. Its suitability is based on the fact that it is very fast growing, it is tough, lightweight and resistant to splintering.

Law 6 of the Laws of Cricket state that the bat shall be no more than 38 inches (96.5 cm) in length and that the blade of the bat shall be made solely of wood and not exceed $4\frac{1}{4}$ inches (10.8 cm) at the widest point. There is no restriction on either the handle or the weight of the bat, and an optional protecting film may be placed over the front face of the bat in order to repair, strengthen or otherwise protect the surface. Kookaburra introduced a bat with just such a protective coating of graphite. Ricky Ponting has used the bat to good effect although the ICC insisted that the bat contravened Law 6 and Kookaburra have now voluntarily withdrawn the design.

The wood is selected when the tree is between 15–30 years old. The 'rounds', or sections of trunk, are split into 'clefts', and then rough-sawn to the approximate shape of the bat blade. The grain ends are then wax-sealed and blades are kiln or air dried. Next comes the most skilled phase of the process: choosing the blades and then machining them to shape. The decision is based on the variability of the wood quality, both between different blades, and within each blade. The integrity of the final product depends, more than any other aspect, on this selection.

The blades are then hydraulically compressed, up to four times with pressures up to 1.5 MPa. The choice of pressure is a delicate balance between creating a bat which is hard enough to maintain its strength over a reasonable lifetime, and yet still soft enough to maintain the coefficient of restitution from the striking ball. The handle is a laminated Asian (Manau or Sarawak) cane/rubber strip construction which is glued to the blade using standard wood glue. The handle is generally positioned slightly forward of the blade to aid balance and to create a good ball 'pick-up'. The blade is shaped, first using a plane, and then a shaver. The surface is finished with course and fine sandpapers. For greater strength, the handles are 'whipped' with twine held in place by glue, and the rubber grip cover applied. The blade is then finely burnished using a compound wax which polishes and flattens the wood leaving a satin finish, prior to the transfer labels being applied. Of course, the first thing the proud owner of a new bat will do is 'knock it in' by repeatedly hitting it with a ball or equivalent hammer device. Knocking in a bat for 4–6 hours may harden the bat's surface by as much as a factor of two, while not influencing the core hardness. This creates a more elastic response from the bat. The resulting bat will typically possess a Young's modulus, E, of 6600 MPa and a density, ρ, of 417 kg m^{-3}.

This construction process has essentially remained unaltered for some considerable time, although in Sir Donald Bradman's day, bats typically weighed in the region of 2 lb 4 oz (1.02 kg) compared with today's weights which may be nearer 3 lb (1.36 kg). Furthermore, modern blades are longer, with a larger profile and thicker edges and the center of percussion is positioned nearer the toe of the bat. This allows modern bats to hit balls harder and further, but they do rather reduce the range of shot styles available to the batsman.

12.2.2 Baseball bats

The earliest bats were simply sticks, later improved by the players by, either whittling, or better still, turning the bat to shape. Modern baseball rules limit bat lengths to 42 inches

(107 cm) with a diameter up to 2.75 inches (7 cm). There is no weight restriction. The bats must be made of wood with no metal, cork (something of a contradiction here, as cork actually comes from the cork oak tree!), or other type of reinforcement inserted into the bat's center.

Traditionally, the bats are made from ash trees from Pennsylvania and upstate New York. Like the cricket bat willow, this tree is valued for its strength, light weight and flexibility. It can take between 40 to 50 years for the trees to reach the maturity required to ensure trunks wide enough to yield approximately 60 bats per trunk. The trunks are cut to length and hydraulically split to width using wedges, producing the raw material known as splits. Each split is crudely turned on a lathe into what are called billets. These are seasoned for between six months and two years, which removes the sap and gum from the wood, thereby strengthening it.

The billets are then turned on a lathe for the second time into the recognizable shape of the baseball bat, and then course and fine sanded to the final shape. The bat may then be stained if required and branded with the manufacturing company's trademark and players name if required, prior to varnishing.

As with cricket, the game is steeped in tradition and the governing bodies resist change, at least in respect of the bat and ball. However, the supply of ash trees is not limitless, and alternatives to natural wood are continually sought. Lanxide ceramic-reinforced material may be a suitable replacement in the coming years. Already, at the lower ranking college level of the game, plastic reinforced bats are legally used, consisting of a plastic foam core surrounded by resin-impregnated synthetic fibers.

12.3 The balls: design, construction and performance

12.3.1 Cricket balls

Law 5 of the Laws of Cricket state that 'The ball, when new, shall weigh not less than $5\frac{1}{2}$ oz (155.9 g) and not more than $5\frac{3}{4}$ oz (163 g) and shall measure not less than 22.4 cm nor more than 22.9 cm in circumference' (MCC, 2000b). However, in 1975, a consortium of cricket authorities together with the British Standards Institute worked to create a series of tests on ball quality, listed below. Only the first two tests address Law 5.

1   Circumference
2   Mass
3   Seam width
4   Seam height
5   Shape
6   Height of bounce
7   Hardness
8   Impact resistance
9   Wear resistance

The Australian Cricket Board run a completely different set of standards since the core of Australian balls are different.

Ball manufacture begins with a 2 cm diameter cork center, onto which wet wool is layered. This dries, shrinking back and compressing the cork. Another layer of cork is added and further wool yarn. The process continues until, typically, five layers of cork are laminated with six layers of wool. The leather hide cover, dyed red or white as appropriate, is sewn as four quarters, thus creating two seams; the primary (or visible) seam which consists of six rows of stitching (80–90 stitches per row, which protrude 1 mm above the surface), and the quarter, or invisible, seam which joins at right angles to the primary seam.

The differences in ball quality and characteristic derive, to a large extent, from the different types of core used, and much debate continues on the best compromise between quality, repeatability and cost. Broadly, cricket ball cores fall into three categories: the traditional solid core, a rolled core – either 'quilted' or layered, or the composite core pressed from cork and rubber fragments. The ball which is most different from the others in terms of play characteristic is the rolled core variety, which exhibits a significant variation in hardness with choice of axis. This adds an element of unpredictability to the impact aspects of the flight (ball/ground and ball/bat interaction). The composite core is thought to be the safer ball in respect of body strikes, but this has not been rigorously tested.

It should be noted that the finish of the white ball is quite different from that of the traditional red ball. The red ball finish is dyed, greased and polished with shellac. The shellac soon wears in play, leaving a greasy surface suitable for polishing. However, the white ball is painted with polyurethane based white paint and then heat-treated to bond the paint to the leather like a hard skin. It is then further coated with a polyurethane varnish. This additional cover smooths out the quarter seam and slows the roughening of the seam-side (with respect to an in-swinging ball). However, Mehta (2000) has not really noted significant and consistent differences in play between the two balls.

12.3.2 Baseballs

The earliest baseballs consisted of a solid core of virtually any material, but typically wood, wound with yarn and then covered in a one-piece brown leather in a fashion known as 'lemon peel' or 'rose petal'. These balls were both soft and light and behaved in a particularly 'dead' fashion on impact. Furthermore, their impact characteristic could be modified by varying the tightness of the wrapping yarn. Standards were such that the home team could gain advantage by choosing balls (it was their prerogative) which they knew would suite their team; tightly wrapped lively balls for strong hitters, looser wrapping for the more defensive teams.

The two-piece leather figure-eight stitch ball was introduced in 1872 at the professional level. In 1910, the simpler wood or cork core was replaced by a vulcanized rubber-coated cork core, which created balls with a much higher coefficient of restitution, leading to a more interesting game. However, the use of rubber strips covering the core were banned in favour of moulded shells as this design resulted in greater consistency of performance.

Today, the specifications demand a cork nucleus of precisely 0.5 oz and a diameter between 1.94 in and 2.86 in. This is encased in two thin rubber layers; one black and one red, weighing a total of $\frac{7}{8}$ oz. This 'pill' is machine-wound under high tension with consecutively, three different grades of wool yarn of differing lengths, topped with 150 yards of fine white polyester–cotton blend yarn. The ball is then coated with rubber cement prior to the two piece cover being attached and stitched. The cover, made from an elongated figure-of-eight shaped white cowhide, is dampened to allow stretching and manipulation. It is hand-stitched with exactly 216 raised stitches using 2.235 m (88 in) of the familiar waxed red cotton. The last stage of the process is to roll the balls for 15 seconds while still slightly damp, so that the seams are even and reasonably flat, thus reducing air resistance.

12.4 Launch techniques

12.4.1 Cricket: the deliveries

The basic action

The basic action consists of five distinct phases;

1 *The run-up and jump into the gather.* The front arm extends high as the body turns sideways. The back foot is parallel to the crease, the head over the front shoulder, with the ball tucked into, or near to, the chin, and pointing towards the target.
2 *The set-up.* With the gather complete, the body starts to uncoil, the front knees brought upwards as the body rocks back. The bowling arm now begins to extend downwards and backwards.
3 *The unfold.* The bowler unwinds as the front arm comes down and forward, both arms being approximately parallel to the ground. The front leg now straightens and is braced. The eyes are now fixed on the target.
4 *The delivery.* The head is level, and the bowling arm swings up and through the full extent, with the ball being released at the top of the arc as the arm is about to swing down. A full follow-through is essential for a good delivery.
5 *The follow-through.* The body completes its action with a turn of the hips and the bowling arm continues to follow the line of the front arm. It is essential to allow sufficient momentum to drain away steadily following the release, to prevent injury.

The process described above relates to the classic side-on action which the vast majority of bowlers adopt. A less common alternative is the front-on action in which the bowler's chest faces the target throughout the bowling stages. This was a much maligned technique originally; coaches would never allow this method to become entrenched into the bowler's style. The method may restrict the range of swing deliveries available. Furthermore, because the power in the throw originates to a large extent from the rotating hips of the side-on action, a fast delivery is difficult to achieve using the front-on approach. However, with the advent of front-on adoption by some great bowlers such as Andrew Flintoff, the technique has now become accepted. The main advantage appears to be that, with less complex body movements occurring between run-up and follow-through, there is simply less to go wrong, and so consistency in accuracy is maintained.

Delivery accuracy

Bowling accuracy is usually defined in terms of what is commonly called 'line and length'.
Line relates to the direction in which the bowler delivers the ball, and is determined by the target they are aiming for. It should be noted that, for a whole variety of complex strategic reasons, that target may not actual be the wicket, and is dependent on such aspects as; the amount of swing, spin and speed of the ball and, indeed what the bowler is trying to achieve in their delivery.
Length defines how far up the wicket the ball bounces. Within the 22 yard range, the length options disaggregate into about six naturally strategic choices. Figure 12.1 illustrates the approximate scale of these lengths in respect to the full length:

a *Long hop.* This gives the batter plenty of time to pick up the delivery. Best avoided!
b *Bouncer.* This bounces off the pitch and often heads for the upper body of the batter who has to avoid or fend off the errant ball. Often used to intimidate the batter! Only one bouncer is allowed per over per batter.

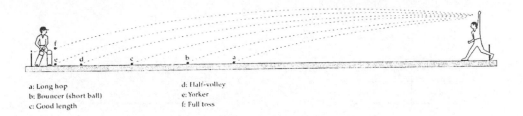

a: Long hop
b: Bouncer (short ball)
c: Good length

d: Half-volley
e: Yorker
f: Full toss

Figure 12.1 Different length deliveries as bowled by a medium fast bowler.

c   *Good length.* The ball pitches about 4 m from the batter's crease, leading to the batter having little time to react; in particular to decide if they will play off the front, or the back foot.

d   *Half-volley.* Often driven as the ball rises from the bounce for four runs! There is little lateral movement for the batter to contend with, and the ball often meets the bat's sweet spot nicely. However, in the complex double-strategy employed by players, the occasional perceived 'gift' of the half-volley may lead to a reckless mistake by the batter.

e   *Yorker.* A deadly ball, especially for the tall batter who has to 'dig the ball out' from under their toes, when it is possibly only a centimeter or so off the ground, and dangerously close to the wicket. The less experienced tail-end batters are often caught by this, difficult to achieve, delivery.

f   *Full toss.* This is considered a batter's 'free hit', often for the boundary, although a low full-toss arriving close to the batter's ankle can again be difficult to dig out. This delivery may be thought of as a 'failed' yorker.

Types of delivery

There are many different types of delivery available to the bowler. These vary by bowling technique and tactical intent.

The 'family tree' shown in Figure 12.2 broadly illustrates the relationship between the most common styles, although it must be stated that there are a number of 'grey areas' and 'fuzzy links' in some parts of the chart. The following descriptions assume a right-handed bowler.

1   *Fast bowling*

   1.1   *Seam bowling*

      The aim here is to make the ball contact with the ground directly on the seam. This creates a degree of bounce uncertainty which can beat the batsman. It follows that the bowler and fielders also do not know the position of the ball after the bounce. The ball is gripped with first and second finger directly above the ball, and with the thumb directly below it. The ball is released with a perfect vertical seam and strong backspin (typical measured values in the region of 14 rev s^{-1}) which, while ensuring vertical seam stability, will also modify the vertical trajectory in accordance with the Magnus effect. However, the ball should not move sideways through the air to any great degree.

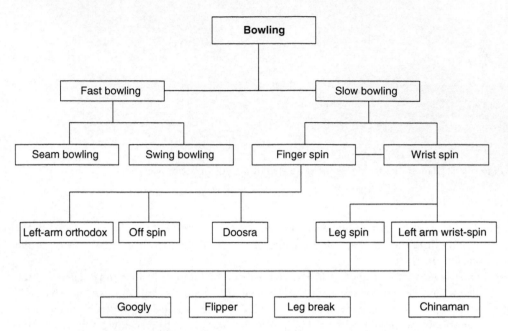

Figure 12.2 The bowler's family tree.

1.2 *Swing bowling*

The ball grip is similar to the seam bowling grip except that the seam is held at an angle to the direction of the release. The swing is best explained by imagining the seam as the prow of a boat which splits the air asymmetrically, pulling the ball to one side, or the other, thereby creating either the in-swinger (seam points slightly to the right), or the out-swinger (seam points slightly to the left). It is most prominent when the ball is new and the seam protrudes. However, the swing is enhanced by having a roughened surface on the inside of the swing and a highly polished surface on the outside, created by spit and wiping of the side of the ball on the bowler's thigh. An aspiration of the bowler is to create the perception of what is termed the 'late-swing'; a somewhat contentious issue which is discussed later in this chapter.

Reverse swing, first created by Pakistani bowlers in the late 1980s, is created using older balls, especially in dry, hot and humid conditions. The older ball creates a turbulent airflow on both sides of the ball but the flow separation is asymmetric, creating a late boundary separation on the polished side, causing it to swing in that direction. It is further considered that the spit and polish, continually applied to the smooth side of the ball over a number of overs, adds weight to that side of the ball which supports the swing in the direction of the smooth side.

2 *Slow bowling*

2.1 *Finger-spin*

Spin is imparted by first flicking the wrist and then tweaking the ball at the point of launch. For the right-handed bowler this is called the 'off-spin', for the left-handed bowler it is known as the 'left-arm orthodox spin'. The correct grip is created by first spreading the first and second finger across the seam of the ball, and then folding the ball down so that it is gripped comfortably below the ball with the third finger and side of the thumb. A ball released in this manner spins simultaneously around both horizontal axes of rotation, i.e. both backspin and lateral side-spin. Obviously the amount of each of these spin components will dictate the resulting ball's flight trajectory.

A variation on this style is known as the doosra (Urdu for 'the other one', or more precisely, 'the second one'). Saqlain Mushtaq is credited with its invention. The action is similar to that of the off-spin style, except that the wrist is completely (or as nearly as is practicable) reversed so that the back of the hand, rather than the palm, faces the batsman. As the wrist unwinds into the throw, the ball is spun in the opposite direction (it is as if its whole symmetry has been turned over), causing the ball to move from the leg side to the off side over the duration of its trajectory.

2.2   *Wrist-spin*

An anti-clockwise spin (with respect to the right-handed bowler) is imparted by releasing the ball from the back of the hand, so that it passes over the little finger. In fact, both the so-called finger and wrist spin styles involve both fingers and wrists; it is just the order of movement which is reversed. In the case of the wrist spinner, the movement starts in the fingers and finishes in the wrist, while for the finger spin, it starts with the wrist and finishes with the fingers.

2.2.1   *Leg spin*

2.2.1.1   *The leg break*

The ball is held in the palm of the hand with the seam running under all the fingers. On release, the wrist is rotated from right to left, causing an anti-clockwise spin. As the ball makes contact with the pitch, it will deviate to the left, or off-side, for the right-handed batsman.

2.2.1.2   *The flipper*

The ball is squeezed out of the front of the hand with the thumb, first and second fingers creating significant backspin. The ball will fly deceptively low and may drop late, but it will pitch high on contact with the ground, hopefully catching the batsman unaware.

2.2.1.3   *The googly*

This is a delivery which looks like a standard leg-break, but spins in the opposite sense towards the batsman. Ideally, the direction of spin is hidden from the batsman. The ball is held as for a normal leg-break, with the index and middle fingers' top joints above the seam, and the ball resting between the bent third finger and the thumb. Immediately prior to the point of release, the palm should be facing skywards, with the back of the hand towards the batsman. On release, the ball is turned anti-clockwise with the third finger doing most of the work. The deception is created because, hopefully, the batsman only sees the back of the bowler' hand, and so does not pick up a sense of spin direction.

2.2.2   *Left arm wrist-spin*

The delivery styles as delineated in 2.2.1 above apply equally to the left-handed bowler, with the directions and senses appropriately reversed. It is, however, worth noting that few good 'left-armers' achieve Test cricket status. There is, however, one additional delivery available to only the left-arm bowler.

2.2.2.1 *The chinaman* (also known as the left-arm unorthodox spin or left handed leg break)
The direction of pitch is the same as a standard right-handed bowler's off-spin, i.e. from right to left as seen from the bowler's end. This would normally be easy for a batsman to 'answer', except that it is an unexpected spin from a left-hander, and, furthermore, because the spin is imparted from the wrist rather than the finger, more turn and bounce is created. Grip and action will be the mirror of the right-handed bowler's leg-break.

12.4.2 Baseball – the pitches

Figure 12.3 illustrates the common types of baseball pitch available to the pitcher, with the similar proviso to that stated in the previous section, that there is a degree of overlap between the different styles of pitch.

1   *Fastballs.* The most common type of pitch which most pitchers use. There is usually some degree of backspin on the ball which adds stability and maintains a straight trajectory against the force of gravity.

1.1 *The four-seam fastball*
The fastest pitch, often incorporating a little topspin to maintain height throughout the trajectory. Speeds of up to 100 mph are achievable. It is so named because all four seams rotate in the air ensuring as straight a trajectory line as possible.

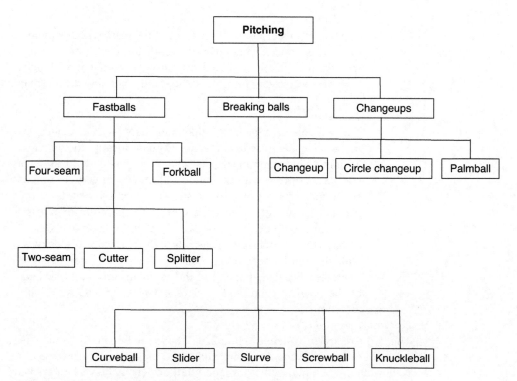

Figure 12.3 The pitcher's family tree.

1.2   *The two-seam fastball*

The 'two-seamer' has the speed of the four-seamer, at least at launch. There will however be more movement on the ball as, ideally, it curves a little in flight towards the target. The throw mechanics are the same as the four-seamer. The difference lies only in the way the ball is gripped for the pitch.

1.3   *The cutter*

The cutter is basically a two-seam fast ball, but with a little more spin created by applying more pressure from the middle finger on release. Because there is still considerable speed on launch, the ball movement (or break as it is called) appears to occur late in the trajectory when the ball is in the vicinity of the batter, allowing them little time to react. This type of pitch is more responsible than any other for damaging bats.

1.4   *The splitter*

So named because the ball is grasped with the index and middle finger stretched and split around the ball. As a consequence the ball is actually released slower from the hand, even though, to the batter, it looks like a standard fastball. The trajectory again appears to break late, but downwards towards the batter's knees, rather than to the side, as in the case of the cutter.

1.5   *The forkball*

Similar to the splitter, although to throw a forkball, the ball is held deep between the first two fingers, and thrown hard with a considerable snap of the wrist. The result is a slower (and therefore arguably this should be in the 'breaking balls' section) and more tumbling action. This type of pitch is known to cause significant damage to the shoulder and elbow. For this reason it is not recommended for younger pitchers, where the bone/muscle system is not yet fully developed.

2   *Breaking balls.* These are balls which are slower than fast balls and do not travel straight, but will have some sideways or additional downward movement. If the ball drops close to the plate, it may defeat both the batter and the catcher. Therefore, these types of balls are only attempted if the pitcher has good confidence in the proficiency of his catcher.

2.1   *The curveball*

Typically, the curveball is about 15 mph slower than a fastball and possesses topspin and side spin, which breaks the ball both downwards and horizontally. One of the most difficult types of curveball is known as the '12–6' which signifies the ball's movement with respect to a clock face. A 12–6 will have little sideways movement, but will drop down, hopefully surprising the batter, as it approaches the plate. However, a ball which breaks downwards too early in flight will be a gift to the batter, as they have more time to react to the slower ball and a clearly predictable trajectory.

2.2   *The slider*

Also known as a nickel curve, this is a pitch considered to be about half way between a fastball and a standard curveball. The batter thinks it is a fastball, but the slower speed and curve, both downwards and sideways, deceives them. The break does occur earlier than a curveball – indeed about halfway through its trajectory – but hopefully the batter is expecting a straighter flight path, and so fails to connect. This is another style of throw that can injure the pitcher's arm.

2.3   *The slurve*

The name derives from a combination of the curveball and the slider. This is not a particularly popular variant, but is created by throwing a slider in the same manner as they might throw a curveball.

2.4   *The screwball*

The screwball is basically a slider which breaks in the opposite direction. So, for a right handed pitcher, the ball will break from left to right as viewed from the pitcher. The ball is gripped firmly, deep in the hand as for the fastball pitch, but released out of the back of the hand with the rotation almost totally provided by the twisting of the arm.

2.5   *The knuckleball*

The ball is released, ideally, with no spin at all, although it is difficult to completely avoid backspin when the ball is released with such force. The result is a ball which will behave unpredictably; indeed it will break randomly for the batter, the catcher and, indeed, the pitcher himself. The erratic behaviour is caused by the boundary layer separation or vortices occurring, in the absence of spin, at particular points on the ball's surface dictated by the orientation of the stitching to the airflow direction. The throw is achieved by holding the ball with the first knuckles (hence the name), or even the fingernails. As the ball is released, the fingers are extended in a manner which attempts to kill the tendency to spin, especially in the backspin sense. Section 12.5.2 provides further details of this type of pitch.

3   *Changeups*

3.1   *The standard changeup*

The standard changeup is thrown with the same arm action as a fastball, but because a particular type of grip is used and the ball is thrown slower, the batter is deceived into mistiming their swing. Even if a strike is avoided and contact with the ball is made, the resulting hit may be a foul ball. The grip is such that the ball is held with three fingers rather than two, and it is held closer to the palm. This grip is designed to kill some of the launch speed usually created by the wrist and fingers.

3.2   *The circle changeup*

With this pitch, the grip is made by forming a circle around the ball with the index finger and thumb along the line of the stitching. On release, the wrist is twisted downwards causing the ball to break in a similar manner to the screwball.

3.3   *The palmball*

This is another method of reducing the release velocity and creating a changeup. This time, the ball is held tightly in the palm of the hand with the fingers merely providing the absolute minimum grip. The ball is thrown in a fastball manner but, without the whipping power from the fingers, the velocity of the pitch is considerably reduced.

12.5 Overview of trajectory models

12.5.1 Cricket

Conventional swing

As stated in Section 12.4.1 (Page 229, Section 1.1), fast bowlers make use of the ball's primary seam to swing the ball. The ball is released with the seam at an angle to the initial line of flight, and above a certain critical velocity, the laminar flow is tripped into turbulence (chaotic flow), but only on the side of the ball which contains the leading edge of the seam (Figure 12.4). The non-seam side remains in laminar flow, aided by a polished surface on that side. The side with the turbulent boundary layer separates further back along the ball due to its increased energy, and it is this that sets up the asymmetric pressure conditions which create the swing.

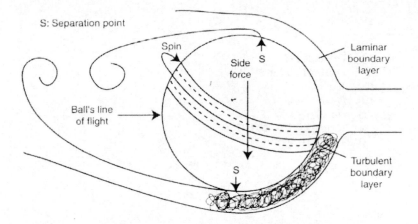

Figure 12.4 Airflow over a cricket ball undergoing conventional swing.

Mehta (2005) discovered that ball velocities in excess of $25\,\mathrm{m\,s^{-1}}$ are enough to trigger turbulence, while the maximum pressure differential on either side of the ball occurs at $29\,\mathrm{m\,s^{-1}}$. The differential then reduced, but was still measurable with speeds up to $37\,\mathrm{m\,s^{-1}}$. The critical velocities, or Reynold's numbers, at which the asymmetry appears and disappears, are found to be functions of seam angle, surface roughness and the spin rate of the ball. Due to the bowling action, the ball will always be launched with some backspin which serves to maintain a stable seam orientation in flight.

In summary, Mehta found that optimum swing occurred when the ball speed was $29\,\mathrm{m\,s^{-1}}$ (corresponding to $67\,\mathrm{m\,h^{-1}}$), with a seam angle of $20°$, and a backspin of $11.4\,\mathrm{rev\,s^{-1}}$. The seam angle of $20°$ ensures that the Reynold's number, R_e (in the region of 1.5×10^5 to 2.0×10^5 for a cricket ball), based on a seam height of 1 mm, is about optimal for effective tripping of the laminar boundary layer. However, for lower speeds, the laminar boundary may still be tripped into turbulence by increasing the seam angle. As the air flows around the ball, it accelerates. Therefore, placing the seam further back around the curve again creates sufficient speed to trip the airflow on the back of the seam. Mehta (2005) found that the deflection remains fairly constant in the range $24 < v < 32\,\mathrm{m\,s^{-1}}$.

It may be assumed (and it has been proved experimentally) that the asymmetric pressure creates a constant side force which acts perpendicular to the initial trajectory. The resulting swing trajectory is therefore that of a parabola.

For the purposes of creating a simple trajectory model of a cricket ball trajectory including swing, De Mestre (1991) suggests that the horizontal displacement in the direction of flight to the wicket, to a first approximation, may be given by:

$$s_x = \frac{m}{k}\,\ln\frac{v_0}{v}$$

with a value of $k/m = 0.003$. The time of flight may be obtained from:

$$t = \frac{m}{k}\left\{\frac{1}{v} - \frac{1}{v_0}\right\}$$

Bentley *et al.* (1982) used experimentally measured values of side force as a function of seam angle and drag coefficients to compute the deviation of a cricket ball over a range of

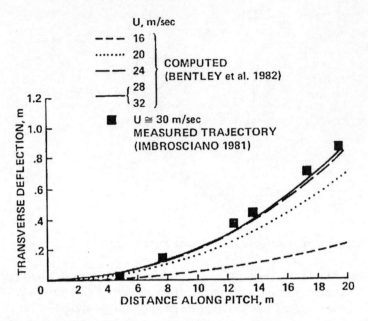

Figure 12.5 Computed flight paths using measured forces for the cricket ball with best swing properties. Seam angle = 20°, spin rate = 14 rev s⁻¹ (Bentley *et al.*, 1982, as cited in Mehta, 1985).

release velocities, $v_0 = 16, 20, 24, 28$ and $32\,\mathrm{m\,s^{-1}}$. The results are shown in Figure 12.5 and indicate that ball deviation is almost identical in the range $24 < v_0 < 32\,\mathrm{m\,s^{-1}}$.

Thus far, the phenomenon of ball swing has been theoretically described and diagrams such as Figure 12.4 can be validated by smoke-flow visualization techniques (see Mehta, 2005, for some excellent examples). However, it is not possible to theoretically *quantify* the magnitude of the side force created by such studies. To perform this type of activity, a full computational flow dynamic (CFD) analysis must be performed. Penrose *et al.* (1996) carried out such an analysis using a CFD software package called FLOTRAN.

The process the software undertakes is to create a suitable mesh around the ball which represents the surrounding atmosphere. It then performs a finite element analysis which distorts the mesh in proportion to the forces created as the object flows through the fluid of the air. The advantage of this method is that the body can be of any general shape, such as a racing car or a yacht hull. In this case, the object is described, in software, as a spherical body with a raised circumferential area: the seam. Account can be taken of the differential surface roughness on either side of the ball. Figure 12.6 indicates a typical finite element mesh, distorted by a seam-created swing. The asymmetry is clearly evident and this can be converted by a further software procedure into pressure contour patterns which more closely relate to the side force and the resultant ball deviation.

Ball tampering

It is clear that, to achieve efficient ball swing it is important to have a contrast between the surface roughness on either side of the seam. For an in-swinger, the ball needs to be polished on the outside, and rough on the seam side. Although the rule clearly states that only natural substances such as sweat or saliva may be used to polish the ball, it is certain that furtive use

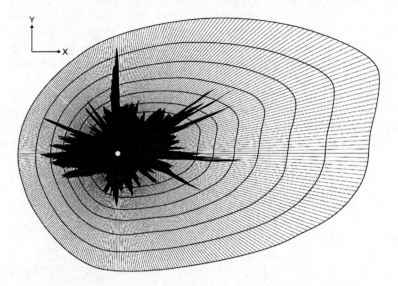

Figure 12.6 The finite element mesh for a seam-swinging ball.

of hair-gel or Vaseline has been used to shine the ball. Conversely, the rule states that the rough side of the ball must be allowed to roughen naturally over the duration of the game. It has similarly been known that unscrupulous players accelerate the process using finger nails, bottle caps (which just happened to be in the player's pocket!) or soil or sand from the pitch (which may also collect at the bottom of a player's pocket).

The late swing

A number of theories exist which attempt to explain the late swing: the ball that swings in flight, but swings much greater as it comes in close proximity to the batsman.

One explanation, expounded in a number of early papers, but now discredited, is that it does not actually exist; it was thought to be nothing more than a figment of the imagination of the batsman, bowler and observers. A further explanation advanced was that the ball was bowled at such a speed that the airflow trips into turbulence on both sides of the ball, ensuring a similar pressure on either side, even though the seam is at an angle. However, some way into its flight, the ball slows down, and then the side which is away from the seam will trip into laminar flow, allowing the conditions for swing to commence.

There is a problem with this theory.

Consider the simple drag equation:

$$m\frac{\mathrm{d}v}{\mathrm{d}t} = -\tfrac{1}{2}C_D \rho A v^2 \tag{12.1}$$

This may be integrated and with v_0 being the ball's speed:

$$v = \frac{v_0}{\left(1 + C_D(1/2m)\rho A v_0 t\right)} \tag{12.2}$$

And for typical values for a cricket ball of $C_D = 0.25$, $m = 0.156\,\mathrm{kg}$, $\rho = 1.2\,\mathrm{kg\,m^{-3}}$ and $A = \pi(0.0355)^2\,\mathrm{m^2}$, Equation 12.2 shows that, after 15 m of flight, $v = 0.967v_0$. The ball slows by just over 3 per cent of its initial velocity. This slight velocity drop just may account for some of the unpredictability in terms of the more remarkable cases of late swing. Although C_D is known to vary considerably with speed as expected, it is in fact found to vary little between old worn balls and new ones.

Another theory of late swing suggests that there may be a change in the ball's orientation due to the gyroscopic precession effect, but this has not been supported by experiment. It has further been suggested that maybe sudden changes in wind direction may cause the seam to rotate and lead to unpredictable late swing. However, this is unlikely to occur in practice.

It is far more likely that the perception of late swing is caused by the projectile forming a section of the trajectory parabola as described earlier in this section. By way of example, consider a ball which is in flight for 0.5 s and is thrown such as to create a significant swing – say, 40 per cent of its weight.

The transverse deflection will be given by, approximately:

$$s_y = \tfrac{1}{2}at^2 = \tfrac{1}{2}(0.4 \times 9.81)t \tag{12.3}$$

Looking at the deflection in 0.1 second intervals for the time of the flight (0.5 s), we find:

$t(s)$	0.1	0.2	0.3	0.4	0.5
s_y (m)	0.020	0.078	0.177	0.314	0.491

It is clear that the ball is deflecting throughout the flight, and yet 75 per cent of the deflection occurs over the latter half of the flight; indeed, half the deflection takes place in the last 30 per cent of the flight-time. The general consensus is, therefore, that late swing is a natural, built-in part of the cricket ball's parabolic swing, and not evidence of some other additional phenomenon.

Reverse swing

The phenomenon of reverse swing has only been in evidence since the early 1980s when, primarily, the Pakistani bowlers such as Imran Kahn, Wasim Akram and Waqar Younis bowled in an apparently standard fast swing bowling style, and yet the ball was deflected in the opposite direction to that expected of a standard swing ball; i.e. in a direction away from the leading edge seam. Perhaps most surprising was the fact that they achieved this type of swing using an older ball which tends to have a rough surface on both sides and a flatter seam.

We commence the explanation of the aerodynamics of reverse swing by first considering the traditional swing, in which the boundary layer on the seam side of the ball is tripped, at the critical R_e value, into turbulence by the seam, as indicated by Figure 12.4. The non-seam side remains in laminar flow, aided by the polished smooth surface on that side.

Now, as the ball speed increases, the critical R_e value will also be activated on the non-seam side of the ball. Hence both sides are tripped into turbulent (chaotic) flow and the result is an equal pressure on either side of the ball. At such velocities the ball will exhibit no deflection at all and fly straight to target. However, if the ball speed is increases still further, the turbulent region begins to move upstream towards the front of the ball, because now the air speed even in this area is such that the critical R_e value is reached. We now have a condition of symmetrical airflow, except that the flow on the seam side still has to negotiate the unevenness of the seam. Now the seam has a detrimental effect; the boundary layer is weakened and thickened as a consequence of the airflow across the 'bumpy' surface of the seam, and this creates an early separation on the seam side, compared to the thinner turbulent boundary layer on the non-seam side. We now have a scenario exactly the opposite of the traditional swing depicted by Figure 12.4, with a chaotic flow evident on both sides of the ball, but separation occurring earlier (i.e. further upstream) on the seam side, leading to a force acting at right angles to the flight direction and pointing *away* from the non-seam side. These are the conditions necessary to elicit the phenomenon of reverse swing.

Effects of humidity

Rather like the reverse swing, this is a much discussed and often misunderstood phenomenon. It is generally acknowledged that a humid day produces the best conditions for swing. Cricket commentators refer to such conditions as 'heavy', and imply that the ball has to do more work in cutting through this type of atmosphere, and they might further suggest that, for this reason (somehow), the swing will be enhanced.

It is true that humidity will affect the flight. After all the side force is directly proportional to the air density through equations similar to Equations 5.1 or 5.4. However, the more humid the air, the *less* dense it is. So, from that argument, in humid conditions, you would expect less swing, not more. It is true that the dynamic viscosity of humid air increases slightly and this will impact on R_e, but detailed calculations, backed by empirical data, confirm that this effect is negligible.

Of course, on a damp day, when the ground is soft with green wet grass, the new ball will retain its shine for longer and so maintain a laminar boundary layer on the non-seam side, but this does not satisfy those who are convinced of the considerable efficacy of humid conditions for the swing bowler. It was also thought that the damp conditions might allow the seam to swell which would naturally produce greater swing, but seam swelling has never been observed, despite a number of experiments. Several investigators have proposed that humid days may create less atmospheric turbulence which might, in turn, create better conditions for swing, while yet others have suggested the opposite; that humid conditions creates more turbulence which helps swing! Not surprisingly, there is no real evidence for the basis of either of these conjectures.

A phenomenon known as condensation shock has been proposed as the cause. This is a sheet of discontinuity associated with a sudden condensation or fog formation in a field of airflow. It occurs, for example, on an aero wing where a rapid drop in pressure causes the temperature to drop considerably below the dew point. This is a scenario worthy of serious consideration. However, condensation shock only occurs when the humidity is near to 100 per cent (essentially, conditions of thick fog). Furthermore, the primary seams on all new balls are adequately prominent to trip the boundary layer in the Reynold's number range of interest.

So there seems to be no scientific evidence (Binnie, 1976) which supports the view that humid conditions are more advantageous to swing bowlers. Notwithstanding, it may be possible for bowlers to create greater spin in humid conditions without even realizing it, especially with a new ball. When the varnish gets wet, it tends to exhibit a sticky characteristic. This may allow the spin finger to acquire more grip in the final contact with the ball on release, thus leading to greater spin. And it has been noted that a relatively small increase in spin rate can result in a significant increase in swing deflection.

Impact of ball with ground

Two key parameters describe the ball–ground interaction: pace and bounce. They express the velocity and trajectory direction, respectively, of the ball following the ground impact. However, it is not just suitable values of pace and bounce that are important, but the *consistency* of the parameters over the duration of the match. Studies by Baker *et al.* (2003) have indicated a positive correlation between pace and the dry bulk density and sand content of the soil, as well as a negative correlation between pace and the moisture content, silt content and organic matter.

To achieve surface-area and match duration uniformity, the soil is compacted with a steel smooth-wheeled roller which increases the soil's bulk density and shear strength. However, if these parameters are increased beyond certain critical values, grass growth will be prohibited, which is essential to allow transpiration of moisture from the deeper soil levels. Shipton *et al.* (2006) have shown that 20 passes with a 4.75 kN roller increases the soil's dry bulk density, ρ_b, from 1200 to a maximum value of 1470 kg m^{-3}, while a larger roller of 7.51 kN can raise the bulk density to 1540 kg m^{-3} in only 10 passes.

James *et al.* (2006) have analysed and experimented in the area of cricket ball impact with the pitch. Initially, they considered the normal impact model and assumed that both ball and pitch deform as a series connected, parallel spring/damper model, as indicated in Figure 12.7.

The study proves, that the deformation of the pitch at any time, t, following the moment of first impact of the ball (at $t = 0$), is given by the finite difference equation:

$$\left(y_p\right)_t = \frac{(k_b)_{t-\Delta t}\,(y_b)_t + (c_b)_{t-\Delta t} + \left(y_p\right)_{t-\Delta t}\left[(c_b)_{t-\Delta t} + \left(c_p\right)_{t-\Delta t}\right]\Delta t^{-1}}{\left(k_p\right)_{t-\Delta t} + (k_b)_{t-\Delta t} + \left[(c_b)_{t-\Delta t} + \left(c_p\right)_{t-\Delta t}\right]\Delta t^{-1}} \tag{12.4}$$

The stiffness and damping parameters of both ball and pitch can be determined experimentally. James *et al.* (2006) determined their values by dropping a cricket ball onto a rigid load cell, followed by dropping a rigid impact hammer onto a cricket pitch. It is worthy of note that, following the impact, the ball decompresses fully and quickly, while the pitch will not quite fully reform to its previous flat surface, and its decompression is a comparatively slow process.

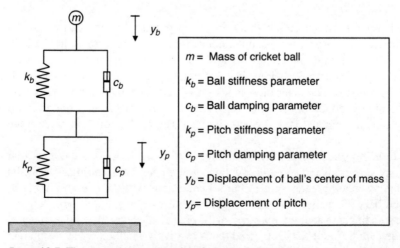

m = Mass of cricket ball

k_b = Ball stiffness parameter

c_b = Ball damping parameter

k_p = Pitch stiffness parameter

c_p = Pitch damping parameter

y_b = Displacement of ball's center of mass

y_p = Displacement of pitch

Figure 12.7 The spring/damper model of a normal ball impact.

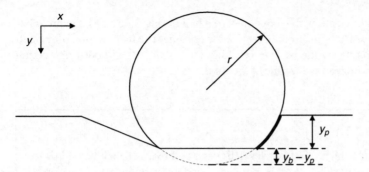

Figure 12.8 Oblique impact of cricket ball with pitch.

In the case of the oblique impact with the pitch, the assumption is made that the ball deforms in a similar manner to the normal impact, i.e. a truncated base of a sphere. The pitch, however, will form a flat bottomed crater with a shallower slope into it on the incoming side of the ball impact (see Figure 12.8).

James *et al.* (2006) calculated that the angle the oblique reaction force makes with the horizontal, γ, is given by:

$$\gamma = \tan^{-1}\left[\frac{\pi}{2}\tan\lambda\right]$$

where:

$$\lambda = \tan^{-1}\left[\frac{r - y_b + 0.5y_p}{\sqrt{r^2\left(r - y_b + 0.5y_p\right)^2}}\right]$$

From these equations and with the coefficient of friction and the direction and magnitude of the normal and tangential forces all known (measured empirically), the ball's angular velocity at time, t, over the progression of the impact, is given by:

$$(\omega)_t = \frac{(F_x)_t\left[r - (y_b)_t\right]\Delta t}{(I)_t} - (\omega)_{t-\Delta t} \tag{12.5}$$

where $(F_x)_t$ is the time-varying horizontal force on the ball and $(I)_t$ is the moment of inertia of the ball, which is assumed to vary due to the deformation and is approximated to a solid sphere with a reduced radius such that:

$$(I)_t = \frac{2}{5}m\left[r - \frac{(y_b)_t}{2}\right]^2 \tag{12.6}$$

To conclude, if one is planning to model a cricket ball trajectory from release to bat impact, including the pitch bounce, the value of the coefficient of restitution, e, may be taken as anything from 0.6 (a hard dry wicket) to 0.1 (a 'sticky' wet pitch). $e = 0.32$ is often taken as a typical average value.

The motion of a ball hit for six

Having dealt with the bowling action, the movement of a ball through the air as it approaches the batsman and the bounce from the ground, let us now consider the final motion of the ball (hopefully!); the ball being hit by the bat and reaching the boundary, without first making contact with the ground, which adds six runs to the total.

Worked Example 12.1

Consider the scenario in which a fast bowler delivers a ball to a batsman who aims to hit the ball in a straight drive back over the bowler's head to the boundary for six runs. Assume the ball is bowled at a typical fast ball speed of $40\,\mathrm{m\,s^{-1}}$ to the batsman, some 18 m away. Calculate the velocity with which the bat must hit the ball.

Assuming no deceleration of the ball in flight, the batsman has only 0.46 s to react and play the shot. From Equation 5.10:

$$s_x = \frac{v_0^2}{v_0 k \sec\theta_0 + g\tan\theta_0}$$

We see that, if we choose $\theta = 30°$ and $v_0 = 30\,\mathrm{m\,s^{-1}}$ and $k = 0.184$ (drag coefficient taken from Coutis, 1998), the range $s_x = 75\,\mathrm{m}$, which would clear most boundaries.

Now, the equation for a bat striking a moving ball, is given by a variant of Equation 4.22:

$$v_1 = \frac{(m_1 - em_2)u_1 + (m_2 + em_2)u_2}{m_1 + m_2}$$

where:

$v_1 = $ ball speed following impact $= -30\,\mathrm{m\,s^{-1}}$

$u_1 = $ ball speed before impact $= 40\,\mathrm{m\,s^{-1}}$

$u_2 = $ bat speed before impact

$m_1 = $ mass of cricket ball $= 0.156\,\mathrm{kg}$ (regulation ball)

$m_2 = $ mass of bat $= 1.7\,\mathrm{kg}$ (typical value)

$e = $ coefficient of restitution $= 0.56$. (typical value)

So:

$$-30 = \frac{(0.156 - 0.56 \times 1.7) \times 40 + (1.7 + 0.56 \times 1.7) \times u_2}{0.156 + 1.7}$$

For this scenario, then, the bat must impact with the ball at $u_2 = -9\,\mathrm{m\,s^{-1}}$.

This speed is attained through a combination of linear and rotational movements of the hips, shoulders, elbows and wrists. In summary, the batsman has 0.46 s to pick up the flight of the ball out of the bowler's hand, choose their shot and accelerate the bat from rest to about $9 \, \text{m s}^{-1}$. Furthermore, they must ensure the ball strikes the bat normal to it, as well as near the bat's center of percussion or 'sweet spot'.

12.5.2 Baseball

A baseball has virtually the same size and weight as a cricket ball. However, the different cover design (which is constructed from two hour-glass shaped segments of white leather seamed together by a single row of about 216 stitches), together with the different throw methods, produce a quite different range of trajectory patterns to that created in the game of cricket.

The curveball

The ball is released with some degree of topspin and possibly side spin as well, which, because the roughness of the seam serves to reduce the critical Reynold's number, leads to a faster drop of the ball than expected, together with a possible curve deflection to left or right. Pitchers may achieve spin rates of up to $30 \, \text{rev s}^{-1}$ and speeds of up to $45 \, \text{m s}^{-1}$.

Briggs (1959) measured the lateral deflections, and hence the lateral force, of a baseball at varying speeds and spins. He found that, for $20 < v < 40 \, \text{m s}^{-1}$ and $20 < \omega < 30 \, \text{rev s}^{-1}$ the lateral deflection of a baseball is directly proportional to both the spin rate and the square of the force magnitude. By a simple application of Pythagoras to the velocity vector, it can be seen that the ball deflection is, therefore, independent of the speed. Figure 12.9a,b shows these two results graphically.

As an indication of the sensitivity of the spin variable, it is worth noting that, for a baseball deflection over a distance of 18.3 m and a throw velocity of $v = 23 \, \text{m s}^{-1}$, Briggs (1959) measured a lateral deflection of 0.28 m for a spin rate, $\omega = 20 \, \text{rev s}^{-1}$, while the deflection was 0.43 m when the spin rate was raised to $\omega = 30 \, \text{rev s}^{-1}$.

As in the case of cricket ball spin, the trajectory of a curveball is parabolic, which may lead the catcher to wrongly conclude that the ball is making a 'late break'.

The knuckleball

The knuckleball is pitched by grasping the ball with the first knuckles or fingernails, and is thrown with the normal backspin and, ideally, with no sidespin. In practice, some sidespin will be created in an indeterminate sense which will affect the trajectory. However, with or without the addition of sidespin, the baseball trajectory will be uncertain as the seams cut the air at some random orientation which may vary slowly in flight.

Watts and Sawyer (1975) carried out force measurements on fixed baseballs in a wind tunnel with varying orientations of seam. For the datum taken in the symmetrical seam position ($\phi = 0°$), the normalized side force, F/mg, varied from 0 at $\phi = 0°$, to 0.6 at $\phi = 50°$. In between those two extreme angles, however, the normalized side force fluctuated rapidly with angle, by as much as ± 0.3. They called this an oscillating wake phenomenon. The largest jump in side force was found to occur where the boundary-layer separation point coincided with the seam. Here, the separation point was observed to suddenly jump from the front of the stitching to the back, causing the rapid step-change in side force. This can be verified because similar rapid jumps were observed at $\phi = 52, 140, 220$ and $310°$, which coincide with the four critical positions for the stitching as the ball rotates around a vertical axis.

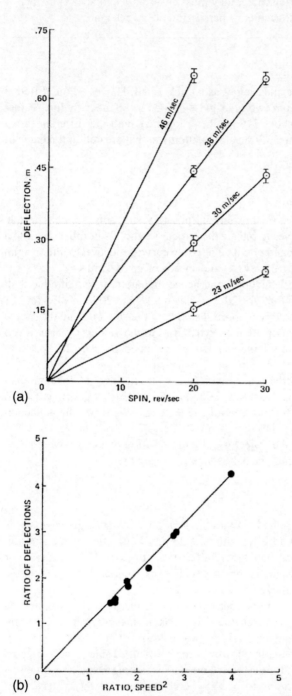

(a)

(b)

Figure 12.9 (a) Lateral deflection of a baseball, spinning about a vertical axis at indicated speeds (Briggs, 1959), (b) Lateral deflection of baseball showing independence with speed (Briggs, 1959).

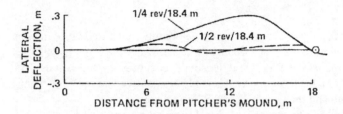

Figure 12.10 Typical computed trajectories for a slowly spinning baseball with $v = 21 \, \mathrm{m \, s^{-1}}$ (Watts and Sawyer, 1975).

Similar to Briggs (1959), Watts and Sawyer (1975) found that, within the range $12 < v < 20 \, \mathrm{m \, s^{-1}}$, the deflection of the knuckleball was independent of flow speed, as was the force frequency, as the ball slowly rotate through the air.

Figure 12.10 shows the computed trajectories for two, slowly turning knuckleballs, indicating the range of trajectories available. Notice how the deflection is much greater for the slower rotating ball; this would be the harder ball for the batter to attack.

12.6 Trajectory diversions: are there no boundaries to the game of cricket?

The peoples of this planet split into two very distinct groups: those who don't appreciate cricket, and those who do. And those who do value the sport, do so with a fervour biased somewhat towards the right end of a hypothetical line drawn between 'Fanatic' and 'Certified Obsessive Compulsive Disorder'. We have previously emphasized the importance of equipment and pitch to the quality of the game. However, wherever and whenever a group of cricket enthusiasts gather, by some mean or other, they will don their whites and play. Consequently, games have been played from latitudes as high as the Langjokull Glacier in Iceland (using rubber matting), all the way down to Antarctica, where the late Harry Thompson took his team, The Captain Scott XI, and out of which he wrote his book *Penguins Stopped Play* (2006). Almost understandably, both these games were played under their respective midnight suns.

However, between the extreme latitudes of the North and South Poles, many impromptu games are reliably reported to have been played. Examples include:

- over a frozen Lake Geneva (rubber matting again!);
- atop a mountain in the Dordogne, France;
- in a crowded chemical engineering laboratory at the University of Connecticut (surrounded by beakers and a tank of liquid nitrogen – allegedly!);
- the Falkland Isles, San Carlos, on the grass airstrip (during active combat);
- the Gaza Strip (similar dangers to above, the boundary patrolled by armed militia);
- the foot of Mount Fuji, Japan;
- inside a volcanic crater near Naples, Italy;
- Death Valley located between California and Nevada, USA (45° in the shade);
- on the deck of the *Canberra* (game was apparently delayed to allow whale watching);
- under the English Channel (inside a car transporter carriage on the channel tunnel train between London and Paris);

- numerous temporary sandbanks including Margate Sands and the Goodwin Sands, on the east coast of England.

However, perhaps the strangest *regular* fixture occurs seemingly in the middle of the sea, right in the path of one of the world's busiest shipping lanes known as the Solent; a channel of water which runs between the Isle of Wight and the south coast of mainland Britain. The match is the Annual Brambles Cricket Match, which is held between two Isle of Wight teams: the Royal Southern Yacht Club and the Cowes Island Sailing Club.

The Brambles is a notorious sandbank, responsible for many a boat's demise, as they run aground. However, on or around the Spring Tide nearest the August Bank Holiday, the tide hits a particularly low point and the sandbank is revealed as a raised island of about $100\,m \times 150\,m$ for anything between half an hour and two hours. Just as soon as the island appears, players, spectators, scoreboard and the all-important drinks bar arrive by mean of a flotilla of boats, and the game commences. This particularly damp, limited-overs match continues until the ball gets lost in the depths of the returning tide. Although the scoreboard is used, traditionally the teams take turns each year to 'win' the match. The losers buy the winning team either breakfast or dinner depending on the time of day of the low water.

One can only imagine the thoughts of any passing ferry passengers if they see the teams dressed in white, earnestly competing and completely surrounded by sea for miles in all directions. Maybe the international passengers simply think 'We must be getting close to England now'! Their second thought could well be, 'If they're *that* keen, why do they seldom win the Ashes'?

Summary

In this chapter we looked at the origins of the game of cricket and its history up to the present day, including the Ashes series and the infamous bodyline tour. Baseball's history is also discussed, and it is interesting to look at the similarities and differences in the two sports' advances, especially with regard to the manner in which commercial aspects took hold at different stages in their respective development.

Cricket and baseball bat manufacture is described. It is reported that the construction, in both cases requires highly skilled labour and good quality natural materials. Next, the balls are addressed; with particular regard to how judicious material choices, as well as correct construction methods, ensure high quality and consistent playing performance.

The wide assortment of bowling and pitching techniques, which utilize the sewn seams on the surface leather, create a range of deliveries with which to confuse the batsmen. It is no surprise that trajectory analysis is dominated by the phenomenon of swing in its multifarious forms. The late swing and the reverse swing are two of the most difficult styles of delivery to both master and deal with at the batting end. Both of these deliveries are described qualitatively and mathematically. In terms of the less sensitive parameters, variations in flight caused by humidity, which is widely thought to exist, is reported neither to be explained analytically, nor measured empirically. Following the flight, the reaction of the ball with the ground is examined, both from a rebound perspective, and in terms of spin changes. Finally, following the impact of ball with the bat, the flight motion to the boundary is investigated.

The final section of this chapter deals with a style of delivery unique to baseball called the knuckleball. This is shown to be a form of non-spinning delivery, which creates an unpredictable trajectory solely as a consequence of the seam orientation.

Problems and questions

1   In cricket, investigate the following types of deliveries (not discussed in this chapter); the slider, the carom ball, a teesra and the zooter.

2   In baseball, investigate the following types of pitches (not discussed in this chapter); the sinker, the shuuto, the fosh, the gyroball, the spitball and the Eephus pitch.

3   In cricket, how do you define an illegal ball which has been 'thrown'?

4   In cricket, find and study a video clip of 'The Ball of the Century' and consider its implications in terms of general trajectory modelling. (Actually this is just an excuse to watch a moment of sporting genius!)

5   In cricket an illegal activity called 'ball tampering' sometimes occurs in which bowlers roughen one side of the ball artificially, using some form of abrasive substance such as earth from the ground (even bottle tops have been reported). Explain what they are trying to achieve in terms of the critical Reynold's number, R_e, and the degree of the swing that may be produced.

6   It has been stated in this chapter (Section 12.5.2) that the side force on both the curveball and the knuckleball is proportional to the square of the speed. This means that the deflection is independent of the ball's speed. Show this to be true.

7   Calculate the return speed of a baseball, of mass 0.145 kg, when it is hit by a bat of mass 0.9 kg and a speed of $20\,\mathrm{m\,s^{-1}}$, when the incoming ball's speed is: 30, 40 and $60\,\mathrm{m\,s^{-1}}$. Assume the bat ball coefficient of restitution to be 0.5. If the balls are all returned at an angle of 30° to the horizontal, ignoring drag effects, calculate the distances the balls would travel.

Chapter 13

Soccer (Association Football)

13.1 Introduction: the game

Although there is some evidence that a game involving foot-dribbling a leather ball took place in China as far back as 200 BC, the first solid scientific evidence of such an event points to the Chinese game of 'Cuju', dating between the third and second century BC. It is also known that the Greeks and Romans played a variant of the game which permitted carrying the ball.

A somewhat gruesome version of the game began in Mediaeval Europe, and more specifically England, where 'mob soccer' was played following public executions; the victim's head being used as the 'ball'. As such, the first ball was hence documented as the head of a Danish brigand. Competitions followed, with two or three parishes taking part, entering teams of up to a hundred per side. The goals would reside within each of the competing parishes which may have been 3 or 4 miles apart. An inflated animal's bladder, usually belonging to a pig, was used as the ball,while the goal posts were vertical sticks.

By the start of the nineteenth century, the game evolved into two teams of equal numbers playing on a field with goals set apart by between 80 and 100 yards. The goal posts were now joined at the top by a crossbar, usually a length of tape. The rules were gradually standardized throughout the nineteenth century, beginning with the Cambridge Rules drawn up at Trinity College, Cambridge University in 1848 by committee representatives from Eton, Harrow, Rugby, Winchester and Shrewsbury schools. The rules were not, however, universally adopted, and in 1857, Sheffield Football Club was formed and they used their own Football Association version of the rules. This eventually developed into the formation of the Football Association, a national body, in 1863. As a matter of interest, when the detailed rules were being drawn up, a committee member from Blackheath, London withdrew his club as a protest against the removal of two specific rules. These rules would have allowed players to (a) run with the ball and (b) obstruct such runners by kicking them in the shins, or tripping or holding them (Chapter 14 covers this variant in more detail!).

The International Football Association Board (IFAB) was formed in 1886, and the first rules of 'association football', or 'soccer' for short, was ratified on 8th December of that year. Although IFAB still determines the laws of the game to this day, the 17 laws currently in place are now published by the Fédération Internationale de Football Association (FIFA), a body formed in Paris in 1904.

The first major soccer competition, the annual FA Cup, was established in 1872, and in 1888, the first football league was founded in Birmingham by Aston Villa director, William McGregor. The Football World Cup, which takes place every four years, was first played in Uruguay in 1930, although in those early days most teams taking part were from the Americas. However, by 1950, international soccer was well established with European teams participating enthusiastically. Today, over 190 international clubs compete for the 32 qualifying places and a chance to win the Cup. Surprisingly, only seven nations have won the 20 World Cups that have taken place so far; these being Uruguay, Argentina, Brazil, Germany, England, Italy and France. FIFA Women's World Cup was initiated in 1991 and hosted by China with only 12 teams represented. However, by the 2003 Women's World Cup, 16 teams competed and 1 billion viewers from 70 countries followed their progress on television.

Another major international soccer challenge is the Olympic Games and there has been a competition at every Olympic event since 1900, except for the 1932 Games. It has to be said that the established professional status of the game makes its incorporation into the largely amateur Olympic Games restrictive, to say the least. Various rule changes regarding permissible players have taken place over the years in order to accommodate realistic quality teams, while at the same time ensuring that nations do not field their strongest squads. Women first competed in the Olympic Games soccer event in 1996 but they are not subject to the same team restrictions as the men's tournament.

Today, Association Football is generally considered to be the most popular sport worldwide. Millions of fans follow their team passionately each week at the stadia, while billions more watch games on television. According to a FIFA survey conducted in 2001, more than 240 million people from more than 200 nations regularly play soccer. It has been said that the game was in fact responsible for securing a truce in the civil war in Côte d'Ivoire, while also exacerbating tensions in the Yugoslav wars of the 1990s, following a riot of fans at a Zagreb versus Belgrade match.

13.2 The ball

Although it is widely known that an inflated pig's bladder was the first ball to be used competitively in a soccer-like game, its playing characteristic was so remote from that of any typical ball that the game's dynamic must have been severely hampered by its performance. Even though the bladder was resilient and may have even been usable for more than one game, it was also extremely light and bouncy, possessing a performance closer to that of a balloon than a traditional ball. Furthermore, its lack of sphericity would have led to an unpredictable bounce. In fact, this type of ball was probably used only for the games in which the aim was to hold the ball in the air for as long as possible.

An alternative ball in use at the time was the leather ball, stuffed with wool, skin and feathers. These would have been heavy, hard and durable balls, possessing little bounce. However, they were more spherical which would have led to greater in-game control of the ball. In 1836, Charles Goodyear, who was a fan of the game, patented vulcanized rubber and the first rubber ball was played in 1855. Although it had many good characteristics such as bounce, weight and uniformity of manufacture, it still played in a somewhat unpredictable manner which tended to slow down the pace of the game.

The FA first defined the rules pertaining to the ball in 1863, stating that its circumference must lie between 27 and 28 in (68.6–71.1 cm), and its weight must be between 13 and 15 oz. In 1872, the weight was slightly modified to 14–16 oz (410–450 g). These specifications remain in force to this day for Size 5 balls. Mitre and Tomlinson were the first companies to mass produce balls for the newly created Football League. By now it was recognized that the ball's shape retention throughout the game was important, and this could be ensured by using only good quality leather, together with the correct type of stitching.

In 1900, the first brown leather balls with inflated heavy duty rubber inner tubes (bladders) were developed. The covering was made from 18 individual leather sections sewn together to produce three panels of six pieces. These were then stitched into the spherical shape, with a slit remaining for the insertion of the bladder. Once inflated, the slit was stitched up and the ball was then ready for play. Overall performance of these balls was quite good, but they were not without their shortcomings. They were quite heavy, and became even heavier when wet from rain or a damp pitch. In fact, their weight, together with their roughened surface, led to many head injuries from heading the ball. Furthermore, these balls were prone to deflation, even over the duration of a single game. Following the hiatus caused by the Second World War, further developments included the insertion of a layer of strong material between the leather outer and the bladder, and the application of a waterproof synthetic paint to protect the leather. The balls of this era were easily recognized as they were often white or orange in color.

From 1970 onward, ball development has occurred in stages corresponding with the World Cup years; generally a new and revolutionary ball is unveiled prior to each scheduled competition. Perhaps the greatest soccer technology advance was created for the 1970 World Cup, when Adidas launched their Buckheimer model ball. The outer covering was constructed from 32 hand-stitched pentagon-shaped panels of leather. This design was said to be the 'roundest ball ever'. The ball had alternate black and white panels which made it clearly visible on the black and white television sets of the day. It was named the 'Telstar' because of its visual similarity to a satellite in flight.

For the 1978 Argentinean World Cup, 12 circular patches were added to the ball's surface which created an interesting optical effect when in flight. The 1982 World Cup heralded the application of waterproof seams which created a ball with much better wet weather performance. The first fully synthetic ball using reinforced composite polymer panels over a latex bladder was created for the 1986 World Cup. This ball was reported to perform better on hard and wet surfaces, which are often found at altitude. In 1990, Adidas confirmed that their World Cup ball for that year would, for the first time, be guaranteed 100 per cent waterproof. This ball was also found to be faster and livelier than its contenders.

It was discovered that a layer of polyurethane foam added under the synthetic covering made the ball easier to control at high speed, as well as giving it a softer touch. By the 1998 World Cup, this layer had evolved into individual micro-balloons of foam placed under each of the ball's segments. This made the ball more durable as well as improving its energy return characteristics and its responsiveness. The balls used in the 2002 Korean World Cup incorporated an advanced syntactic foam sheet supported by a three-layer knitted chassis, which was reported to create a more precise and consistent flight path.

The Adidas 'Teamgeist' ball introduced for the 2006 German World Cup was constructed from 14 panels, machine pressed and thermally bonded into the shape of the spherical polyhedron. This latest football was again stated as the roundest ball ever, with a guaranteed diameter variation of less than 1 per cent. In fact, the improved flight dynamics of this ball has led to a change in game tactics, with wingers being utilized more frequently. This is as a direct result of more accurate, longer passes being attainable with balls which possess precision flight characteristics, rather than having to limit kicks to shorter punts down the middle of the pitch. However, it has to be said that goalkeepers felt disadvantaged by the adoption of these innovative balls. They believed that these balls clearly favoured the striker, since the ball appeared to stay close to the ground and fly straight to target without deviation.

The latest developments in the field have concentrated on performance test beds for footballs, which utilize robotic legs to create repeatable and realistic impact characteristics. Using these kicking robots, parameters such as; weight, shape, dimensions, internal pressure and trajectory consistency may be measured over recurring, realistic, and yet controlled, simulated typical use (Grund and Senner, 2006).

13.3 The boots

The boots originally used to play Association Football were heavy, high-topped and made of hard leather, the emphasis being on foot protection. Following the end of the Second World War, international fixtures became more commonplace and the lighter, more flexible boot favoured by the South American countries became established world-wide.

In the 1950s, the German shoemaker Adi Dassler was the first to create boots with screw-in studs which could be altered depending on the conditions of the pitch. His boots were used to great effect by the German national squad of the day. Dassler continued his burgeoning boot development under his company's name Adidas; reducing the boot height still further and making them lighter to allow greater player mobility. The trend has continued to this day, although the current minimal amount of ankle support, together with the use of the thinnest and most flexible material for the uppers, have undoubtedly been responsible for numerous metatarsal and other foot-related injuries prevalent in the modern game.

The surface traction provided by the studs is crucial to overall boot performance. The amount of grip created is proportional to the depth of pitch penetration which, in turn is governed by the stud profile and the wetness of the turf. Short studs on a wet surface will not be effective as they will fail to penetrate to the firmer ground beneath. On the other hand, long studs on a hard surface may inhibit good penetration and lead to painful pressure areas on the underpart of the foot at the stud points. The situation is further complicated by the fact that players require just the correct level of grip, and this varies both with the team positional roles, and for each individual player. Obviously too little grip will cause a player to slip or fall. However, too much grip can cause knee or ankle injuries as their feet become locked while executing tight turns and manoeuvres. Modern boots provide a choice of places and angles to insert studs, and for certain playing conditions, blades rather than studs may be chosen. It is claimed that blades provide a more stable base without loss of grip, compared to studs. By covering a larger surface area, the blades reduce the pressure felt by the player through the insole board, thus reducing the instances of chronic injuries such as shin splints. Blades are, however, thought to be more dangerous than studs to the opponents during play due to the blade's comparative sharpness.

The hard outsole of the boot must be thin enough to provide the required flexibility, while being thick enough to allow the studs to be attached to it. The heel counter is non-existent in the modern boot; the heel being held by a cup. Ground impact shock absorption is kept to a minimum, and only in the region of the midsole. A foam sock is placed inside this minimal structure for the player's comfort.

The Adidas 'Predator' range of boots, introduced in 1994, utilized low profile liquid-rubber elements on the top of the boots to aid boot–ball friction and to facilitate curve shots. Today, boot uppers are invariably modelled from kangaroo leather, which provides good foot support and protection, low water ingress, lightness, as well as a high abrasion property for ball feel and grip. Compared with calf hide, kangaroo leather retains considerably more tensile strength when the material is split down into the very thin laminar sheets required for boot construction.

13.4 Kicking: the boot–ball impact

In soccer, kicking the ball is arguably the most important competence required of a player and takes precedence over dribbling and other skills. Variations of kicking style arise due to ball type, ball speed and position as well as the nature and intent of the kick. However, the most important variant, and the one which has been most widely researched, is the maximum velocity instep kick of a stationary ball; essentially the style used most often in penalties and free kicks.

Equation 4.22 relates the velocity of a ball following impact. It is rewritten below for the particular case of a boot hitting a soccer ball:

$$v_{ball} = \frac{Mv_{foot}(1+e)}{M+m} \tag{13.1}$$

where v_{ball} and v_{foot} are the velocities of the ball and foot respectively, M is the effective striking mass of the leg, m is the mass of the ball, and e is the coefficient of restitution (boot/ball).

The effective striking mass of the leg relates to the rigidity of the limb and is the mass of an equivalent rigid implement hitting the ball. The term $M/(M+m)$ gives an indication of the rigidity of the impact and relates to the muscles involved in the kick and their strength of impact. The term $(1+e)$ relates to the firmness of the foot on impact and, of course the pressure of the ball. A higher value of e can be achieved by striking the ball closer to the ankle such as occurs, for instance, with a drop kick.

For a typical maximal kick, the value of $M/(M+m)$ may be taken as 0.8, while the value of e has been found to be in the region of 1.5.

Equation 13.1 now simplifies to:

$$v_{ball} = 1.2v_{foot} \tag{13.2}$$

Competent players will be able to strike the ball between 16 and 22 m s^{-1} which leads to a ball velocity in the range 24–30 m s^{-1}.

For the submaximal kicking case, Zernicke and Roberts (1978) derived a regression equation for ball speeds in the range of 16–27 m s^{-1}, which is similar to Equation 13.2.

$$v_{ball} = 1.23v_{foot} + 2.72 \tag{13.3}$$

Perhaps the most detailed boot–ball impact computer simulation designed and empirically tested for accuracy is that carried out by Asai et al. (2002) and Carré et al. (2002). In their model, the leg and the ball's surface were described using a Lagrangian frame of reference, in which the object of interest is set at a fixed point in space with all movement referenced to that point. By contrast, the pressured air within the ball was described using the more common Eulerian frame of reference, where the object of interest moves through a fixed reference frame (such as the case of the position of a ball in flight which is given, at any instant of time, as a function of $x(t)$, $y(t)$). The mesh structure required to carry out the finite element analysis was a hexagonal arrangement for the leg and boot section, while shell elements were used for the ball's surface. Figure 13.1 shows the components appropriately 'meshed-up'.

Critical to the model dynamic is the behaviour of the air within the ball under conditions of impact. The gamma law equation relates the enclosed air pressure to stored energy and was used in the model to predict the outcome from the impact conditions:

$$p = (\gamma - 1)\rho E$$

where p is air pressure, γ is the ideal gas ratio of specific heats, ρ is the overall air density, and E is the specific internal energy.

Asai et al. (2002) and Carré et al. (2002) modelled kicks with varying degrees of off-set and calculated the resultant spin. Validation was carried out using high speed video

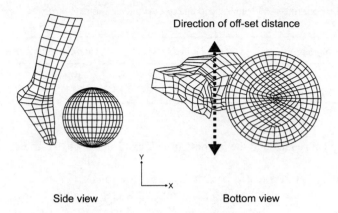

Direction of off-set distance

Side view

Bottom view

Figure 13.1 Foot and ball simulation models (from Asai *et al.*, 2002).

recordings of real kicked footballs and the theory-measured results variance was found to be impressively small.

Interestingly, their model showed that spin would occur even if the coefficient of friction between boot and ball was zero. Although this may initially seem surprising, one must remember that, on impact, the ball deforms inwards creating a surface indentation, with the boot lying within it. When moments later, boot–ball contact is lost, the ball launches with a spin created by the torque of the forces from the offset boot, which was effectively 'locked' into the offset position on the ball's surface during the contact phase. No friction is therefore necessary. Nevertheless Asai *et al.* (2002) and Carré *et al.* (2002) did find that spin increases with boot–ball coefficient of friction, but this is not as sensitive a parameter as the offset distance. Figure 13.2 shows the effect of changing the offset on both the spin rate and the velocity of launch. As expected, the optimum offset distance is a trade-off between the ball rotation and the ball speed.

Obviously, when a striker is trying to maximize spin, the direction of the boot's approach has an influence. The model described by Asai *et al.* (2002) and Carré *et al.* (2002) ascertains this optimum force vector in terms of the magnitudes of horizontal, vertical and

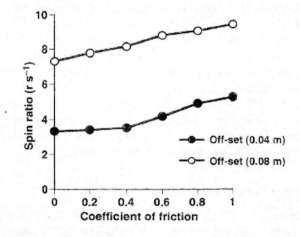

Figure 13.2 The relation of the offset distance, spin ratio and ball speed (from Asai *et al.*, 2002).

Table 13.1 Impact forces required to optimize ball spin for the instep curve kick and the infront curve kick (from Asai *et al.*, 2002)

Curve kick type	Horizontal impact (N)	Vertical impact (N)	Lateral impact (N)
Instep	2439	853	−452
Infront	2206	1221	−1143

lateral forces, for both the instep curve kick and the infront curve kick. Table 13.1 lists their results.

The vertical and lateral forces generate a net force that increases the rotation of the ball as well as the flight vector of the ball in a direction different from the swing direction of the kicking leg.

In previous chapters we have discussed the vibration modes of striking implements with particular reference to the so-called sweet spot. Any object which is struck will vibrate with preferential vibration frequencies at their resonance points which are governed by the various dimensional lengths that make up the object's shape. Now, as it happens, the shape which is most prone to a single severe resonance is the sphere. The sphere has only one major dimension; its diameter, and it is that length which captures this single, strong resonance. The effect on a flexible sphere which is struck is for it to vibrate, increasing and decreasing its diameter at the one resonant frequency. For an efficiently resonating sphere, the vibration will be locked in, leaking out slowly, rather as a tuning fork's vibrations decay with time. While the sphere is vibrating, it possesses stored, vibrational potential energy. This energy, which has been created by the impact is, to all intents and purposes, wasted, as it slowly dissipates its vibrational energy away into the surrounding air. It would be far better if the ball did not vibrate, so that all the energy from the boot's impact was transferred into the ball's kinetic energy, and thereby contribute to its launch velocity.

For this reason, a paper by Ronkainen and Harland (2006) is significant. They utilized a piece of measuring equipment known as a Scanning Laser Doppler Vibronometer (SLDV) to measure the resonances from soccer balls. Of course, any resonances found represent areas of possible energy loss in the boot–ball impact event. The SLDV allows resonance measurements to take place remotely, non-invasively and without having to put location marks on the ball. The alternative resonance measurement techniques require the positioning of accelerometers on the surface of the ball or coloured tape markers, both of which must interfere with the measurements taken.

The results of the measurements showed a small number of large resonances at less than 100 Hz and one at 150 Hz. As expected, these were due to the vibrations of the overall structure of the ball. There were further, smaller resonances measured in the region of 1 kHz and 1.4 kHz which were due to the individual panel oscillation. The signature of these resonances varied depending on the size of the panels and the style of stitching utilized in the ball's construction. The SLDV plots indicate that the ball is much stiffer in the region of the valve, vibrating only 86 μ m, whereas the rest of the ball had a vibration displacement of around 210 μ m. Clearly, stiffness in the region of the valve could be a source of unpredictable ball bounce in play.

13.5 Bouncing: soccer pitch characteristics

There is no doubt that pitch condition has a significant bearing on the style of play. Harder and dryer pitches offer greater ball bounce, while soft and wet pitches allow the ball to embed into the ground where energy is lost into the top layers of the pitch, creating a preponderance

Table 13.2 Traction regression equations for sand and clay loam soil types (from Jennings-Temple et al., 2006)

Sand	Wet	Traction = 0.31 (grass cover) − 0.29 (grass length) − 9.43 (50–100 mm bulk density) + 37.80
	Dry	Traction = 0.23 (grass cover) − 0.22 (moisture content) + 1.98 (log silt content) + 28.79
Clay loam	Wet	Traction = 0.17 (grass cover) − 0.54 (grass length) +0.86 (evenness) + 31.66
	Dry	Traction = − 0.94 (moisture content) + 7.19

Table 13.3 Hardness regression equations for sand and clay loam soil types (from Jennings-Temple et al., 2006)

Sand	Wet	Hardness = 0.002 (grass cover) + 0.56 (0–50 mm bulk density) + 4.34
	Dry	Hardness = − 0.01 (moisture content) + 0.23 (50–100 mm bulk density) + 4.95
Clay loam	Wet	Hardness = − 0.01 (grass length) + 0.04 (evenness) + 0.71 (0–50 mm bulk density) − 0.03 (medium sand) − 0.27 (log fine sand) − 0.98 (log silt) + 8.64
	Dry	Hardness = 0.004 (grass length) − 0.04 (moisture content) − 0.01 (medium sand) + 6.58

of low balls in play. Such bounces on soft wet pitches also generate considerable topspin (or at any rate reduce any backspin the ball possessed prior to the bounce) as the ball clings to the sticky ground throughout its contact phase. However, it would seem that these inconsistent pitch parameters can be factored in to the game play, with elite players having the ability to adequately accommodate pitch conditions with regard to deducing future ball locations from visual observations of the trajectories.

A more important element is the way in which players' boots interact with the pitch under varying conditions. Not only does this impact on the range of allowable players' manoeuvres in the game, but more importantly, serious player injuries can result from poor or misinterpreted pitch conditions. Jennings-Temple et al. (2006) investigated the links between soil physical conditions and playing quality in terms of hardness and boot traction. They first identified six soil types in use over 25 different pitches: clay, clay loam, loamy sand, sand, sandy loam and sandy silt loam. They then conducted a series of hardness and traction tests at several points on each pitch, and hence derived a series of regression models, from which hardness and traction figures can be calculated *for any pitch*. To calculate the values, groundsmen only need to deduce the following properties; surface type (one of the six listed above), length of grass (mm), moisture content (%), grass cover (%), surface evenness (mm), soil water tension (kPa) and bulk density (Mg m^{-3}). Fortunately, only some of these measurable values are required to calculate traction and hardness figures for each of the soil types. Tables 13.2 and 13.3 shows an extract of the regression equations for traction and hardness respectively for the two most common soccer pitch types; sand and clay loam pitches.

13.6 Ball trajectory analysis

13.6.1 Flight through the air

From a trajectory analysis perspective, soccer balls behave in an archetypical manner, although its large drag-to-weight ratio (≈ 1) tends to leave it more vulnerable to deviations caused by the wind gusts. Notwithstanding, the trajectory analysis described in Chapter 6 will apply and certainly Equation 6.1 is a good starting point for the analysis. Asai *et al.* (2006) have experimentally determined the important soccer ball parameters for use in simulation models. Physical parameters were taken as, area, $A = \pi \times 0.11^2 = 0.038\,\mathrm{m}^2$ and mass, $m = 0.42\,\mathrm{kg}$. They determine the critical Reynold's number, $R_{ecrt} = 2.2 \times 10^5$, at which point the drag coefficient drops from a subcritical value of 0.43 to a supercritical value of 0.15. For typical powerful kicking speeds of between 25 and $35\,\mathrm{m\,s}^{-1}$, the ball will easily reach beyond the critical Reynold's number and the airflow will be of a turbulent nature. Finally, the Asai *et al.* (2006) wind tunnel photographs indicated that the airflow around a slow ball ($5\,\mathrm{m\,s}^{-1}$) has a separation point which is about 90° around from the front stagnation point on either side of the ball's velocity vector. However, for a fast ball ($29\,\mathrm{m\,s}^{-1}$), the separation recedes to about 120° around from the stagnation point.

Asai *et al.* (2002) and Carré *et al.* (2002) performed similar experiments but concentrated on the trajectory of *spinning* soccer balls. Their 3D simulation considered a ball aimed at the top right hand corner of the goal from a free kick taken just outside the penalty area. Their model calculated trajectories for kicks with differing offsets, and the results clearly showed that launch velocity and spin rate are inversely and directly dependent on the offset respectively.

One particularly interesting case study highlights the difference between executing two identical kicks, first with a dry ball, and then a wet one. They found that the reduction in ball–boot coefficient of friction caused by a wet ball reduced the spin rate by only $2\,\mathrm{rev\,s}^{-1}$ ($13\,\mathrm{rad\,s}^{-1}$), which in turn reduced the Magnus force and the swing. A dry ball kicked at $18.5\,\mathrm{m\,s}^{-1}$ would result in a goal, by swinging into the left of the right-hand post by half a ball's radius (i.e. just 'kissing' the post on the inside). On the other hand, a wet, slower spinning (but otherwise identically) kicked ball, would not swing in sufficiently, and would just miss the post, again by half a ball's radius. This demonstrates the importance of drying the ball before free kicks or penalties are taken.

Finally, in this section, a paper by Bray and Kerwin (2003a) will be briefly described. They took the standard three-dimensional equations of motion (which, as we know, have no closed-form solutions, and can only be solved by numerical techniques such as Runge–Kutta), and simplified them for the two-dimensional case. That is, they assumed that the ball was spinning around a truly vertical axis, and that there was no interest in the vertical motion; only its lateral deflection. If the ball is launched horizontally towards the goal in the y direction (and it stays horizontal for the period of interest), and the ball then swings in the x direction due to spin and the Magnus effect, Bray and Kerwin (2003a) found that the positional coordinates of the ball at intervals of time t, are given by:

$$x = \frac{1}{k_L}\left\{1 - \cos\left(\frac{k_L}{k_D}\ln\left(1 + k_D v_0 t\right)\right)\right\} \qquad (13.4)$$

$$y = \frac{1}{k_L}\sin\left(\frac{k_L}{k_D}\ln\left(1 + k_D v_0 t\right)\right) \qquad (13.5)$$

In these equations:

$$k_D = \frac{\rho A C_D}{2m} \quad \text{and} \quad k_L = \frac{\rho A C_L}{2m}$$

Equation 13.4 gives the deflection of the ball at each point in time, while deflections can be calculated as a function of distance by determining the corresponding y values from Equation 13.5. These equations are modelled and a plan view of the ball's deflection is given in Model 13.1 – 2D Deflection model of soccer ball. Bray and Kerwin (2003a) claim that there is very good correlation in deviation between the full 3D analysis and their simplified 2D analysis.

13.6.2 The penalty kick

The penalty kick is an important set play in competitive soccer which often determines the game's outcome. However, some simple three-dimensional geometry will reveal that, when a penalty kick is taken, large areas of the goal cannot be defended by the keeper.

With reference to Figure 13.3, let us assume that the ball is kicked with a velocity v_0 at the goal with an elevation angle of θ to the horizontal, and an azimuth (horizontal offset) angle of ϕ, then:

$$v_{0x} = v_0 \cos \theta \quad \text{and} \quad v_{0y} = v_0 \sin \theta$$

where v_{0x} is the horizontal component of the ball's velocity and v_{0y} is the vertical component of the ball's velocity.

The horizontal distance the ball travels from penalty spot to goal line is given by:

$$d = v_0 t \cos \theta$$

If the penalty spot is taken as 11 m from the center of the goal line (it is actually 10.97 m), then:

$$d = 11/\cos \phi$$

The time of flight will be given from:

$$\frac{11}{\cos \phi} = v_0 t \cos \theta$$

So

$$t = \frac{11}{v_0 \cos \phi \cos \theta} \qquad (13.6)$$

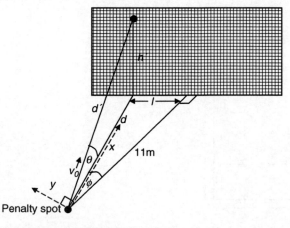

Figure 13.3 Coordinate system for penalty save.

The height of the ball as it crosses the goal line, h' will be given by:

$$h' = v_{0y} t - \tfrac{1}{2}gt^2$$

Substituting for t, gives:

$$h' = v_0 \sin \theta \left(\frac{11}{v_0 \cos \theta \cos \phi} \right) - \frac{1}{2}g \left(\frac{11}{v_0 \cos \theta \cos \phi} \right)^2 \qquad (13.7)$$

The horizontal offset of the ball as it crosses the goal line, l, will be simply given by:

$$l = 11 \times \tan \phi \qquad (13.8)$$

So the goalkeeper must get his hand(s) to coordinates l, h', in time t, given by Equations 13.6–13.8, in order to save the ball.

Now, if the goalkeeper is of height h, and has a center of mass located at $0.5\,h$, prior to the kick, the goalkeeper will take up the pre-impulse position with legs semi-flexed and torso curved forwards and downwards. In which case, we can assume that his center of mass at the start of the save will be $\tfrac{3}{8}h$ above the goal line (see Figure 13.4).

If the goalkeeper has arms of length b, then his center of mass must move a distance l' to the left (say), and a distance h'' upwards.

So $h'' = h' - \tfrac{3}{8}h$ and by similar triangles:

$$\frac{h'}{l} = \frac{h''}{l'} = \tan \alpha$$

So, the center of mass movements is given by:

$$l' = \frac{h''l}{h'} = \frac{\left(h' - \tfrac{3}{8}h \right)l}{h'} \quad \text{horizontally}$$

and

$$h'' = v_0 \sin \theta \left(\frac{11}{v_0 \cos \theta \cos \phi} \right) - \frac{1}{2}g \left(\frac{11}{v_0 \cos \theta \cos \phi} \right)^2 - \frac{3}{8}h \quad \text{vertically.}$$

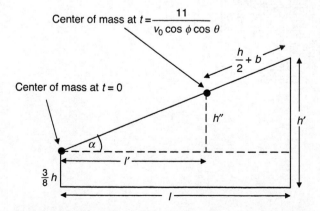

Figure 13.4 Goalkeeper center of mass movement to save penalty.

A model of total center of mass movement ($= \sqrt{l'^2 + h''^2}$) versus ball velocity may now be easily constructed.

Finally, the explosive impulse, I, required of the goalkeeper to save the ball is given by:

$$I = -mgt = -\frac{11\,mg}{v_0 \cos \phi \cos \theta}$$

Kerwin and Bray (2006) carried out an extensive study of penalty kicks, both within the game, and for end-game penalty shoot-outs. They reported that 85 per cent of in-game penalty kicks were successful, while only 75 per cent of penalty shoot-outs scored. An electrically driven ball launcher was used to deliver balls at experienced goalkeepers. They concluded, both from their observations, and from their model, that 28 per cent of the goal area cannot be protected by the goalkeeper for a typical launch velocity of $21\,\mathrm{m\,s^{-1}}$. The 'unsaveable' zones lie roughly outside two adjacent semicircular regions whose apexes lie about 1 m on either side of the goal center, and rise above the bar. Crucially, however, the regions do not quite reach the post unless the goalkeeper moves a step before diving. However, even allowing for such adjustments before diving, there remains a pair of large, unsaveable areas at each of the top corners of the goal, amounting to 12 per cent of the total goal area on each side.

Finally, Suzuki *et al.* (1988) identified seven different styles of goalkeeper saves, listed in Table 13.4 which should be used when saving in each of the 12 areas of the goal mouth, as illustrated in Figure 13.5.

13.6.3 The free kick

If a free kick is taken in proximity to the goal (i.e outside the penalty area, but within about 18–25 m of the goal), the opposition will construct a defensive wall of players between the striker and the goal mouth. It is clear that the nearer the wall is to the striker, the more restricted is the range of angles the striker can shoot to score. However, the game rules state that the wall must be formed at least 10 yards (9.14 m) from the kick. The perceived wisdom

Table 13.4 Goalkeeping saving styles (Suzuki *et al.*, 1988, cited Reilly and Williams, 2003)

Goalkeeping action	Cross-over step	Side step
Type 1 area A Collapse both legs and drop to ground	0	0
Type 2 areas E and I Single or two legged jump to get in line with ball	0	0
Type 3 areas B and F Right leg comes under body and dive off left leg	0	1
Type 4 area C Small step to right followed by low dive driving off left leg	0	1
Type 5 areas G, J and K Small step followed by drive upwards of right leg	0	1
Type 6 areas D and H Cross-over step before diving off left leg	1	0
Type 7 area L Two cross-over steps and dive off right leg	2	0

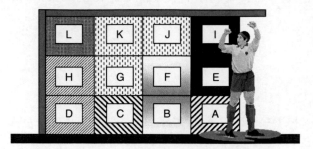

Figure 13.5 Twelve goalkeeping saving areas (Suzuki *et al.*, 1988, cited Reilly and Williams, 2003).

is that the defender's wall should extend from the far post towards the near post but leaving a gap close to the near post to allow the goalkeeper a clear sight of the kick,

Bray and Kerwin (2003b) have extended on the studies of Bray and Kerwin (2003a) by placing a rectangle representing the wall at an appropriate location 10 yards from the kick point and then investigating the available range of angles for a shot to arrive inside the goal area.

They added to the usual three equations of motion for a ball subject to gravity, lift, drag and spin, the geometry of the wall's rectangle given by:

$$w = d \sin \alpha \left[\frac{p}{(R - p \cos \alpha)} + \frac{q}{(R + q \cos \alpha)} \right]$$

where $2p$ is the width of the goal, w is the required width of the wall, d is the compulsory 10 yard (9.14 m) distance from kick spot to defensive wall, R and α are the distance and angle of the free kick respectively, and q is the implied coverage of the goal line by the wall.

A value of $q = p$ would completely cover the line, so $0 < q < p$, and a value of $q = p/2$ seems to be a reasonable compromise. The height of the rectangle, h, is taken as the average height of the players. So, in summary, the blocking rectangle has an area given by $w \times h$. The constraints on the free kick presented by walls of specified sizes and distances may now be calculated.

Bray and Kerwin (2003b) used this model to carry out a number of case studies. They used their own measured values of $C_D = 0.28$ and $C_L = 0.26$ for the ball. They took $\alpha = 90°$ and $R = 20$ yards (18.28 m) which places the ball symmetrically along the center line and in the middle of the 'D' outside the penalty area, as in Figure 13.6. The wall height was taken as 1.83 m and q was taken as 1.83, which represents 75 per cent of the goal protected. The positions of the wall and the goalkeeper are as in Figure 13.6.

The first case study considers a ball launched at $25 \, \mathrm{m \, s^{-1}}$ with an elevation of 16.5°, directed straight down the center line to the goal ($\alpha = 90°$). The spin angle is taken to be vertical (pure sidespin). The model indicates that such a ball just clears the wall and crosses under the bar, about 4 m from the goalkeeper's guard position. However, by altering the horizontal angle by just 3.5° from the center line, the ball will now enter the net just inside the far post. The model, therefore, exhibits a very sensitive angular variable. This is further supported by increasing the elevation angle by just 1° to 17.5°. Now, the ball clears the wall but only just passes under the bar and into the goal. In fact, the same result would be achieved if the launch angle was held at 16.5°, but the launch velocity was increased by just $1 \, \mathrm{m \, s^{-1}}$ to $26 \, \mathrm{m \, s^{-1}}$.

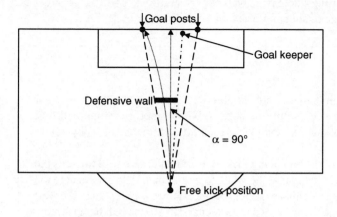

Figure 13.6 The free kick and defensive wall position used in the kick simulation (from Bray and Kerwin, 2003b).

Another case study involves reducing the spin axis from 90° (vertical) by 7° to 83°. With the same launch speed and elevation angle of $25 \, \text{m s}^{-1}$ and 16.5°, again the ball enters the net just below the goal bar. In summary from the initial launch parameters, the model shows three ways in which the ball can enter the net at the highest allowable point: (a) increasing the speed by $1 \, \text{m s}^{-1}$, (b) increasing the angle by 1° and (c) applying backspin by decreasing the spin angle by 7°.

The results illustrate how closely the striker must control the parameters of their kick to score a goal. From the goalkeeper's perspective, he has about 0.45 s to react to save the ball from the moment it appears over the defensive wall.

Summary

This chapter commences with a description of the historical development of the game from the gruesome early versions of the sport practiced in the middle ages. The history, construction and manufacture of both the ball and the boot follow, with much consideration given to the advantages of using modern, man-made materials to create exciting and consistently predictable play, irrespective of the weather.

The science of the dynamics of the sport commences with a description of the ball–boot impact process, in terms of momentum and energy conservation. A complex 3-D mesh simulation analysis is presented with the results given in graphical form. The vibration modes set up in the ball following its sharp impact with the boot is addressed, and the effect of varying ball pressure is stated. Bounce off the pitch is described and the importance of pitch type and moistness is highlighted. Charts show the resulting bounce for a range of pitch types.

The trajectory dynamics of a soccer ball flight is treated in the usual format of addressing, simultaneously drag and lift, albeit for a ball with a higher Reynold's number, R_e, than any other type of sporting projectile. It is shown that the equations of motion simplify considerably if the trajectory path is considered to be limited to only two dimensions; assuming the height to be constant and movement limited to the horizontal plane in a spin induced curve. By this means, the effects of curving balls on the goal are examined in some detail, as is the consequential defensive moves available to the goalkeeper. It is shown that the goal area breaks down to 24 areas which correspond to the different styles of save. The trajectory analysis is then extended by inserting a defensive wall between the kicker and the goalkeeper, showing how that limits the range of shot angles available to the kicker.

Problems and questions

1   For submaximal kicks the ball velocity is given by Equation 13.3, whereas for maximal kicks, Equation 13.2 would be used. It would appear that the ball/foot velocity ratio is greater for submaximal kicks than for maximal kicks. Why would this be?

2   In the second case study of the ball being kicked over a defensive wall towards the goal in Section 13.6.3, it is stated that the goalkeeper has only 0.45 s to react to the ball, from the moment the ball appears over the wall. However, as stated, the wall and keeper usually arrange themselves so the keeper can see the ball being struck. For how long is the ball out of sight of the goalkeeper after it has been kicked. Use the angles and velocities given in the case study and the standard basic trajectory equations.

3   One of the world's most feared throw-in exponents is Rory Delap, currently of Stoke City. A former schoolboy javelin champion he uses the run-up method and can achieve ranges of up to 40 m. If, in a particular throw-in, Delap, releases the ball at an angle of 30° and at a height of 2.3 m from the pitch, what will the release velocity of the ball be if he achieves a range is 40 m? Ignore the effect of backspin.

4   Determine which has the greater hardness factor, a wet, sand-based pitch or a dry, sand-based pitch, if the grass cover is 90 per cent, the bulk density is 1.5 Mg m^{-3} and the moisture content is 50 per cent.

Chapter 14

Rugby and American football

14.1 Introduction: the games

14.1.1 Rugby

Legend has it that the game of rugby was instantly 'invented' in 1823 by a student of Rugby School, England, a certain William Webb Ellis. In flagrant disregard of the conventional rules of football, he picked up the ball in his arms and ran with it to the amazement of his classmates. It is said that his teammates were so enthralled that they instantly followed suit and so the game of rugby was invented.

The truth of the game's inception is not, however, so simple, clear cut, or indeed as instantaneous as the fable would have us believe. Historians agree that games involving running with balls were known to have been imported to England by the Norwegian Vikings around AD 800. Some experts even believe that they can trace the origins back to the sixth century, with the Roman sport of Harpastum. This game involved both hands and feet in ball movement, and was thought to consist of teams of 27 a side, with the aim being to throw the ball over a designated goal area on the perimeter of the field. Also worthy of note during this early period is the Greek game of Phaininda. The name itself means 'to pretend', and may allude to the practice of avoiding ball interceptions by setting up fake or dummy passes.

So the game, widely thought to have been initiated at Rugby School was, in fact, nothing more than the natural evolution of a variety of games that had already existed for centuries, restructured to enhance the sense of fair play and competition. Nevertheless, ball running exploits certainly did occur at Rugby School in the period 1820–1830. In 1838, Mr Jem Mackie, a distinguished powerful runner, certainly helped to popularize this element of game play which was legalized in 1842, and which was incorporated into the first written set of rules in 1845. In fact, many schools created their own rule-sets at this time. However, the Rugby rules rapidly became the most popular, probably due to nothing more than the reputation of the school.

As the schoolboy players matured and left their scholastic institutions, the appeal of the game naturally spread and proliferated. In 1863, two meetings took place; one at Cambridge and the other at Blackheath in London, with the aim of drawing together the different rules in use around the country into one, common set of rules of play. As it turned out, the governing groups split into two distinct factions: the so-called Cambridge Rules group, which precluded running with the ball and from which Association Football evolved; and the Rugby rules group which, in 1871, became the Rugby Football Union (RFU), and which obviously allowed running with the ball, although apparently 'hacking' and 'tripping' were prohibited.

The RFU believed strongly in the amateur status of the game. However, in 1893, some North of England clubs broke away from the RFU in order to develop the game on a professional basis, so creating the Rugby League. The club members felt compelled to change the rules to make the game more attractive to spectators and to ensure financial viability with regard to securing their player salaries. For this reason, they reduced the team from 15 players per side to 13, and they created a 'multiple downs style of play'. Even though rugby union also achieved professional status in 1995, there remains a very distinct difference in both appearance and play strategies between the two games.

In summary, regardless of the rumoured contribution of William Webb Ellis (which incidentally has resulted in numerous statues and a trophy in his name), rugby and association football essentially developed alongside each other, until the meetings in 1863 when the separate rules of the two distinct sports were formulated. The laws have changed a great deal since then and spawned other, related games, notably American football, Australian rules football and Gaelic football.

14.1.2 American football

American football was obviously a derivative of English rugby. A group of students from Princetown University played a game they called 'Ballown', the aim of which was to force the ball past the opposing team using hands and feet. At Harvard, students played a particularly violent version of this game which they called 'Bloody Monday'. Around the mid-nineteenth century, various colleges adopted similar games but, in order to play each other, some commonality of rules was required. In 1873, the first intercollegiate set of rules was created by representatives of Columbia, Rutgers, Princetown and Yale Universities, and the Intercollegiate Football Association (IFA) was established.

The Yale coach, Walter Camp, is acknowledged as the one man who was responsible for the rule changes which made the game what it is today (and so distinctive from British rugby). Known as 'the Father of American football' he dissented in so many ways from the IFA establishment. Over his years on the Massasoit House Conventions Committee, where rules were debated and changed, he was responsible for:

- reducing the team from 15 to 11 players, which opened up and increased the tempo of the game (1880);
- establishing the series of downs, the line of scrimmage and the snap, which turned the game into one of incremental progress towards the end zone (1880);
- reducing the size of the field to the current dimensions of $120 \times 53\frac{1}{3}$ yards (1881);
- various alterations to the scoring rules and game time (1887);
- introducing of paid referees and umpires (1887), and shortly after, the employing whistles and stopwatches (1889);
- allowing tackling below the waist (1888).

The result was a unique, exciting game which, as more sophisticated strategies emerged, became a seriously dangerous sport. Within one decade, nearly 180 players suffered serious injury and 18 deaths were reported. Even President Roosevelt intervened to save the sport from a seemingly inevitable demise. The National Collegiate Athletic Association (NCAA) was set up to oversee rule changes in an attempt to make the game acceptable from a safety perspective. Legalization of the forward pass in 1906 opened up the game and so reduced injuries, as did the banning of mass team formation strategies such as the 'flying wedge'. In 1910, pushing and pulling of opponents were outlawed, as was the interlocking of hands in arms, belts or 'uniforms' to create group attacks as a single unit.

Numerous less significant rule changes have continually improved the game to its current format. A hundred years after the formation of the NCAA, the sport of American football thrives as the most popular collegiate game in the US. There are currently three divisions under NCAA guidelines. Seasonal and Conference games then lead to the post-seasonal 'Bowl games', where the Division champions play, watched by world-wide television audiences.

Since 1921, professional football has been played outside the college circuit under the auspices of the National Football League (NFL). Over the ensuing years, there have been numerous alterations in the number of professional teams that make up the League, as well as a merger with the American football Association in 1970. There are currently 32 teams arranged into eight divisions of four teams each; an unusual structure for the organization of a major professional sports league, but this arrangement purports to offer greater parity between the teams, as well as a realistic opportunity for any of the teams to win the championship.

14.2 The balls

14.2.1 Rugby

There were two cobblers located close to the entrance to Rugby School, one owned by Richard Lindon and the other by William Gilbert. Both independently saw the business opportunity of rugby ball manufacture as the game's popularity increased. By 1850, these two companies were the main suppliers of pigs bladder/leather clad balls for the game.

In these early days, the balls were more plum shaped than oval, the shape being primarily governed by the shape of the bladders which were inflated directly by mouth. In fact, Richard Lindon's wife, who had the unenviable task of inflating the bladders, died from a lung related disease, thought to be contracted by inhaling from diseased bladders. It was Lindon who first utilized India rubber for the bladder and he designed a hand pump to inflate the new bladders based on the ear syringe mechanism in use at the time. Charles Macintosh, of raincoat fame, was the chief supplier of bladders for the balls. Lindon and Macintosh were partners for many years, and following further mergers, the company we now know as Dunlop emerged from the collaborations.

Gilbert has succeeded in maintaining the company name since those earliest days, with the last, and fourth generation member of the Gilbert family, James, taking over the Company in 1917 on the death of his father in the First World War. In 1946, Gilbert merged with the Tomlinson Soccer Ball Company of Glasgow and, since 1995, has been the official ball in use at the Rugby World Cups.

In *Tom Brown's School Days*, Thomas Hughes describes the shape of the ball as oval for the first time. It is thought that Gilbert was the first to make the leather casings from four hand-stitched sections, cut so that, when assembled, the finished shape was an oval, and for the first time, the ball's dimensions were reasonably constant. The rugby union balls of today are the only balls which are truly oval; a shape which, in mathematical parlance, is called a prolate spheroid. The dimensions are 290 mm long with a 600 mm minor axis girth. They are manufactured from a hybrid of polyurethane, synthetic leather, foiled polyester, latex and adhesive to keep the ball intact. Rugby league balls are slightly more pointed and American football, Canadian football and Australian rules football even more so, to accommodate their single hand passing technique. It is clear that these shape variances are important when studying the ball trajectory dynamics as applied to the different games.

14.2.2 American football

The American football is still referred to as a pigskin in recognition of its initial construction material. The current regulation size is 11 inches (28 cm) in length and 22 inches (56 cm) in circumference. Modern balls are assembled from leather panels tanned to a natural brown colour. The tanning process increases the essential grip properties in both wet and dry conditions. For a short time, white balls were adopted for nighttime games as they were thought to present greater televisual contrast under floodlights. However, white balls were soon abolished as players did not take to their performance in a variety of respects.

Good grip characteristics of the ball are essential for play, and so to improve the grip, the leather is stamped with a pebble-grain pattern prior to assembly. Four leather panels are required for each ball. The pre-tanned hide is cut into the panel shapes using a hydraulic metal die. Two panels are perforated along adjacent edges, so that they can later be laced together. One of these panels is given further reinforcement and perforations which hold the valve in place. The individual panels are first attached to the interior lining which is composed of three layers of cross-laid fabric firmly cemented together. The four panels are then stitched together in an 'inside-out' manner leaving a slit at the edges of the lacing holes. The ball is then turned inside out through the lacing slit. A polyurethane or two-ply butyl bladder is inserted and inflated to 12.5–13.5 psi (860–893 kPa). Finally, the polyvinyl chloride laces are threaded and laced up and the ball, now weighing 14–15 oz (397–425 g), is ready for play.

It would seem that American footballs are still manufactured using top quality leather in much the same way that soccer balls used to be. The ball manufacturer, Spalding are currently experimenting with a range of composite materials for use in American footballs which, they hope, will be an improvement in terms of moisture ingress, as well as maintaining a constant weight and flexibility during damp playing conditions.

14.3 Ball trajectory dynamics

14.3.1 Rugby

Examples which ignore drag effects

When a try is scored in rugby, the kicker attempts to 'convert' that try by successfully kicking the ball over the goal bar. They must choose a kicking point on a locus parallel with the sideline which runs through the point where the try was scored, as indicated in Figure 14.1. A try in rugby league is worth four points, while in rugby union it is worth five points. A successful conversion will add two to the score in either form of the game.

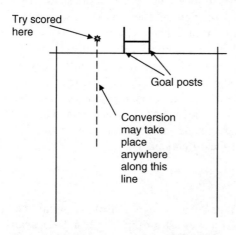

Figure 14.1 Rugby pitch indicating allowable points from which a try may be converted.

For the dragless case, we can use Equation 3.6 to calculate the minimum velocity required, and the optimum angle which will allow the ball to just clear the goal bar at height, h. For both rugby league and union, $h = 3$ m:

$$s_y = s_x \tan \theta_0 - \frac{as_x^2 \left(1 + \tan^2 \theta_0\right)}{2v_0^2} \tag{3.6}$$

If the kicker chooses to convert from a distance d from the goal, then:

$$h = d \tan \theta_0 - \frac{ad^2 \left(1 + \tan^2 \theta_0\right)}{2v_0^2}$$

Therefore, the minimum kick velocity will be given by:

$$v_0 = \sqrt{\frac{as_x^2 \left(1 + \tan^2 \theta_0\right)}{2 \left(d \tan \theta_0 - h\right)}} \tag{14.1}$$

If we now differentiate v_0 with respect to θ_0 and equate to zero to obtain the turning point, we find:

$$\theta_0 = \tan^{-1} \left[\frac{h + \sqrt{\left(h^2 + d^2\right)}}{d} \right] \tag{14.2}$$

Kicking the ball at this angle will require the least possible speed.

An interesting very simple model may be constructed to derive the optimum position for conversion along the kicking line. Obviously, the optimum point is where the angle the ball makes between the two goal posts is the greatest. As the ball is positioned further from the goal line, this angle reduces. However, the nearer the ball is to the goal line, the angle will also reduce as the offset cosine closes the window with proximity. This would suggest that there is a single point along the kicking line which will present the maximum angle of opportunity. Going either forwards or backwards from that optimum point will reduce the target angle.

Assuming a simple two-dimensional scenario in which the distance between the two vertical goalposts is W ($= 5.6$ m in both rugby league and union), the offset of the kicking line to the nearest goal post is X and the distance along the kicking line from the goal line to the ball is L. By simple geometry and with reference to Figure 14.2, we have:

$$\tan \alpha = \frac{X}{L}$$

$$\tan \beta = \frac{(X + W)}{L}$$

But since $\theta = \beta - \alpha$:

$$\theta = \tan^{-1} \left[\frac{(X + W)}{L} \right] - \tan^{-1} \left[\frac{X}{L} \right] \tag{14.3}$$

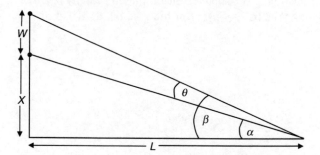

Figure 14.2 Geometry of kicking angles in rugby conversion.

The actual distance which gives the widest angle, L_{max}, may be found by differentiating Equation 14.3 and equating to zero. We thus find:

$$L_{max} = \sqrt{WX + X^2} \tag{14.4}$$

Equations 14.3 and 14.4 can be simply modelled in MS Excel with the offset, X, and distance from goal, L, as a variables and the angle, θ, derived. The distance between the goalposts, W, is fixed at 5.6 m. The angle versus distance can then be plotted and the optimum distance calculated. This model has been constructed and is presented as Model 14.1 – Rugby angle to goal model. Figure 14.3 shows one instance of the model where the offset distance is 20 m.

Even with this most basic of two-dimensional models, one can see two very important properties which may be of use to conversion kickers. First, the distance which gives the optimum angle is fairly close to the offset distance. Try varying the offset angle in the model and see how the optimum distance closely follows it. Put another way, the kicking angle

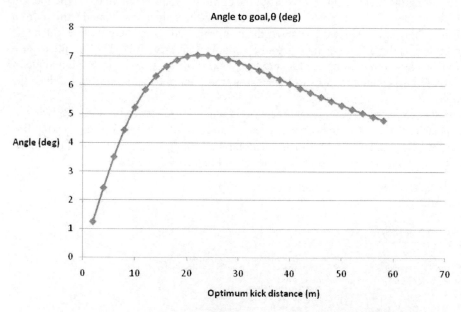

Figure 14.3 Output from Model 14.1 – Rugby angle to goal model. Distance of kick from goal line against angle between goal posts.

should be around the 45° mark *regardless of offset*. Second, the function is not linear or symmetrical around the optimum kicking point. The angle drops off faster towards the goal line, compared with moving away from the goal line. The moral here is: if the kicker is unsure of their optimum kicking point, it is safer to move slightly away from the goal than towards it, as the angle closes off much faster in the direction of the goal.

Before leaving this section, let us consider the forward pass in rugby. The rules state that the ball must not be passed forward to a teammate; the ball can only be passed backwards or laterally. What follows is a proof that, even if the receiver remains backward of the passer over the whole duration of the throw/catch manoeuvre, technically, relative to the pitch, the ball may have travelled forwards, which if spotted by the referee, would constitute a foul pass. Let us assume that the passer and receiver are running towards the try-line at $6\,\mathrm{m\,s}^{-1}$, and the receiver is positioned 4 m to the right of the passer and 1 m behind him. Now, if the speed of the pass is $10\,\mathrm{m\,s}^{-1}$ relative to the passer, then the ball's speed relative to the pitch will be the vector addition of ball and player speed $\sqrt{10^2 + 6^2} = 11.66\,\mathrm{m\,s}^{-1}$. This would make the time of flight $4/11.66 = 0.34\,\mathrm{s}$. During this time, the passer may have slowed down in order to avoid being tackled, but the receiver will have continued to move forward a distance of $6 \times 0.34 = 2\,\mathrm{m}$. The receiver is now 1 m ahead of the point where the ball was released when they catch the ball. The ball has, therefore, clearly been passed forward relative to the pitch. Of course, it does not seem that way to the players involved, or indeed to the referee, if they are running with the game. However, to a static spectator on the sideline or in the side stand, it appears that a clear infringement has taken place.

Examples involving drag effects

Although there are many ways to launch a rugby ball into flight by kicking, the methods broadly fall into two categories; the torpedo kick in which the ball is launched nose-first and rotates about its major axis, rather like a rifling bullet, and the tumbling kick in which the ball rotates about its minor axis in a tumbling, end-over-end motion. The torpedo kick is capable of significantly greater range, not only because of the reduced drag of the lower cross-sectional area ($0.028\,\mathrm{m}^2$ end-on, compared with $0.080\,\mathrm{m}^2$, side on), but also, like a javelin, lift can be generated by setting the attack angle slightly above the direction vector. The tumbling kick, which is usually achieved by an in-step boot contact with the ball, can produce more accurate targeted kicks which are less susceptible to crosswind deviations.

A particularly interesting style of kick, from a trajectory analysis perspective, is known as the torpedo swerve kick. In this manoeuvre, the ball can be made to swerve to the right by swiping the boot across the lower part of the ball (i.e. in close proximity to the ground) in a left to right direction. Initially, the ball is launched so that its major axis is aligned with the direction vector and so, even though the ball is spinning along its major axis, there will be a negligible Magnus effect and, consequently, little or no swerve. However, later in the flight, as the ball begins to fall under the influence of gravity, while still maintaining a fixed-axis orientation, its sense of spin will be lying above the direction vector. The Magnus effect and gyroscopic precession effect (a phenomenon explained in greater detail later in this chapter) will now cause the ball to swerve to the right. This can be a particularly effective kick when performed at the extreme right-hand side of the field. Here, the ball can be made to fly forwards and upwards in a direction parallel to the touchline and then suddenly swerve into touch as it begins to drop. The late deviation is especially enhanced when kicking into a prevailing wind.

Daish (1972) calculated the range for both the torpedo and the tumbling kick using the military ballistic table given in Table 5.3. Wind tunnel tests on rugby balls indicate that the drag force is proportional to the square of the velocity for speeds up to 30 m s^{-1}. Daish (1972) measured the forces as follows:

Angle between major axis and air stream	0°	45°	90°
Force on ball (Newtons)	3.3	5.4	8.8

From Equation 5.1 with $m = 0.4$ kg, $\rho = 1.22$ kg m^{-3} and $v_0 = 30$ m s^{-1}, we find:

$$F_D = C_D \rho A \frac{v_0^2}{2} = 1.22 \times C_D A \frac{30^2}{2}$$

and:

Angle between major axis and air stream	0°	45°	90°
AC_D (m^2)	0.0060	0.0098	0.0160

Equation 5.15 can be written in terms of the cross-sectional area, A, rather than the diameter, d, thus:

$$C = \frac{1}{C_D} \frac{m}{d^2} = \frac{\pi m}{4 A C_D}$$

So, now the ballistic coefficients, C, can be calculated for the three angles using the appropriate values of AC_D:

Angle between major axis and air stream	0°	45°	90°
C	52	32	20

Now, with a launch velocity of $v_0 = 30$ m s^{-1}, the ballistic range table, Table 5.3, can be used to calculate the range for the torpedo kick, where the angle of 0° may be used, and also the tumbling kick, which obviously rotates through all angles, but a mean angle of 45° would give an approximate indication of the attainable range.

	Torpedo kick	Tumbling kick
C	52	32
A	17.3	28.1
Z	1.10	1.52
Range	**57 m**	**49 m**

It should be recognized that these are older, relatively basic calculated values which are not supported empirically.

Holmes *et al.* (2006) carried out accurate video analyses of three types of kick, as well as a pass, using match balls, and they measured the release velocity, the ball spin, the launch angle and the distance achieved. Table 14.1 shows their results for the number of samples, $n = 14$.

Table 14.1 Mean ($\pm$ SD) test results for kicking and passing data ($n = 14$) (from Holmes *et al.*, 2006)

Kick/pass	Velocity (m s^{-1})	Ball spin (rpm)	Launch angle ($^\circ$)	Distance (m)
Place	26.44 $\pm$ 2.97	238.10 $\pm$ 44.92	30.22 $\pm$ 4.41	53.74 $\pm$ 5.72
Drop	25.60 $\pm$ 3.77	234.25 $\pm$ 66.57	35.76 $\pm$ 4.28	51.30 $\pm$ 5.70
Spiral	28.06 $\pm$ 3.70	216.41 $\pm$ 46.11	43.91 $\pm$ 4.55	55.42 $\pm$ 7.22
Spin pass	13.79 $\pm$ 1.48	219.09 $\pm$ 32.98	12.19 $\pm$ 5.26	

These values may be used with confidence when modelling rugby ball trajectories.

Alam *et al.* (2006) carried out both experimental measurements [using Experimental Fluid Dynamics (EFD)] and computational methods [using Computational Fluid Dynamics (CFD)] to ascertain the drag coefficients at varying yaw angles and velocities. They found a degree of correlation between the values derived from both methods, especially at lower velocities, and yaw angles close to 0° (major axis aligned with the flow direction). The average drag coefficients at zero yaw angle for experimental and computational methods were found to be $C_D = 0.18$ and 0.15 respectively. However, the values maximally diverge between the two methods when the yaw angle is around 70° and with a velocity of $120\,\text{km h}^{-1}$. Here the EFD methods gives $C_D = 0.67$, while, for the CFD method, $C_D = 0.30$.

The most detailed modelling of rugby ball flight has been described in two papers by Seo *et al.* (2006a, b). They have calculated the drag, lift, side force and pitching moment coefficients, respectively, and fitted the values to polynomial functions in angle of attack and angular velocity vector as follows:

$$C_D(\alpha) = 2.44 \times 10^{-3} - 1.09 \times 10^{-3}\alpha + 3.03 \times 10^{-4}\alpha^2 + 3.59 \times 10^{-7}\alpha^3$$
$$- 2.50 \times 10^{-8}\alpha^4 \tag{14.5}$$

$$C_L(\alpha) = 6.25 \times 10^{-3}\alpha + 2.41 \times 10^{-4}\alpha^2 - 3.44 \times 10^{-6}\alpha^3 \tag{14.6}$$

$$C_Y(\alpha, \omega) = \left(-1.50 \times 10^{-3} + 6.49 \times 10^{-4}\omega - 8.35 \times 10^{-5}\omega^2\right)\alpha$$
$$+ \left(-4.11 \times 10^{-5} - 3.82 \times 10^{-5}\omega + 2.64 \times 10^{-6}\omega^2\right)\alpha^2 \tag{14.7}$$
$$+ \left(4.94 \times 10^{-7} + 2.74 \times 10^{-7}\omega - 1.77 \times 10^{-8}\omega^2\right)\alpha^3$$

$$C_m(\alpha) = 1.51 \times 10^{-2}\alpha - 1.69 \times 10^{-4}\alpha^2 \tag{14.8}$$

where C_D is the drag coefficient, C_L is the lift coefficient, C_Y is the side force coefficient, C_m is the pitching moment coefficient, α is the angle of attack ($^\circ$) and ω is the magnitude of the angular velocity vector (rev s^{-1}).

Model 14.2 – C_D, C_m, C_g and C_L calculations for rugby ball, solves this group of equations. The polynomial functions are plotted in Figure 14.4.

A full analysis is complicated by the fact that, unlike the balls discussed in previous chapters, a rugby ball is not a spherically symmetric object. Seo *et al.* (2006a), assumed a prolate spheroid in their calculations which, although only strictly true for rugby union balls, is a close enough approximation for rugby league, American football, Australian rules football and the other games involving roughly ovoid shaped balls. To handle this extra anisotropy, the analysis is carried out using a vector matrix notation in which the direction of the ball's major axis as well as the ball's direction vector are carried through the analysis in matrix form.

a)

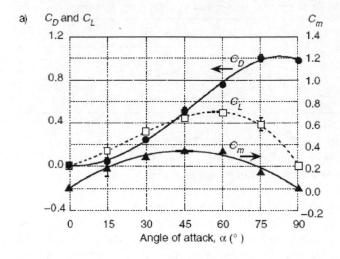

b)

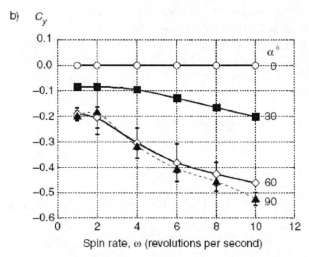

Figure 14.4 (a) Aerodynamic coefficients C_D, C_L and C_m as a function of the angle of attack, α (b) The side force coefficient C_Y as a function of the spin rate, ω. The angle of attack α is taken as a parameter (taken from Seo *et al.* 2006a).

The longitudinal and transverse moments of inertia of a prolate spheroid have been given respectively by Brancazio (1987) as:

$$I_L = \frac{b^2(4a+b)m_b\sqrt{(a^2-b^2)}}{15a\left[b\frac{\sqrt{(a^2-b^2)}}{a} + a\sin^{-1}\frac{\sqrt{(a^2-b^2)}}{a}\right]} \tag{14.9}$$

$$I_T = \frac{m_b\sqrt{(a^2-b^2)}\left[(a+b)^3 + a(a^2+b^2)\right]}{30a\left[b\frac{\sqrt{(a^2-b^2)}}{a} + a\sin^{-1}\frac{\sqrt{(a^2-b^2)}}{a}\right]} \tag{14.10}$$

where a and b are the lengths measured along the longitudinal and transverse axes, respectively, and m_b is the mass of the ball.

For a rugby ball, $a = 0.25\,\text{m}$, $b = 0.19\,\text{m}$ and $m_b = 0.42\,\text{kg}$, which gives values of $I_L = 0.0026\,\text{kg}\,\text{m}^2$ and $I_T = 0.0033\,\text{kg}\,\text{m}^2$.

Now the flight trajectory may be obtained by numerically integrating the following six equations of motion; three linear and three angular:

$$\dot{U} = \frac{1}{m_b}\left[X_a - m_b g \sin \Theta\right] - QW + RV \tag{14.11}$$

$$\dot{V} = \frac{1}{m_b}\left[Y_a + m_b g \cos \Theta \sin \Phi\right] - RU + PW \tag{14.12}$$

$$\dot{W} = \frac{1}{m_b}\left[Z_a + m_b g \cos \Theta \cos \Phi\right] - PV + QU \tag{14.13}$$

$$\dot{P} = \frac{L_a}{I_L} \tag{14.14}$$

$$\dot{Q} = \frac{M_a}{I_T} - PR\left(\frac{I_L}{I_T} - 1\right) \tag{14.15}$$

$$\dot{R} = \frac{N_a}{I_T} + PQ\left(\frac{I_L}{I_T} - 1\right) \tag{14.16}$$

where U, V and W are the direction vectors of the ball, so $\dot{U}$, $\dot{V}$ and $\dot{W}$ are the velocity vectors (in $\text{m}\,\text{s}^{-1}$), P, Q and R are the components of the angular velocity vector around the ball's principle axes (in s^{-1}), X_a, Y_a and Z_a are the components of the aerodynamic force, L_a, M_a and N_a are the components of the aerodynamic moment, m_b is the mass of the ball, g is the acceleration due to gravity, Θ is the pitch angle of the ball and Φ is the roll angle of the ball.

Figure 14.5b shows the output of the model for a spiral screw kick, launched from a position 2 m behind the 22 m line and 20 m in from the right-hand touch line. With reference to the pitch dimensions and coordinate system origin indicated in Figure 14.5a, $(X_{E0}, Y_{E0}, Z_{E0}) = (20, 50, -0.5)$. Its launch velocity is taken to be $25\,\text{m}\,\text{s}^{-1}$, the launch angle is $30°$ and the pitch angle is also $30°$. Its azimuth angle is $0°$, as is its initial yaw angle. Finally, its initial angular velocity is taken as $10\,\text{rev}\,\text{s}^{-1}$.

The complete launch parameter set is summarized as follows:

$$\begin{bmatrix} X_{E0} \\ Y_{E0} \\ Z_{E0} \\ U_0 \\ V_0 \\ W_0 \\ P_0 \\ Q_0 \\ R_0 \\ \Psi_0 \\ \Theta_0 \\ \Phi_0 \end{bmatrix} = \begin{bmatrix} 20 \\ 50 \\ -0.5 \\ 25 \\ 0 \\ 0 \\ 10 \\ 0 \\ 0 \\ 0 \\ 30 \\ 0 \end{bmatrix}$$

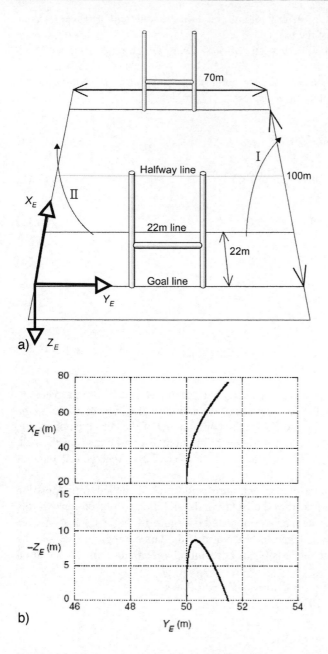

Figure 14.5 (a) Pitch dimensions and coordinate origin for implementation of the Seo et al. (2006a) model (b) Example of the flight trajectory of the screw kick.

It may be noted that (for a right-footed kicker) the ball tends to hook to the right even though initially the direction of spin is aligned with the launch direction vector. This is explained by the fact that, over the duration of the flight, the flight trajectory curves downwards under the influence of gravity and this tends to open up an angle between the direction vector and the spin vector. As the ball advances, the center of pressure shifts forward from the center of mass. The ball now tries to equalize the pressure by orientating its axis along its new direction.

However, because the ball is spinning through that axis, the only way the nose of the ball can correct its direction is by precessing in a gyroscopic action and veering to the right from the kicker's point of view. This causes the ball to swerve in that direction. Indeed, Brancazio (1985b), using the theory of torque-free motion demonstrated that an American football, in a vacuum, will precess three times for every five spins, which suggests a gyroscopic ratio of 1.67.

In the second paper from Seo *et al.* (2006b), this model was applied in order to optimize three types of kick:

1   *The punted kick*
The aim here was to optimize the number of extreme positions of X_E and Y_E and also the hang time. The optimum initial conditions were found to be:

$$
\begin{bmatrix}
\text{launch velocity} \\
\text{flight path angle} \\
\text{azimuth angle} \\
\text{angular velocity vector} \\
\text{elevation angle of } \omega_0 \\
\text{azimuth angle of } \omega_0 \\
\text{yaw angle} \\
\text{pitch angle} \\
\text{roll angle}
\end{bmatrix}
=
\begin{bmatrix}
\vec{v}_0 \\
\gamma_0 \\
\chi_0 \\
\vec{\omega}_0 \\
\iota_0 \\
\kappa_0 \\
\Psi_0 \\
\Theta_0 \\
\Phi_0
\end{bmatrix}
=
\begin{bmatrix}
25\,\text{m s}^{-1} \\
80° \\
-2.7° \\
0.94\,\text{rps} \\
99° \\
-128° \\
85° \\
-84° \\
7.7°
\end{bmatrix}
$$

For these initial conditions, the hang time was found to be 5.4 s. The number of extrema (or oscillations back and forth in that plane) were found to be; 3 in the X_E-axis, and no less than 13 in the Y_E-axis. The magnitudes of the extrema varied from a few centimeters up to a maximum of 10 cm. It is clear that such a punt would be difficult for the opposition to catch.

2   *The kick into touch*
Here two possibilities are considered; both involving a right-hand swerve created by a normal, left-footed kicker. One ball was kicked from a position I $= (X_{E0}, Y_{E0}, Z_{E0}) = (20, 50, -0.5)$, and the other from a position II $= (X_{F0}, Y_{F0}, Z_{F0}) = (20, 20, -0.5)$. The two positions labeled I and II are shown in Figure 14.5a. The key optimizing element is how far the ball has travelled in the forward direction when it crosses the sideline.

It was found that the optimum kick into touch would be made using the leg nearest the touchline. However, the range of optimum conditions for II was narrower.

3   *Kick for goal*
The kick chosen to be optimized was taken to be on the 22 metre line with a launch angle of 45°. The launch position was given by $(X_{E0}, Y_{E0}, Z_{E0}) = (78, 13, -0.145)$. The two objectives stated are: (i) a minimal lateral deviation at the point at which the ball crosses the goal bar, and (ii) a minimum clearance of the ball above the goal bar as it crosses. In fact, these two objectives cannot be simultaneously satisfied since, as the lateral deviation reduces, the height above bar increases. So, the optimization chosen was a minimum lateral deviation conducive to the ball safely clearing the bar.

The initial conditions were given by:

$$
\begin{bmatrix}
\text{launch velocity} \\
\text{flight path angle} \\
\text{azimuth angle} \\
\text{angular velocity vector} \\
\text{elevation angle of } \omega_0 \\
\text{azimuth angle of } \omega_0 \\
\text{yaw angle} \\
\text{pitch angle} \\
\text{roll angle}
\end{bmatrix}
=
\begin{bmatrix}
\vec{v}_0 \\
\gamma_0 \\
\chi_0 \\
\vec{\omega}_0 \\
\iota_0 \\
\kappa_0 \\
\Psi_0 \\
\Theta_0 \\
\Phi_0
\end{bmatrix}
=
\begin{bmatrix}
25\,\text{m s}^{-1} \\
41° \\
45° \\
10\,\text{rps} \\
41° \\
45° \\
51° \\
90° \\
6°
\end{bmatrix}
$$

The optimized trajectory was found to be a tumbling kick with the axis of rotation perpendicular both to the ball's major axis and to the direction vector. With these conditions, the deviation was found to be negligible with a height 12 m from the ground, or a clearance of 9 m above the goal bar.

14.3.2 American football

The 'kicker's dilemma'

Brancazio (1985a) highlighted what he called the 'kicker's dilemma' which he stated as the simultaneous requirement to kick the ball up the field as far as possible and also for it to hang in the air for as long as possible, allowing team-mates to reconfigure themselves ready to tackle the receiver and prevent a substantial runback.

Equations 3.10 and 3.11 define the range and hang time of a ball launched under dragless conditions. As discussed in Chapter 3, the greatest range on a level pitch can be achieved with a launch angle of 45°, while the greatest hang time can be achieved when launched at the totally impractical angle of 90°. Indeed, it can easily be shown that launching for maximum range reduces the hang time by nearly 30 per cent of the maximum. Brancazio (1985a) collected data from 26 NFL games consisting of 206 kick-offs and 238 punts. He measured and collected data on launch angle range and hang time and presented the two charts reproduced in Figure 14.6a,b. The chart of Figure 14.6a will provide information on any one missing variable from: distance, hang time, launch angle and drag coefficient for any ball launched at 26.82 m s^{-1} (60 mph). The chart of Figure 14.6b fixes the drag factor at $k = 0.01$ m^{-1} (0.003 ft^{-1}), but allows launch velocity as a variable. Remember: k, the drag resistance per unit mass $= \rho A C_D / 2m$.

Brancazio (1985a) concludes:

- For the kick-off, in which the launch velocity may be augmented by a 10 m or so run in, launch velocities of 31.3 m s^{-1} (70 mph) may be attained. Kickers tend to go for maximum range, with a launch angle around 45°. However, it would appear that players are aware of the importance of the launch velocity variable in that they go all out for the strongest boot impact, with less emphasis on launch angle accuracy. In so doing, they are maximizing both the range and hang time.
- Punters seem happier to sacrifice range for a longer hang time. Typical launch angles are in the region of 55° to 70°. Launching at 55° will reduce the range by about 5 m from the optimum value, but gain an extra 0.55 s of hang time.

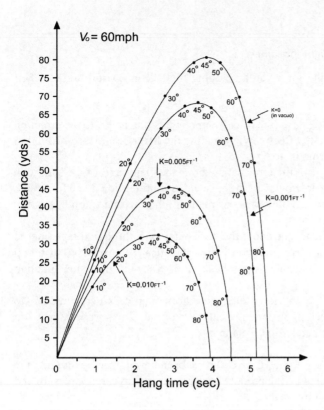

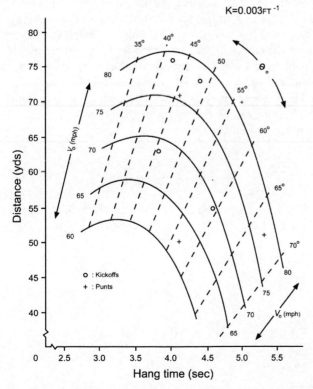

Figure 14.6 (a) Distance–hang time curves for a football launched at 60 mph as a function of launching angle and drag factor (b) Distance–hang time curves for $k = 0.003\,\text{ft}^{-1}$. The launch speed ($v_0$) and launching angle ($\theta_0$) for a given kick may be interpolated from the chart. (Taken from Brancazio (1985a) – note all values are in imperial units.)

Measuring American football flight parameters

Rae and Streit (2002) measured the forces on a spinning football in a wind tunnel. Their results are briefly delineated below:

- The normal force coefficient, C_{normal}, (the force component acting at 90° to the direction of the major axis of the ball and, effectively, providing the lift) rises linearly from 0 to 0.8 for pitch angles ranging from 0° to 65°.
- The axial force coefficient, C_{axial}, (the force acting along the major axis of the ball, with the positive direction pointing downwind of the ball) drops from 1.5 down to 0 for a pitch angle ranging from 0° to 55° in an almost linear manner albeit with a smooth, slightly negative second-order differential bias.
- The pitching moment, C_M (the torque about the horizontal axis, through the geometric center of the ball, with the positive direction being nose up) rises from 0.0 at 0° up to a maximum of 2.0 at 40° and then drops back to 0.18 at 55° in a sine function manner.
- The side force coefficient, C_Y, (the force along the horizontal axis perpendicular to C_{normal}, with the positive direction pushing the ball to the right) drops from 0.0 at 0° to −0.5 at 55° in a very approximately linear manner.

All the above coefficients (C_{normal}, C_{axial}, C_M, C_Y) have been normalized by dividing the actual measured forces and moments by the product of the measured dynamic pressure and the cross-sectional area of the axial midplane.

Cunningham and Dowell (1976) considered that punts may be partitioned into three main types: the type I punt, in which the ball travels with its major axis aligned with its velocity vector so that it lands on the forward tip of the ball; the type II punt, in which the major axis points above the velocity vector on the downward section of the flight such that the ball lands horizontally; and the type III punt, in which the angle between the major axis and the velocity vector on the downward flight is so great that it lands on the rear tip of the ball. The range was measured for a large number of Varsity players' kicks. The average launch angle was found to be 50°, players clearly choosing to sacrifice range for hang time.

The range for the dragless case is given by Equation 3.11:

$$s_m = \frac{v_0^2}{a} \sin 2\theta_0 \qquad (3.11)$$

Experimentally, Cunningham and Dowell (1976) determined reduction factors, K, for the three types of punt: Type I punt, $K = 0.48$; Type II punt, $K = 0.46$; Type III punt, $K = 0.44$.

Type I punt	$s_m = 0.48\dfrac{v_0^2}{a} \sin 2\theta_0$
Type II punt	$s_m = 0.46\dfrac{v_0^2}{a} \sin 2\theta_0$
Type III punt	$s_m = 0.44\dfrac{v_0^2}{a} \sin 2\theta_0$

However, for an elite punter who may be able to launch at 40 m s^{-1}, the authors thought that a value of $K = 0.5$ may be nearer the correct reduction factor.

The final paper in this section, by Rae (2003), is a detailed analysis of the forward pass. This paper is broadly similar to the treatment carried out by Seo *et al.* (2006a) on the rugby ball, as discussed in some detail in Section 14.3.1. Indeed, the equations of motions, both linear and angular, are identical in both papers, apart from the use of different symbols for the variables. In the paper by Rae (2003), however, there is a greater emphasis on the ball's precessing motion (or 'wobbling', as the paper terms it) and the plots clearly show the precessing spirals superimposed on the standard trajectory shapes.

The paper also discusses the limits of stability where, if the spin rate is too low, the precessing motion builds up and becomes so large that an unstable overturning moment is created. This is particularly prone to occur on the longer flight durations in which the precession period is nearly twice that of the spin rate, and its amplitude grows slowly, yet exponentially, with time.

The paper further explains how the gyroscopic force generates a sideways deviation of the ball in a similar manner to that of a rugby ball. Again, as the ball's nose attempts to follow the trajectory path downwards, it creates a gyroscopic force acting laterally, which pulls the nose sideways as it dips. This gyroscopic effect of a spinning football creates a significant lateral force which can considerably affect the flight trajectory path. By contrast, the Magnus force is insignificant on American footballs and rugby balls which are spinning at normal rates, where as for spinning spherical balls, such as soccer balls and tennis balls, the Magnus force is of great importance in computing trajectories.

Summary

In this chapter we chronicle the short (compared to many sports) history of rugby and American football and consider how commercial enticement has, in differing ways, influenced the sports' progress over the years. We then look at the way in which balls differ in physical attributes and manufacture.

Clearly, trajectory dynamics of these balls holds a fascination with the researchers due to their inherent axis asymmetry. Analysis is carried out with the rotation axis as a variable: the differences in trajectory path for (a) kicks directed with the long axis along the line of ball's velocity vector, and (b) at 90° to it in a tumbling motion, are calculated, and the main differences in flight path are outlined. As before, the dragless case introduces the analysis, and this is then advanced by the inclusion of drag and lift forces.

A number of papers on rugby and football trajectories are contrasted, from the early Daish (1972) method, up to full computation fluid dynamic analysis carried out most recently, using the full power of modern, high level computers.

A problem rather unique to American football is discussed: kickers need to achieve two mutually exclusive aims; first to have the ball hang in the air for as long as possible, and second for it to have as long a range as possible. Optimization strategies are discussed to provide informed compromises. The penultimate paper examined in this chapter concludes with three simple equations that derive the range of a football kick, dependent on the three common types of kick encountered in the game.

Problems and questions

1   A rugby kicker attempts to convert a try by placing the ball 30 m from the goal line. Ignoring drag effects, calculate the minimum velocity and optimum angle for a successful conversion if the goal bar height is 3 m.

2   Using Model 14.1 – Rugby angle to goal model, alter the offset distance and study how the angle versus distance function alters. What do you notice?

3   Calculate the drag coefficient, C_D, the lift coefficient, C_L, the side-force coefficient, C_y, and the pitching moment coefficient, C_m, for a rugby ball, kicked with an angle of attack of 50° and an angular velocity vector of 7 rev s⁻¹.

4   Using the charts of Figure 14.6:

   (a) Estimate the hang time for a ball kicked with a launch angle of 50° at a speed of 60 mph, which travels a distance 60 m (65 yd). Assume the drag factor, k, is 0.0033 m⁻¹ (= 0.001 ft⁻¹ as indicated on the chart).

   (b) Estimate the distance a ball travels if it is kicked with a velocity of 31.3 m s⁻¹ (70 mph), a launch angle of 45° and a hang time of 3.8 s, if the drag factor, k is 0.1 m⁻¹ (0.003 ft⁻¹ as indicated on the chart).

5   Calculate the range of a football punt, if the launch velocity is 30 m s⁻¹ and the angle of launch is 55° for: (a) the dragless case; (b) the case when the ball lands nose first; (c) the case when the ball lands with its major axis horizontal with the pitch and (d) when the ball lands tail first.

Chapter 15

Some assorted sporting projectiles

15.1 Projectiles that fly though the air

15.1.1 Basketball

Brancazio (1981) published a detailed and wide ranging paper covering the physics of differing aspects of basketball. Much of the game involves bouncing the ball off the ground, and Brancazio dealt with this in much the same manner as in Section 4.4.4, Case 2 Page 62 (the non-skidding case). However, he only deals with the ball's velocity component *parallel* to the court, so that the coefficient of restitution, e, is assumed to be equal to 1. He focuses on the kinetic energy changes on impact with the court and derives the formulas delineated in Table 15.1 for the translational, rotational and combined kinetic energy changes for balls possessing, initially: no spin, forward spin and backward spin (where v_0 is the initial horizontal velocity, ω_0 is the initial spin with topspin being positive, and all other symbols have their usual meaning).

Stronge and Ashcroft (2006) studied the bounce of large, thin walled, inflated balls, such as basketballs. They stated that such a ball, bouncing on a firm surface will have its trajectory modified by virtue of the fact that the contact area will be a circle of comparatively large radius; a dimension which is dictated by both the radius of the ball itself and its inflation pressure. They presented graphs, including that of rebound angle versus angle of incidence for their 'real' ball, and compared it with the traditional 'rigid body' calculations which assume impact occurs at a singular point of contact. They showed a considerable variation in the rebound angle for the larger angles of incidence.

Brancazio (1981) and others derived equations for optimum velocity and angle of shot to basket in broadly the same manner as Section 3.5, assuming zero drag. If the difference in height between the throw and the basket is h, and the distance between the launch and basket is L, then:

$$v_0^2 = \frac{gL}{2 \cos^2 \theta_0 \left(\tan \theta_0 - (h/L) \right)} \tag{15.1}$$

Figure 15.1 is an output plot taken from Model 15.1 – Graph of v_0 versus θ_0, based on Equation 15.1, showing the variation in velocity with angle for a ball released 4 m horizontally from the basket and 0.6 m below the basket height. The model allows you to adjust both L and h.

Equation 15.1 represents a family of parabolic curves and shows that there is a unique velocity for any angle which will allow the ball to cross the point representing the middle of the hoop. Note that there is a minimum to the curve implying that there is a minimum value

Table 15.1 Energy changes for a bouncing ball (no skidding). Taken from Brancazio (1981)

	$\Delta KE_{translation}$	$\Delta KE_{rotation}$	ΔKE_{total}
No initial spin	$-\dfrac{12m}{49}v_0^2$	$\dfrac{5m}{49}v_0^2$	$-\dfrac{m}{7}v_0^2$
Initial forward spin	$-\dfrac{2m}{49}\left(6v_0^2 - 5R\omega_0 v_0 - R^2\omega_0^2\right)$	$\dfrac{m}{49}\left(5v_0^2 + 4R\omega_0 v_0 - 9R^2\omega_0^2\right)$	$\dfrac{m}{7}\left(v_0 - R\omega_0\right)$
Initial backward spin	$-\dfrac{2m}{49}\left(6v_0^2 + 5R\omega_0 v_0 - R^2\omega_0^2\right)$	$\dfrac{m}{49}\left(5v_0^2 - 4R\omega_0 v_0 - 9R^2\omega_0^2\right)$	$\dfrac{m}{7}\left(v_0 + R\omega_0\right)$

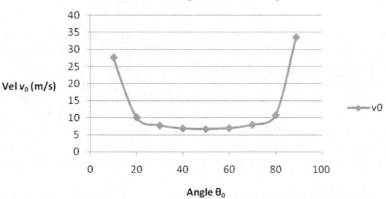

Figure 15.1 Output of Model 15.1 – Graph of v_0 versus θ_0, based on Equation 15.1, a plot of velocity versus angle for $h = 0.6$ m and $L = 4$ m.

of release velocity for every combination of (h, L). Differentiating Equation 15.1 gives us the minimum velocity, v_{0m}, and the angle, θ_{0m}, at which that minimum occurs:

$$v_{0m}^2 = g\left[h + \sqrt{(h^2 + L^2)}\right] \tag{15.2}$$

$$\theta_{0m} = 45° + \left(\tfrac{1}{2}\right)\arctan(h/L) \tag{15.3}$$

It is clear that we must place limits on the allowable angle as the expressions above simply state the conditions for the ball to pass through the dead center of the ring. Obviously one condition is that $\theta_0 < 90°$ which ensures that the ball is at least heading in the right direction towards the basket. Furthermore, the ball must be on a downward trajectory at the point where it crosses the center of the ring. This leads to the further condition that $\tan\theta_0 > 2h/L$ for the ball to be on the descendent. Now, one further condition we might want to apply to the launch angle is that the allowable trajectory does not touch either the front or the back of the ring as the ball falls through it. Brancazio (1981) proved by geometric means that, for a size 7 basketball, and a standard basket ring diameter, the angle of approach to the basket θ_e must be greater than 32° to clear both front and back sections of the ring. So, combining all those limitations, the final target window must lie such that the launch angle is given by $\theta_{0L} \le \theta_0 < 90°$, where $\tan\theta_{0L} = 0.62 + 2h/L$.

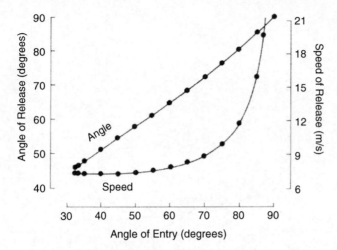

Figure 15.2 Plots of (a) release angle and (b) release speed, versus entry angle for $h = 0.92$ m and $L = 4.57$ m (from Hay, 1993).

The plot shown in Figure 15.2 may be used to ascertain the angle of entry for any given release speed and angle of launch. These curves were calculated for $h = 0.92$ m (i.e. 3 ft below the ring), and $L = 4.57$ m (15 ft).

The margin of error in the lateral direction, ψ, is again calculated by geometric means and, for a size 7 basketball and a standard basket ring diameter, the margin of error, $\Delta\psi$, is given by: $\tan\Delta\psi = 0.11/(h^2 + L^2)^{1/2}$ (N.B. converted from Brancazio's imperial dimensions to metric).

All the above equations show that, the smaller the value of h (i.e. the higher the release point), the lower the value of the minimum release velocity and the greater the allowable margin of error in both horizontal dimensions. As expected, all values decrease with increasing L.

Figure 15.3 shows the relationship between score probability and distance from basket. Note that, to double the probability of a successful basket from, say, 17 to 34 per cent, the range must be reduced by a factor of 3, from 9 m to less than 3 m.

Effects of air resistance

Although many researchers have assumed that basketballs travel so slowly that drag effects are insignificant, Brancazio (1981) fed typical values for a basketball shot, together with the ball's properties of diameter and weight, into the standard drag equation (Equation 5.1), and calculated that a basketball travelling with a velocity of $7.6\,\mathrm{m\,s^{-1}}$, decelerates by $1.2\,\mathrm{m\,s^{-2}}$ due to drag. This drag would cause a typical free-throw shot to fall short of the basket by 30–40 cm. To compensate, the velocity would have to be increased by 5 per cent for a given launch angle. Clearly therefore, drag is not something that can be ignored in basketball.

Basket-rim interactions

Okubo and Hubbard (2004) have modelled the motion of an off-center basketball around the rim of the basket to ascertain, given values of the initial contact location on the rim, the entry angle and velocity, and the spin of the ball, whether it would drop through the basket, or bounce off it. As expected, the equations of motion are complex, involving no less than

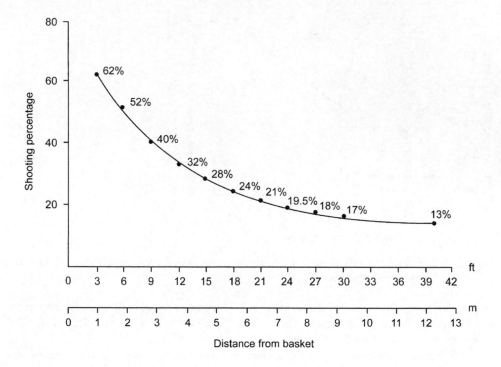

Figure 15.3 The relationship between score probability and distance from basket (from Hay, 1993).

six degrees of freedom. Plots presented in the paper show the effects of adjusting each of the key parameters, with regard to achieving a successful basket, or otherwise.

The free throw

There are two basic styles of free throw in basketball; the so-called overhand push, in which the ball is pushed from shoulder height in a straight line, and the underhand loop, in which the ball is swung up in an arc, from above the knees. Tan and Miller (1981) have analysed these two styles kinematically using broadly the same equations as Brancazio (1981). They concluded that, from a trajectory analysis perspective, the overhand throw is clearly preferable, requiring a lower launch angle and velocity, which, of course, leads to a wider entry angle.

However, from a biomechanical perspective, the underhand loop may be preferable as the ball is balanced more evenly between the more muscularly-relaxed hands which promote accuracy. Furthermore, the underhand loop style results in the apex of the trajectory being nearer the basket which increases the angle of approach. By contrast, the overhand push normally involves a predominant drive from one hand creating an unbalanced dynamic.

In either case, a degree of backspin is beneficial; not only does it stabilize the ball's flight and increase ball release precision, but it may also divert the ball in an advantageous direction should it initially make contact with the back of the rim or the backboard.

15.1.2 Badminton

The game of badminton originated in the 1860s and is an adaptation of the ancient game of 'shuttlecock and battledore' in which players would attempt to keep the shuttle in the air for as long as possible. It is thought that the earliest games were played by the Duke of Beaufort's family at Badminton House in Somerset, England.

There are currently two main types of shuttlecock. Feather shuttlecocks are made of a hemispherical cork or plastic nose, with 16 goose feathers (interestingly, always from the right-hand wing) attached to it. The quills of the feathers are glued into holes in the cork or plastic and fan out behind the nose to form the cone. Feather shuttlecocks tend only to be used at elite level, since their manufacture is labour-intensive and costly, even though they are usually manufactured in the Far East. Synthetic shuttlecocks have the same shape, but replace feathers with an injection moulded plastic skirt which uses a special blend of nylons. Modern manufacturing processes are highly automated, producing a shuttlecock every two seconds. The International Badminton Federation, formed in 1934, issues guidelines on size, mass and shape, but tolerances are so great that the projectiles can differ greatly, both aerodynamically and mechanically.

Both types of shuttlecock weigh between 4.74 and 5.50 g which is about twice that of a table tennis ball, although the shuttle's terminal velocity is the slowest of all the sporting projectiles at $7 \, \text{m s}^{-1}$. The designer's aim for many years now has been to create a synthetic shuttlecock with similar properties and characteristics to a feather one. However, it is still generally considered that the feathered type has better speed and control in the air. This is especially true at high speeds when the synthetic skirt tends to deform and fold inwards. This reduces the drag and distorts the flight path from the desirable 'parachute trajectory', in which the shuttlecock undergoes a steep rising gradient, followed by a sudden, near-vertical drop. Drag coefficient versus velocity measurements support this hypothesis as the plots clearly deviate at $16 \, \text{m s}^{-1}$, with the synthetic drag coefficient dropping below that of the feather shuttle.

When studying shuttlecock trajectories, it is important to consider two types of dynamics. The unsteady state occurs for that short period of time following racket/shuttlecock impact, when the shuttle is turning around and the skirt is still stabilizing. The quasi-steady state describes the flight behaviour once it has left the racket and resumed a constant shape, moving in a linear direction and experiencing spin, pitch and yaw. It is recognized that feather shuttles, because of their natural asymmetry, tend to spin more readily and with greater angular velocity than the synthetic types which are naturally more symmetric in form. The latest attempts to design synthetic shuttlecocks that more closely emulate their feather counterparts involve building into the synthetic models mismatched weight elements to help instigate greater spin, together with the utilization of stronger, carbon-fiber, artificial feathers to reduce deformation at the higher speeds. Despite all the efforts of the manufacturers, elite players still prefer using the feather variety, arguing that its performance is more predictable, especially in the long, court-length attacks.

Cooke (1996, 1999, 2002) has extensively measured, analysed and computer modelled the flight properties of four types of shuttlecocks in common use (Feather, Championship, ProCork and Tournament) over a range of types of shot, including, the serve, the net shot, the smash and the high clear.

Figure 15.4 shows the air flow over (a) a feather shuttlecock and (b) a modern championship synthetic shuttlecock; both travelling at $1 \, \text{m s}^{-1}$ ($R_e = 4400$). Note how, in both cases, air flows through the gaps in the skirt prior to reforming downstream. However, the stagnation area behind the feather variety is more severe than that of the synthetic one, clearly leading to greater drag in the feather shuttlecock.

Cooke (1999) measured the drag coefficient of the feather shuttles at a relatively constant value of $C_D = 0.48$, for $13\,000 < R_e < 200\,000$. The synthetic championship shuttles, on the other hand, had drag coefficients dropping from about $C_D = 4.8$ in an approximately linear fashion down to 4.0 in the range $13\,000 < R_e < 200\,000$, but with a clear reduction occurring at $R_e = 70\,000 \, (= 16 \, \text{m s}^{-1})$.

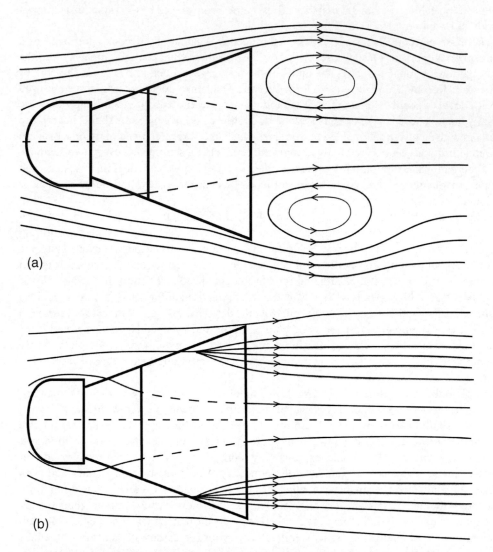

Figure 15.4 Smoke flow diagrams over (a) a feather and (b) a synthetic, shuttlecock, each at a velocity of 1 m s^{-1} (from Cooke, 1999).

Cooke (2002) stated the equations of motion in the usual manner:

$$m\ddot{x} + D\cos\theta + L\sin\theta = 0 \tag{15.4}$$

$$m\ddot{y} + D\sin\theta - L\cos\theta + mg = 0 \tag{15.5}$$

$$I_t\ddot{\beta} + c\dot{\beta} + \left|\frac{\mathrm{d}M}{\mathrm{d}\alpha}\right|\alpha = 0 \tag{15.6}$$

where β is the pitch angle, I_t is the transverse moment of inertia, c is the damping factor, M is the aerodynamic pitching moment, α is the angle of attack, and the other symbols have their usual meaning.

These equations were numerically solved, as before, using the fourth-order Runge–Kutta method. Cooke (2002) presents graphs for each of the four types of shuttlecocks, and each graph has a plot of the modelled trajectory together with a plot of the measured trajectory obtained by video analysis. The predicted and actual trajectories are commendably close to each other in each case. Figure 15.5 shows the modelled graphs for the four main types of shots: the high clear, the smash, the serve and the net shot. Each graph includes four plots depicting the trajectory for each of the four shuttlecock types.

It is evident that the largest discrepancies in trajectory path between the different types of shuttlecock occur where the velocities are greatest. This correlates with the premise of greater skirt deformation of the synthetic shuttles at high velocity.

Finally, some consideration should be given to the axial moment of inertia of the shuttle-cock. It is this property which governs how much the shuttle will spin axially during its flight. Axial spin is essential to maintain flight stability in much the same way that a discus travels in a stable manner because of its spin. However, combined with the spin, there is a pitching moment, since the angle of attack alters throughout its flight. These two rotations lead to a gyroscopic effect which causes a degree of flight precession and nutation (see Section 6.4.2). More significantly though, the precessing shuttle will create a yaw moment causing the pro-jectile to deviate sideways as well as creating greater pitch towards the end of the flight. This phenomenon is more evident in the feather variety of shuttlecock which produces greater spin in flight.

15.1.3 Table tennis

There is arguably no other ball game whose dynamic characteristic is more influenced by ball spin than table tennis. The ball is the lightest used in any sport at $m = 0.002$ kg, and, except for the shuttlecock, has the greatest drag to weight ratio of $\varepsilon = 8.8$ (cf. golf ball, $\varepsilon = 1.2$; men's shot put, $\varepsilon = 0.01$). It is a smooth, gas-filled celluloid shell with a matt surface finish. The diameter of the official ball changed from 38 mm to 40 mm following the 2000 Olympic Games in an effort to slow the game and increase spectator enjoyment. Key to generating spin is the surface of the paddle (also termed the bat or racket) which is a pimple contoured rubber that provides the coefficient of friction. It is backed by a sponge which depresses on impact, which increases the contact area and hence, the overall grip factor.

Spin modifies the ball's trajectory as well as the bounce off the table, and the rules state that volley shots are not allowed, so the ball must bounce once on the opponent's side before the return is attempted. Often one will observe, in a long rally, a succession of topspin and backspin played alternately by each player; the spin mounting with each stroke. And, although the topspin player is perceived as the attacker, with the backspin player forced into playing defensively, the ultimate loser of the point will most likely be the one who is finally beaten by the strength of the spin.

Sidespin is also a powerful tool, providing the means to curve the ball several degrees over the short length of the shot (the table's length is 2.74 m). Indeed a skilled player may circumvent the requirement to send the ball over the 15.25 cm high net, preferring instead to curve the ball around the side of the net, even though the net overhangs the table by some 15.25 cm!

With regard to ball trajectory, little has been written in the academic journals but Danby (1998) describes a model of the simple spun table tennis ball. He states that the terminal velocity has been measured at $7 \, \text{m s}^{-1}$. This gives a value of $k_D = 0.0616$ if $C_D = 0.4$ and the unspun ball's trajectory may be modelled in the usual manner using these figures. The spin, however, creates a lift factor C_L which is a complicated function of the quantity $r\omega/v$ where r is the ball's radius ($= 20$ mm), ω is the angular velocity of the spin and v is the ball's velocity. Danby (1998) presents results in the form of Table 15.2 from which values of C_L may be derived for any spin rate and velocity.

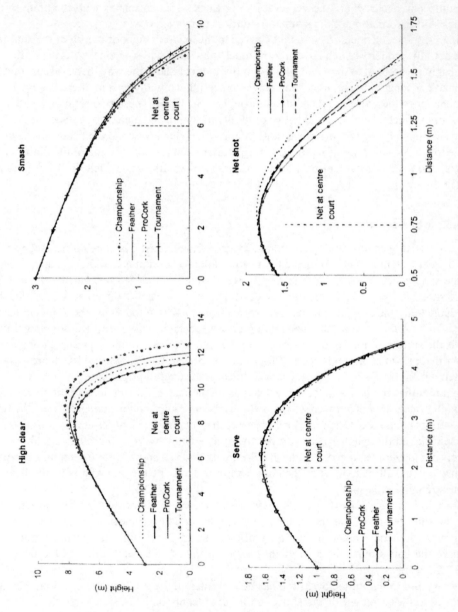

Figure 15.5 Computer modelled trajectories for four badminton shots (from Cooke, 2002).

Table 15.2 Values of C_L for given v and ω

$x = r\omega/v$	C_L
$0 < x < 0.251$	$-0.40x + 0.79x^2$
$0.251 < x < 0.435$	$0.128 - 1.4216x + 2.821x^2$
$0.435 < x < 0.625$	$-0.386 + 1.222x - 0.528x^2$

Values greater than $x = 0.625$ (values greater than 1 are possible) may be found by extrapolation. k_L may now be derived from:

$$k_L = \frac{\rho A}{2m} C_L \frac{v}{\omega}$$

and the trajectory computed in the usual manner using equations of the type:

$$\ddot{x} = -k_D v v_x + k_L \left(\omega_y v_z - \omega_z v_y\right)$$
$$\ddot{y} = -k_D v v_y + k_L \left(\omega_z v_x - \omega_x v_z\right) \qquad (15.7)$$
$$\ddot{z} = -k_D v v_z + k_L \left(\omega_x v_y - \omega_y v_x\right) - g$$

For completeness, the coefficient of restitution varies with impact velocity, but at typical velocities of $6\,\text{m s}^{-1}$, $e = 0.8$, while, for $24\,\text{m s}^{-1}$, $e = 0.7$. The Reynold's number for a table tennis ball is found to be in the region of 0.7×10^{-5}.

15.1.4 Volleyball

Volleyball was invented in Massachusetts, USA, but within a decade the game was established in the Philippines, Japan and Russia. The ball is made of a leather, or leather-like case with an interior rubber bladder. Regulations state that its circumference must be 65–67 cm with a weight of 260–280 g. The internal pressure is 0.30–0.325 kg cm^{-2}. With such specifications its aerodynamic performance understandably resembles that of a soccer ball. It is only the mode of impact which sets this projectile apart from the soccer ball, such that it warrants its own, albeit short, section.

In play, there are fundamentally two types of serve; a fast serve, usually with applied topspin ($v_0 \approx 30\,\text{m s}^{-1}$), and the floater which is served slower ($v_0 \approx 10\,\text{m s}^{-1}$) and with the palm of the hand generating little or no spin. This serve behaves in a similarly unpredictable manner to that of baseball's knuckleball. The volleyball's Reynold's number lies just above that of the golf ball at, $R_e = 2 \times 10^5$. Of course, in the case of the golf ball, the high Reynold's number is due to its surface dimples rather than the large diameter. The range of speeds volleyballs typically encounter will cause the ball to move either side of the critical Reynold's number so, as in cricket and some other sports, late swing may be encountered, depending on the launch speed of the ball and how the speed reduces over the duration of the flight.

Kao *et al.* (1994) have empirically and mathematically studied the most powerful of offensive volleyball shots; the spike. This is where the player jumps high in the air and, impacting the ball as high as possible, slams it down onto the opponent's court, usually aiming towards one of the corners. Kao *et al.* (1994) discovered that, for typical values of spin and velocity created by the spike, the Magnus force, M, may be predicted by the correlation equation:

$$M = 0.000041\omega^{0.8}v^{2.4}$$

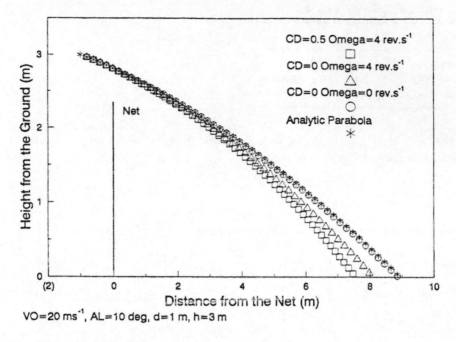

Figure 15.6 Mathematical model of a volleyball spike.

Figure 15.6 depicts the trajectory of a spiked volleyball launched from 3 m high at a declination angle of $10°$ and with a launch velocity of 20 m s^{-1}. The plots show the effects of changing the drag factor from $C_D = 0$, the dragless case (which produces a perfect parabola) to $C_D = 0.5$. Spins of 0 rev s^{-1} and 4 rev s^{-1} are plotted.

15.1.5 Softball

Softball was developed from baseball in the late 1880s and was originally an indoor game, created to maintain baseball practice during the winter months. It is reported that the first games were played using rolled up and tied boxing gloves as the ball, and a broomstick for the bat. Today's balls are larger than a baseball, being usually 12 in (30.5 cm) in circumference in fast pitch games, although sometimes 11 in (29.7 cm) may be used for slow pitch games. In Chicago, 16 in (40.6 cm) 'mushballs' are often used. These are also utilized in wheelchair events. Contrary to its name, softballs are quite hard, being constructed from two figure-of-eight shaped leather pieces, sewn together with a tight filling of long fiber Kapok. Modern balls may be filled with a cork/rubber/polyurethane combination. The bats are constructed from metal, wood or composite materials (carbon fiber, etc.), and sizes may vary. In fast pitch softball, wooden bats are not allowed.

Although there are many differences in the rules between baseball and softball, the main difference is that, in softball, the ball must be pitched underarm. There are two forms of delivery: the windmill, in which the arm swings a full $360°$ before release; and the slingshot, in which the arm is bent back and up until perpendicular with the ground before being swung forward and released.

In aerodynamic terms, the softball has a drag-to-weight ratio about 1.7 times greater than that of the baseball, and a Reynold's number, $R_e \approx 7.6 \times 10^5$, which is similar to that of a baseball at their equivalent pitching speeds. As a consequence, in trajectory calculations,

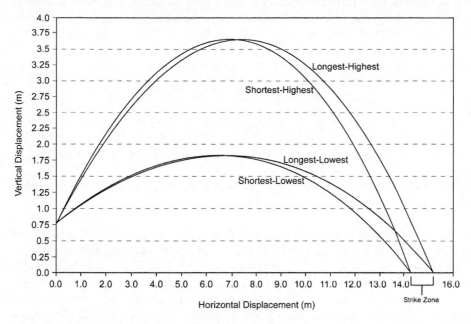

Figure 15.7 Backspin softball range versus vertical displacement for pitches depicting combinations of maximum peak and range and minimum peak and range (from Wu and Gervais, 2008).

similar values of drag and lift coefficients may be used. Wu and Gervais (2008) took the value $C_D = 0.475$, $C_L = 0.230$. With these values they simulated the four main types of desired flight for both backspin and topspin pitches: the shortest range with the highest peak, the longest range with the highest peak, the shortest range with the lowest peak and, finally, the longest range with the lowest peak. Figure 15.7 illustrates the predicted trajectories for just the four backspin cases.

Wu and Gervais (2008) also empirically measured, by means of video analysis, pitching flights for topspin, no spin and back spin. Their overall conclusions were that, contrary to some reports, ball spin did have a significant effect on softball trajectory. They stated that, in nearly all cases, backspin is preferable to topspin, since it creates a steeper slope on the rising flight, and it can then drop onto the front edge of the strike zone, which creates a challenge for the batter. Furthermore, backspin pitches tend to partially counteract the backspin which is usually created by the batter, reducing his strike range, and possibly, creating an easier catch from a slower spinning ball. The optimum angle of release was found to be 39° at a release height of 0.8 m and with a horizontal velocity of 12.1 m s^{-1}. This would result in an approach angle of 45° as seen by the batter; i.e. a gradient of 1.0. By way of contrast, Alaways and Hubbard (2001) estimated the slopes of a fastball or a curveball to be 5° on launch, and 11° on impact with the bat.

15.1.6 Archery

The earliest detailed mechanical and mathematical study of the bow and arrow system was carried out by Klopsteg (1943), included in Armenti (1992). Salient points include a value he calls the virtual mass of the bow, K, given by:

$$Fd = \tfrac{1}{2}(m + K)v^2$$

where F is the horizontal force exerted on the arrow, d is horizontal distance the arrow is pulled back against the elastic force of the string, m is the mass of the arrow (incidentally arrow masses are often measured in grains: 15.43 grains $= 1$ g), and v is the velocity of the arrow.

From this, the figure of merit of the bow may be derived as:

$$v = \sqrt{\frac{Fd}{K}}$$

This represents the limiting velocity that a bow can impart onto an arrow which approaches zero mass.

The trajectory equations for an arrow released from a bow may be simplified by a number of means. These are as follows:

1   If we assume that the flight is near-horizontal so both launch angle and target impact angle are small, then we can make use of small angle approximations in the analysis.
2   If we ignore crosswind in, say, the z direction, we need only concern ourselves with the variations in the horizontal, x, and the vertical, y, directions.
3   For target shooting, we are not particularly interested in the range. What we really need to know is how y varies with drag at the target. The z deviation is a simple tangent function of the offset angle.

First, the dragless case may be taken from Equation 3.6 as:

$$s_y = s_x \tan \theta_0 - \frac{a s_x^2 \left(1 + \tan^2 \theta_0\right)}{2 v_0^2}$$

which, for small θ_0, $\tan^2 \theta_0 \approx 0$, and, if the arrow is released at height s_{y0}, we have:

$$s_y = s_{y0} + s_x \tan \theta_0 - \frac{a s_x^2}{2 v_0^2} \tag{15.8}$$

Now, in the case of an arrow experiencing drag as a function of v^2, and from Equation 5.1:

$$F_D = C_D \rho A \frac{v^2}{2} = k v^2$$

where k is the drag constant.

We find, by using the assumptions delineated above, and with further simplification using the power series, that the trajectory equation reduces to:

$$s_y = s_{y0} + s_x \tan \theta_0 - \frac{a s_x^2}{2 v_0^2} \left(1 + \tfrac{2}{3} k s_x\right) \tag{15.9}$$

So the drag term is simply given by $\left(1 + \tfrac{2}{3} k s_x\right)$. The drag factor for typical arrows have been measured at $k = 0.003\,75$ m^{-1}.

Worked Example 15.1

Calculate the difference in height at a target between the dragless case and the case where the drag factor $k = 0.003\,75$, for a target which is located 50 m from the archer, and for an arrow fired from a height of 1.5 m, at 6° above the horizontal, with a velocity of 48 m s^{-1}.

The dragless case is given by Equation 15.8:

$$s_y = s_{y0} + s_x \tan \theta_0 - \frac{a s_x^2}{2 v_0^2} = 1.5 + 50 \times \tan 6 - \frac{9.81 \times 50^2}{2 \times 48^2} = 1.43\,\text{m}$$

For the case including drag, from Equation 15.9:

$$s_y = s_{y0} + s_x \tan \theta_0 - \frac{a s_x^2}{2 v_0^2}\left(1 + \frac{2}{3} k s_x\right)$$

$$= 1.5 + 50 \times \tan 6 - \frac{9.81 \times 50^2}{2 \times 48^2}\left(1 + \frac{2 \times 0.003\,75 \times 50}{3}\right)$$

$$= 0.76\,\text{m}$$

The difference in height at the target between the drag free case and the case including drag is, therefore, $1.43 - 0.76 = 0.67$ m. Many targets are 36 in (0.915 m) diameter and are placed with their centers 1 m above the ground, so our drag free arrow would hyperthetically strike the target close to the top, while for the flight including drag, the arrow would strike close to the bottom edge of the target.

The archer's paradox

In the case of an archer taking aim at the target, the stressed bow with the string located in the nock (the notched end of the arrow) will have the arrow pointing directly at the target and lying alongside the bow on its arrow-rest. Looking from above, it can be seen that this represents a highly stressed, yet non-symmetric system. For symmetry, the arrow would ideally run through a hole in the center of the bow's thickness. One can imagine, on release of the arrow, the string and arrow nock would accelerate to the equilibrium position and, while doing so, the line of the arrow would progressively skew off at an ever increasing angle, to the side of the bow with the arrow rest. The question must be asked, if this is the direction of the arrow shaft at the point of release, why does the arrow not shoot off at this angle to the plane of the bow/string assembly?

Klopsteg (1943) addressed this issue, supported by a series of speed-flash photographs. He showed that, on release, the arrow actually wraps around the bow's thickness, and continues along the expected route; that being along the original direction of the arrow just prior to release. In fact, the momentum of the arrow is such that, instead of turning through an angle, it is energetically favourable for the arrow to bend at progressive points along its shaft as it works its length past and around the bow's thickness. This is shown diagrammatically in Figure 15.8, although the original Klopsteg (1943) paper shows the actual sequence of flash photographs. The arrow will, however, fly with residual lateral vibration caused by its enforced accommodation of the bow's thickness.

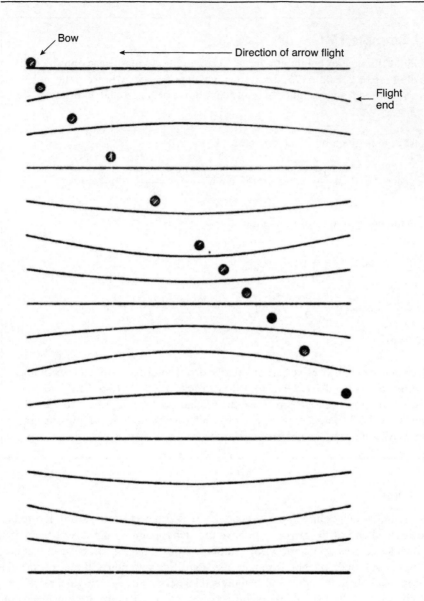

Figure 15.8 Schematic representation of the phases of the arrow in its passage by the bow (from Klopsteg, 1943).

15.2 Projectiles that move along the ground

15.2.1 Bowls and bowling (ten pin)

Bowling balls are played on a grass surface (the green) using balls made from a type of hardwood known as lignum vitae. Their path along the green is complicated by a bias on one side of the ball which causes the ball to curve in the direction of bias. Although in the past, the balls were biased by the addition of weights, these days the bias is created solely by shaping the ball. One can see that, for the case where the ball is rolling along a flat surface, its overall velocity will be the resultant of its larger forward direction velocity and a smaller (and initially constant) velocity at right angles to it, which is caused by the ball's tendency to roll over in that direction. These are the conditions for a parabolic trajectory path which begins, from launch, at the turning point of the parabola. As the ball slows down, the forward velocity reduces, increasing the affect of the biasing side force, allowing the ball to follow the parabolic path into its larger gradient tail.

One variation to this path occurs as the ball slows to a halt. Here, the rolling inertia is not enough to hold the bias directly orthogonal to the direction of travel as the ball rolls over the surface. Hence, as the ball slows, the biased region begins to rotate *under the ball*. The side force therefore reduces, and the last few centimetres of travel may again approach that of a straight line.

A variant of the game of bowls is a game primarily played in the north of England, called crown green bowling. Here, the green is not flat, but consists of a series of undulations, or sometimes a single large hump in the middle of the green. Players mainly utilize the four sides of the green, where they have to balance, not only the ball's bias but the asymmetric effects of the gravity tending to take the ball down the slope. Sadly, the author has been unable to locate any studies of trajectory analyses of such projectiles.

Ten pin bowling, known in America as simply bowling, uses a spherical ball made from urethane, plastic and resin. Their size and weights vary, as do their moments of inertia. Lower inertia value balls carry more weight near the center, compared to the high moment of inertia balls, where the weight is closer to the surface. Choice of ball is down to the player; they simply choose the type they feel will most likely carry, or knock down, all the pins in a strike.

Hopkins and Patterson (1977) have carried out a thorough analysis of the bowling ball trajectory down the lane. Their mathematics is simplified by rotating their coordinate system from one which is aligned down the center of the lane, to one aligned to the initial direction of the throw. The most likely possibility of a strike occurs surprisingly, not when the ball is sent down the center of the lane to strike the first (number 1) pin square on, but rather, to attack the pins slightly from one side or the other, aiming to just graze the number 1 pin with the edge of the ball.

Hopkins and Patterson (1977) characterized the trajectory path in two ways:

a   If the ball is released with topspin, such that there is no initial sliding of the ball on the lane (or bounce), it will, to a first approximation, travel in a straight line to the pins.

b   If the ball is given either no spin at all, or even back spin, it will initially go through a sliding phase before it picks up grip from the lane. The extent of the sliding phase will be governed by the amount of back spin and the amount of wax on the lane. Hopkins and Patterson (1977) calculated that this sliding phase creates a parabolic trajectory. So typically, the ball will leave the bowler's hand and follow a short parabolic arc, followed by a straight route to the pins commencing at the tangent to the parabola at the point where the ball achieves grip and rolls with the surface. This is counter to the perceptions of many bowlers who feel that they can throw bowls with hooks, or late swings, which start their trajectory in a straight line and then curve into the pins late in the trajectory path (cf. cricket balls).

15.2.2 Snooker, billiards, pool, etc.

Cueing the ball to eliminate sliding

For a body pivoted about a point o, with a center of gravity at point G, a distance l away from o, as in Figure 15.9a, it can easily be shown (Daish, 1972: 150) that:

$$x = l + \frac{I}{Ml}$$

where M is the mass of the body, I is its moment of inertia and x is the distance from o to the center of percussion. Now it is clear that, to avoid sliding of a cue ball on impact, it must be hit at the ball's center of percussion. For a sphere, the moment of inertia, I, is given by (Table 2.6):

$$I = \frac{2Ma^2}{5}$$

where a is the radius of the ball (see Figure 15.9b).

With $a = l$, we find:

$$x = a + \frac{2Ma^2}{5Ma} = \frac{7a}{5} = \frac{7}{10}(2a)$$

So, to eliminate the sliding phase, the cue ball should be struck at a height above the table equal to seven-tenths of the diameter of the ball. Indeed, the point of the ball's impact with the table cushions are set at this height to ensure that, on bounce, no rebound energy is lost due to either top or backspin off the cushion.

As in the case of bowling (ten-pin), if the ball is struck below this point, the induced backspin will create an initial slide on the baize which will tend to divert the ball in a parabolic arc, its direction governed by any sidespin imparted. It is commonly thought that the swerving ball is due only to sidespin (i.e. rotating around an axis vertical to the table). However, the backspin also contributes to this diversion. As soon as the ball acquires grip, just as in the case of the bowling ball, it will continue in a straight line tangential to the parabola.

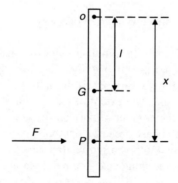

(a) Location of center of percussion for a body hung from a pivot

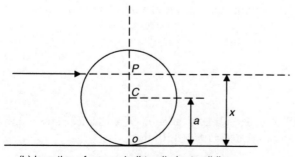

(b) Location of cue on ball to eliminate sliding

Figure 15.9 (a, b) Cueing a ball to eliminate sliding.

Ball-on-ball impact: some further thoughts

Section 4.5 mathematically considers the ball-on-ball impact. There follows some additional notes pertaining to the special case of impact between wooden balls on the green baize of the table.

From the previous section, we can see now that a ball-on-ball impact will create backspin on the second, or object, ball, because it is actually struck at a distance of $0.5 \times (2a)$ above the table, rather than the $0.7 \times (2a)$, which would be the position required for no spin.

The detailed process of the collision is as follows.

Immediately on impact, the cue ball stops in its tracks, and the object ball moves at the initial impact of the cue ball, as would be expected by Newton's second law and conservation of energy. However, this does not take into account the angular velocity possessed by the cue ball as it rolls along the table. The object ball will depart from the impact with backspin, slowing and sliding on the table until its velocity is $\frac{5}{7}$ that of its initial value following impact. At this point, the object ball grips the table and continues as expected.

Meanwhile, although the cue ball has instantaneously halted, it still possesses angular momentum as it skids on the table, with what is now effectively, topspin. The frictional force causes the cue ball to accelerate forward, until it succeeds in gripping the table and starts to roll normally. The cue ball, therefore, follows behind the object ball and with a velocity equal to about $\frac{2}{7}$ its initial pre-impact velocity.

In the case of the cue ball being cued at the center of the ball, rather than the ideal non-sliding height ($\frac{7}{10}$ the diameter above table), the ball will be propelled with backspin. If the cue ball were to collide with the object ball with this amount of backspin, the result would be a collision in which the cue ball stops dead, and the object ball launches in fully-rolling mode and at the same velocity as the cue ball. Newton's second law is clearly shown to apply in this scenario.

The above cases assume, for mathematical simplicity, that the coefficient of restitution, e, of such ball-on-ball impacts is unity. In reality, snooker/billiard ball experiments have shown that figures of $e = 0.9$ are nearer the correct value.

15.2.3 Curling

Curling was developed in Scotland as early as the sixteenth century, although some evidence exists that it was also played in the low countries of Europe at this time. The curling stone is made from granite and weighs between 19.1 and 20 kg. It is 25 cm in diameter and 13 cm high. The handle, bolted onto the stone, is of a tough thermoplastic moulded construction. Interestingly, the shoes worn by curlers on each foot are different. The sliding shoe has a Teflon and steel sole to reduce the coefficient of friction down to $\mu \approx 0.1$. The other shoe has a rubber base for a greater coefficient of friction. Early brooms were of a wooden construction with horsehair heads, or even, off-the-shelf household brushes. The modern broom is constructed from either fiber-glass or carbon-fiber with a sweeping face of fabric. It is not only used to sweep away debris and to polish the ice sheet, but also to help melt the surface to a lower friction at required points in the stone's path. The broom is also used as a tool to aid the balance of the stone deliverer.

The base of the curling stone is concave, so the surface in contact with the ice, known as the running surface, is an annular ring, of inner radius 64 mm and only 5 mm in thickness. When the stone is released with a spin, it will curl in the direction that the front edge of the stone is moving. The process is simple to describe.

Low friction between stone and ice is caused by a lubricating layer of melt-water created by the velocity of the stone over the surface. The coefficient of friction as a function of velocity has been accurately modelled by Stiffler (1984) as:

$$\mu = \frac{2k\left(T_m - T_0\right)}{\sigma\sqrt{\pi\alpha l_c U}} \tag{15.10}$$

where k is the thermal conductivity of the stone, T_m is the melting point of the ice, T_0 is the ambient temperature, σ is the load per unit area, α is the thermal diffusivity of the stone, l_c is the length of contact, and U is the velocity of the stone.

If this equation is plotted for typical physical values of curling stone and ice sheet, and for a range of typical stone velocities, it may be seen that the μ versus U curve is characterized by a highly inversely non-linear function for U values between 0.1 and 0.5 m s^{-1}, followed by a levelling out of the function, especially from 1 m s^{-1} upwards. Figure 15.10 shows a plot of this function for typical curling game values and a range of ice temperatures.

A stone which is launched spinning down the course towards the target will possess both translational and rotational velocities. If we assume the rotation to be anti-clockwise (as viewed from above) then the relative velocity between the stone and ice will be greater on the right-hand side of the stone than on the left. This creates asymmetric friction on either side of the stone, with the right-hand side having a much lower friction coefficient than the left. The stone, therefore, deviates to the left side: the side with the greater friction.

Figure 15.11 shows the model of the trajectory paths for stones travelling at 2 m s^{-1} with varying spin rates.

Friction will slow the stone down as it approaches the target (known as the button), but its inertia almost maintains a constant angular velocity over the whole duration of its travel. Therefore, as the stone slows down and the frictional asymmetry is preserved, its deflection will become greater, spiralling into a halt. This can be seen in the modelled path of the stone in Figure 15.11, and is certainly observed by curlers. Indeed, they make use of this curl in the path in what is known as the come-around shot, in which skilled practitioners bring the stone around the side of any previously planted 'guard stones', to home in on the button.

Another facet of the game is the brushing action of the two sweeping curlers. Their job is to sweep directly in front of the stone as it travels, in order to modify its path and improve its accuracy. The sweeping action warms the ice directly in front of the stone. This decreases its friction coefficient by increasing the ambient ice temperature, T_0, in Equation 15.10, such

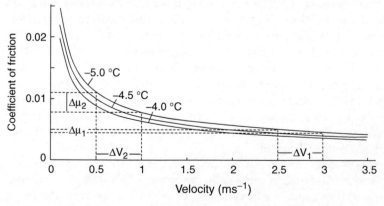

Figure 15.10 Coefficient of friction versus velocity for a curling stone (from Marmo and Blackford, 2004).

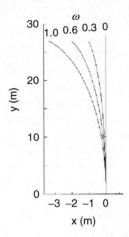

Figure 15.11 Curling stone trajectory for stones released at 2 m s^{-1} at varying angular velocities (from Marmo and Blackford, 2004).

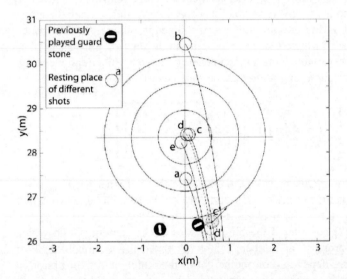

Figure 15.12 End paths and final resting positions for five stones, each released at translational velocity 2.0 m s^{-1} and rotational velocity 0.8 rad s^{-1} (from Marmo and Blackford, 2004). Shot **a** was not swept at all, shot **b** was swept for the entire length, shot **c** was swept from 0–10 m, shot **d** was swept from 10–20 m and shot **e** was swept from 20 m until it stopped.

that the value of the expression $(T_m - T_0)$ is reduced. Marmo *et al.* (2006) have measured temperature increases of T_0 of as much as 2°C as a consequence of brushing. Reduction in the coefficient of friction of the ice sheet has two effects. First, and quite obviously, it enables the stone to travel further. Second, it reduces the frictional asymmetry on either side of the stone which tends to straighten its path. So, accurate stone placement is achieved by close collaborative teamwork between the stone deliverer and the two sweepers.

Figure 15.12 shows the end path and finishing positions of five stones which have been subject to different types of sweep areas.

15.3 The human projectile

15.3.1 The long jump

With a few provisos, the human body may be treated as a projectile located at the point of the body's center of gravity, and the normal projectile equations may, with care, be used. Taking the long jump as a case in point, the landing height (i.e. the distance above the ground of the body's center of gravity at the point of landing) is going to be considerably lower than the launch height, even though the landing area is, essentially, horizontal. This is solely as a consequence of the body's different configurations for take-off and landing.

The next issue is the body's drag coefficient, C_D. This is clearly going to vary dependent on the body's configuration. Frohlich (1985) stated the value of $C_D = 0.9$ for a sprinter. The drag forces, in the case of the long jump, is going to both slow down the run-up, and create a deceleration in flight. Using Frohlich's value of C_D in Equation 5.1, for a man weighing 75 kg a decelerating force of 0.348 m s^{-2} is created. Therefore, for a typical long jump flight time lasting 1 s, the jump length will be reduced by 0.18 m.

We know that, to obtain optimum length, the jumper should aim to launch at an angle of 45°. For typical elite performance lengths, this leads to a center of gravity height at the peak of the jump of 2.55 m. This is closer to the height requirement for a high jumper! The problem is that the human body is not good at converting kinetic energy into vertical potential energy, and the kinetic energy gained during the run-up is better utilized in maintaining horizontal momentum to carry the athlete further. Indeed, even elite high jumpers only succeed in converting 50 per cent of their available launch kinetic energy into vertical potential energy.

The simple ballistic equation for a long jump may be stated as:

$$d_{LJ} = \sqrt{\left(V_0^2 - 2h_{\text{jump}}g\right)} \left\{ \sqrt{\left(\frac{2h_{\text{jump}}}{g}\right)} + \sqrt{\left[\frac{2\left(h_{\text{jump}} + \Delta h_{LJ}\right)}{g}\right]} \right\} \tag{15.11}$$

where d_{LJ} is the long jump jumping distance, V_0 is the launch velocity (5 m s^{-1}), h_{jump} is the rise in center of gravity from launch to peak (1 m), and Δh_{LJ} is the drop in height of center of gravity at launch to landing (0.5 m) (typical values shown in brackets).

Wakai and Linthorne (2002) considered the biomechanics of the standing jump which is used as a measure of leg strength for jumpers. They calculated that a launch angle of around 22° would be optimum, depending on the height of the athlete and their preferred body configuration at launch. However, most athletes launched at around 34° and yet still achieved close to the theoretically calculated maximum range. In other words, the athletes were not significantly disadvantaged by increasing their launch angle.

15.3.2 The pole vault

As stated, the human body is not designed for efficient transfer of running kinetic energy (KE) into vertically raising potential energy (PE). To convert all available sprinting energy into gravitational potential energy we could use an equation of the form:

$$h_{pv} = h_{c.g.} + \Delta h_{c.m.} = h_{c.g.} + \frac{V_0}{2g}$$

where h_{pv} is the available height, $h_{c.g.}$ is the standing center of gravity ($= 1$ m), $\Delta h_{c.m}$ is the change in height, and V_0 is the horizontal velocity.

For a sprinting velocity of $10\,\mathrm{m\,s}^{-1}$ the available height is 6.10 m. Unaided, men can only jump to 2.3 m (only?!), but with the aid of a pole vault, heights of 5.8 m may be achieved. Although the pole may be considered by some to be a form of lever, or even a climbing implement, in actual fact it is most accurately described as a tool to improve the KE $\rightarrow$ PE energy conversion efficiency.

The earliest pole vaults were of wooden construction, usually bamboo. In the 1950s the lighter aluminum poles were introduced which allowed for greater running speeds. However, in the 1960s carbon fiber poles completely reinvented the sport. They were capable of considerable flexing which allowed the athlete's kinetic energy to be temporarily stored as elastic potential energy. This was then released at an optimum instant in time, creating the required potential energy for the athlete, with near 100 per cent efficiency. The introduction of carbon fiber poles corresponded to a steep increase in the world record from 4.5 m to 5.5 m. Since then, further advances in pole material and design, together with improvements in technique, have increased the world record to the current height of 6.14 m.

Danby (1998) has created three energy conversion models of the pole vault; each a refinement on the previous one:

1   The simplest model considers the pole vault to be a linear compressible spring, of original length r_0, pivoting around one end (the origin), which is embedded in the ground. The other end has the athlete's weight and momentum acting to compress the spring, as its angle to the horizontal, θ, swings from near 0° (as the pole embeds in the ground) up to nearly 90° (as the athlete releases the pole). The equations of motion of such a system are:

$$\left.\begin{aligned}\frac{\mathrm{d}^2 r}{\mathrm{d}t^2} &= \left(\frac{\mathrm{d}\theta}{\mathrm{d}t}\right)^2 - g\sin\theta - \frac{kg}{W}(r - r_0)\\[2mm]\frac{\mathrm{d}^2\theta}{\mathrm{d}t^2} &= -\frac{2}{r}\frac{\mathrm{d}r}{\mathrm{d}t}\frac{\mathrm{d}\theta}{\mathrm{d}t} - \frac{g}{r}\cos\theta\end{aligned}\right\} \tag{15.12}$$

where k is the 'spring constant' of the pole, and W is the weight of the athlete.

For typical initial values of $\theta = 0$, $r_0 = x(0) = 4.6\,\mathrm{m}$ (length of pole) and $\mathrm{d}x/\mathrm{d}t = 8.0\,\mathrm{m\,s}^{-1}$ (run in speed) then, for a pendulum:

$$\omega = \sqrt{\frac{kg}{W}} \quad \text{and} \quad x(t) = 4.6 - \frac{8.0}{\omega}\sin\omega t$$

Therefore:

$$\frac{\mathrm{d}x}{\mathrm{d}t} = -8.0\cos\omega t$$

If the 'spring' compresses down to, say, 2 m and $\omega = 4\,\mathrm{s}^{-1}$, then for $W = 72.6\,\mathrm{kg}$, the spring constant, $k = 120\,\mathrm{kg\,m}^{-1}$. We are now in a position to model the motion using Equation 15.12.

2   The next development is to use the theory of forces on a bending rod, known as an elastica. Danby (1998) shows that, the equation of force on such a flexed rod, optimized to a pole vault is:

$$F \approx \frac{4B}{r_0^2}\left[\left(\frac{\pi}{2}\right)^2 + 1.3\left(\frac{r_0 - r}{r_0}\right)\right] \tag{15.13a}$$

where B is the stiffness factor for the rod (a value of 23 000 kg m² is suitable for modelling purposes – but try varying it).

3   Now a passive weight, W, is added to the rod representing the weight of the athlete. Danby (1998) complements Equation 15.13a with the following set:

$$\left.\begin{aligned} r &= \sqrt{x^2 + z^2} \\ \frac{\mathrm{d}^2 x}{\mathrm{d}t^2} &= \frac{g}{W}\frac{x}{r}F \\ \frac{\mathrm{d}^2 z}{\mathrm{d}t^2} &= \frac{g}{W}\frac{z}{r}F - g \end{aligned}\right\} \tag{15.13b}$$

with the coordinates of the athlete at (x, z).

4   And finally, rather than having a dead weight at the end of the pole, the athlete is modelled as a secondary pendulum, swinging from the end of the pole with their center of mass located a distance $\rho(t)$ from the pivot point at the end of the pole, and making an angle $\phi(t)$ to the horizontal. Note that both ρ and ϕ are functions of time. This is to accommodate the athlete starting the manoeuvre at full stretch, curling up through the vaulting process and then releasing the pole when, again, at full stretch. Danby (1998) derived and stated the complete set of equations of motion for this system as follows:

$$\left.\begin{aligned} r &= \sqrt{x^2 + z^2 + \rho^2 + 2x\rho\cos\phi + 2z\rho\sin\phi} \\ F &\approx \frac{4B}{r_0^2}\left[\left(\frac{\pi}{2}\right)^2 + 1.3\left(\frac{r_0 - r}{r_0}\right)\right] \\ \frac{\mathrm{d}^2 x}{\mathrm{d}t^2} &= \frac{g}{W}\frac{(x + \rho\cos\phi)}{r}F \\ \frac{\mathrm{d}^2 z}{\mathrm{d}t^2} &= \frac{g}{W}\frac{(z + \rho\sin\phi)}{r}F - g \\ I\frac{\mathrm{d}^2\theta}{\mathrm{d}t^2} &= -\frac{\mathrm{d}I}{\mathrm{d}t}\frac{\mathrm{d}\theta}{\mathrm{d}t} + F\frac{z + \rho\sin\phi}{r}\rho\cos\phi - F\frac{x + \rho\cos\phi}{r}\rho\sin\phi \end{aligned}\right\} \tag{15.14}$$

15.3.3 The ski jump

There have been numerous studies involving the modelling of this winter sport. Perhaps the most detailed is that of Podgayets et al. (2002) in which the jump is sectioned into four phases: acceleration, take-off, flight and landing. Although each stage is modelled in the paper, we are now only going to look at the most important one: the flight stage.

As ever, once launched, maximum flight length occurs when the mutually opposing variables of drag and lift are optimized. In this case, drag and lift may be continually controlled by the jumper while in flight; the greater the attack angle (within limits), the greater the lift and the greater the drag.

Based on wind tunnel tests, it has been suggested that the relations:

$$C_L = -0.00025\beta^2 + 0.0228\beta - 0.092$$
$$C_D = 0.0103\beta$$

hold true, where β is the skier's attack angle.

The trajectory model may be constructed around that of a javelin using the equations of motion given in Equation 9.4, restated below to reflect the ski jump application, where $\varphi =$ angle of flight to horizontal:

$$
\left.
\begin{aligned}
m\frac{dv}{dt} &= -mg \sin \varphi - \frac{1}{2}\rho A C_D v^2 \\
m\frac{d\varphi}{dt} &= -mg \cos \varphi + \frac{1}{2}\rho A C_L v^2 \\
\frac{dx}{dt} &= v \cos \varphi \\
\frac{dy}{dt} &= v \sin \varphi
\end{aligned}
\right\}
\tag{15.15}
$$

Once this model is constructed for a range of attack angles, β (and consequential values of C_L and C_D), it is found that: to maximize range, the skier should minimize β for the first part of the flight (thereby reducing both drag and lift), and then gradually increase β from about half way through the flight, which will indeed create more drag, but crucially at a lower flight velocity. It will also increase lift, thereby extending the flight time and the range.

Figure 15.13 shows measured trunk angles as a function of flight time. The initially large angle is the forward tilting section which occurs before the onset of steady flight. The angle then steadily rises from about 1.5 s until the final backward tilting stage, ready for the landing.

The optimized flight model solution presented by Seo *et al.* (2002) is summarized as:

1   The body-ski angle is almost 10° during the flight, while the ski opening angle increases in the first half of the flight.
2   The angle of attack oscillates around the ideal angle to accommodate ski tremble and wind gusts.
3   The drag should be small in the first half, while the lift should be large in the latter half of the flight.
4   The flight distance in the horizontal direction is larger in the first half of the flight duration, but the height of the jumper decreases quicker in the latter half.

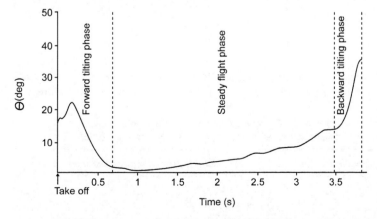

Figure 15.13 Ski jumper's trunk angle versus flight time (from Ohgi et al., 2006).

15.3.4 The sky diver

The terminal velocity of any projectile occurs when the downward force due to gravity is matched by the opposing drag force. At equilibrium, the net force must be zero, so:

$$mg - \tfrac{1}{2}\rho v^2 A C_D = 0$$

and:

$$v_T = \sqrt{\frac{2mg}{\rho A C_D}} \qquad (15.16)$$

where v_T is the terminal velocity.

Two points are worth mentioning. First, the only reason that a terminal velocity, v_T, exists is because the drag force generally increases *as a square of* the velocity. So, as the projectile's velocity increases from zero, the drag force, which starts at a negligible value, increases at such a rate that it eventually catches up with, and matches, the gravitational force at the value of the terminal velocity. Second, strictly speaking, the projectile's velocity approaches v_T *asymptotically*. That is, it accelerates at a slower and slower rate the closer it approaches v_T, but it never actually reaches the value of v_T. So, from that point of view, it could be said that the terminal velocity is never quite achieved.

For the sky diver, Aoyama and Nakashima (2006) measured values of drag coefficient, C_D, for free-falling humans with orientations both normal and tangential to the airflow. Their values are given in Table 15.3, together with the results from Equation 15.16 for the terminal velocity for both orientations. Values for the mass of a typical 85 kg man are used, and the air density at altitude is taken as $\rho = 1.205 \text{ kg m}^{-3}$.

Aoyama and Nakashima (2006) have modelled the displacement and rotation of free-flying, horizontally oriented sky divers as a consequence of various joint movements. Four joint angles are expressed: θ_k is the flexion angle of the knee joint, θ_s is the flexion angle of the shoulder joint, θ_e is the flexion angle of the elbow joint and θ_h is the abduction angle of the hip joint.

The models are stated in terms of four simple linear regression equations:

$$\left.\begin{aligned}
v_x &= -0.38\theta_k + 0.087\theta_s + 7.57 \\
v_z &= -0.10\theta_k - 0.045\theta_s - 50.4 \\
v_y &= -0.043\theta_e - 0.071\theta_h \\
\omega_z &= -0.57\theta_e - 0.82\theta_h
\end{aligned}\right\} \qquad (15.17)$$

where v_x is the forwards/backwards motion (relative to the major axis of body), v_y is the right/left motion (relative to the major axis of body), v_z is the up/down motion (relative to

Table 15.3 Terminal velocity for skydiver falling in two orientations; horizontal to earth, and head or feet first (values for calculation taken from Aoyama and Nakashima, 2006)

Normal orientation (body horizontal to earth)	Tangential orientation (head or feet first)
$C_D = 0.92$	$C_D = 0.80$
$A = 0.5 \text{ m}^2$	$A = 0.2 \text{ m}^2$
$v_T = 55 \text{ m s}^{-1}$	$v_T = 93 \text{ m s}^{-2}$

a default terminal velocity of 54.5 m s^{-1}), and ω_z is the rotating clockwise/anti-clockwise motion (relative to the major axis of body as viewed from above).

15.4 Trajectory diversions: surviving a fall

It is quite simple to prove that a human reaches 99 per cent of their terminal velocity after about 70 m of vertical free fall and yet amazing stories abound of survival from high altitude descents.

Vesna Vulović, a stewardess of the JAT Serbian Airline, survived a fall from 10 km on January 26, 1972 when her plane was brought down by terrorist explosives. Her fall was thought to be broken by the fuselage she was trapped inside following the mid-air explosion. A catering trolley pinned her in place during the fall and subsequent impact. She was, however, badly injured with a fractured skull, three fractured vertebrae and she remained in a coma for 27 days. The man who treated her at the scene was an ex-medical officer from Hitler's Nazi army during the Second World War.

There were numerous reports of Second World War military aircrew surviving falls from planes.

- Alan Magee fell 6.7 km from a Flying Fortress which had been hit by enemy fire over France. Rendered unconscious by the high altitude, his damaged parachute failed to open and he landed in a railway station, breaking his fall as he crashed through the glass roof.
- Flight Sergeant Nicholas Alkemade survived after his Lancaster Bomber was hit and spiralled out of control over Germany. Preferring a quick death to either being burnt to death or captured by the Germans, he decided to jump from an altitude of 5.5 km, even though he knew his parachute had been destroyed. He landed in snow covered pine trees and suffered a sprained ankle. All his crew members died in the crash. As a Prisoner of War, the Germans honoured him with a certificate and looked after him well.
- Ivan Chisov bailed out of his damaged plane at 6.7 km and chose not to open his parachute until near the ground so as to avoid being an enemy target from German fighters. However, he lost consciousness and hit a snowy ravine at something close to 150 mph. He rolled and ploughed his way downwards to the bottom, sustaining spinal injuries and a broken pelvis. He was flying again three months later.

It is thought that two of the Lockerbie bombing victims survived for a short period following ground impact, although they both died before help arrived. Both were found in the nose section of the fuselage.

Finally, there is James Boole from Staffordshire, England. In May 2009, the skydiving cameraman, having jumped from an altitude of 1.8 km, was filming another skydiver in descent. The plan was for the skydiver being filmed to inform Boole by hand signal when he neared the ground. The signal came too late. Boole landed in a deep snow drift, breaking his back and ribs, bruising his lung and … chipping a couple of teeth. Boole is particularly proud of the fact that the incident was recorded for posterity on film.

Summary

In this, our last chapter, we have covered a selection of sports that, though interesting from a trajectory dynamic perspective, did not warrant a dedicated chapter in their own right, or a substantive section of a chapter. This may be due to the sport's relative unpopularity, scarcity of academic material in the field, or that there is limited knowledge to be gained over that which has previously been described. The remaining projectiles have been considered in three sections: projectiles that fly through the air, those that roll along the ground, and the final section deals with the situation where humans may be treated as projectiles, even though their flight may be of a transitory nature.

Basketball is considered in terms of the bounce from the court, the trajectories required to score a basket and the effects of drag on the flight. In badminton, the trajectories of both natural and synthetic shuttlecocks are compared, with particular reference to their performance at speed, and the differences in the characteristics of the skirt. In table tennis, the importance of spin is emphasized and its impact on the character of the game. Volleyball trajectories are considered by highlighting the similarity in performance with the soccer ball. Indeed, it is shown that the main difference lies only in the way the two balls are struck. In a broadly similar manner, softball is compared with baseball. Here the only significant difference is the way in which the ball is pitched, and their consequential pitching speed. Archery is shown to be an interesting sport to analyse, not least because many assumptions may be utilized which simplifies the analysis. Furthermore, the analysis is influenced by the fact that we are not looking to optimize range, but rather, to improve accuracy.

We commenced the section concerning projectiles that roll or slide along the ground with a study of bowls and bowling. We looked at how the bias on a bowling ball creates a curved trajectory. A tenpin bowling ball may also commence its trajectory with a parabolic curved path if it is made to slide, rather than roll, down the lane. However, as soon as it grips onto the lane and rolls, its path becomes a straight line. In snooker, the effect of striking the ball at different heights above the table is examined. It is shown that a ball that is struck seven-tenths up the height of the ball will immediately launch in fully rolling mode. Any other heights will result in a launch with either topspin or backspin. In curling, although the spin mechanism is very different from that of aerodynamic spin and its Magnus force, the spinning stone causes a similar type of curve on the trajectory path. It is shown that the path can be modified by using the brooms to warm the ice immediately in front of the stone.

In the final section of this chapter we looked at the long jump in terms of trajectory modelling theory, and showed how inefficient the body is at converting its kinetic energy into potential energy to raise the body. The long jump is treated as a launched body-particle, landing at a lower height than the launch height, due to the landing posture. The pole vault is treated as a tool to make the KE$\rightarrow$PE conversion more efficient. The section shows the pole vaulting action to be modelled in four different ways, of increasing complexity and accuracy. The concept of the elastica as a means of storing energy and releasing it at the optimum time is introduced. The ski jump is an example of where the varying shape of the body allows the coefficients of drag and lift to be altered in flight. We see that, for optimized range, the athlete requires a low drag at the start of the flight and a high lift factor towards the end. In the final section on sky diving we see how, again, the body can be oriented in flight to alter coefficients of drag, and to alter the body's displacement and rotation by a range of body limb movements.

Problems and questions

1   In basketball, calculate the optimum velocity and angle of release for a ball to pass through the dead center of the basket when it is launched from a point 4.57 m (15 ft) horizontally from the basket and 0.92 m (3 ft) below the basket height.

2   For the case of the basketball trajectory of Question 1, use the plot of Figure 15.2 to calculate the angle of entry into the basket for a ball which passes through the basket's center. Further, calculate the margin of error for such a ball (i.e. the

allowable range of launch angles between the angle where the ball just touches the front of the ring, and the angle where the ball just touches the back of the ring).

3   Calculate the lift force on a table tennis ball which possesses a backspin of $25\,\mathrm{rad\,s^{-1}}$ travelling at $10\,\mathrm{m\,s^{-1}}$. $R = 0.02\,\mathrm{m}$, $m = 0.002\,\mathrm{kg}$. Take the density of air to be $\rho = 1.3\,\mathrm{kg\,m^{-3}}$.

4   Calculate the Magnus force on a volley ball which is spinning at $4\,\mathrm{rev\,s^{-1}}$ and has a velocity of $20\,\mathrm{m\,s^{-1}}$.

5   In a long jump, calculate the length of the jump using the ballistic equation, when the athlete's launch velocity is $5\,\mathrm{m\,s^{-1}}$, the rise in the center of mass from launch to peak is 1 m and the drop in height of the center of gravity at launch to landing is 0.5 m.

6   Calculate the return force on a pole vault of length 4.6 m if it is flexed such that the distance between the two ends reduces to 2.5 m and the stiffness factor, B, is $23\,000\,\mathrm{kg\,m^{-3}}$.

7   Compare the drag and lift coefficients for a ski jumper oriented with angles of attack of both $10°$ and $20°$.

8   It states in the Trajectory Diversions section of this chapter that:

> It is quite simple to prove that a human reaches 99 per cent of their terminal velocity after about 70 m of vertical free fall, and yet amazing stories abound of survival from high altitude descents.

Why did I not simply say '... reaches their terminal velocity ...' (i.e. 100 per cent instead of 99 per cent)?

Prove that, for typical values, the rest of the statement is true, i.e. humans do indeed reach 99 per cent of their terminal velocity after 70 m of free fall.

Appendix I

Table of physical values and drag-to-weight ratio for a variety of sporting projectiles

Physical values and drag-to-weight ratio for a variety of sporting projectiles (taken from De Mestre, 1991)

Projectile	A (m^2)	m (kg)	v_0 $(m\,s^{-1})$	l (m)	$R_e = \frac{v_0 l}{v}$	$\varepsilon = \frac{\rho A C_D v_0^2}{2mg}$
Australian rules football	0.026	0.46	30	0.18	3.8×10^5	0.62
Basketball	0.004	0.15	40	0.07	1.9×10^5	0.53
Baseball	0.045	0.62	15	0.24	2.4×10^5	0.20
Cricket ball	0.004	0.16	35	0.07	1.5×10^5	0.38
Discus (mens)	0.008	2.00	25	0.22	3.7×10^5	0.15
Discus (women)	0.007	1.00	25	0.18	3.0×10^5	0.28
Golf ball	0.001	0.05	70	0.04	1.9×10^5	1.23
Hammer	0.011	7.26	25	0.12	2.0×10^5	0.02
Javelin (mens)	0.001	0.80	30	0.03	0.6×10^5	0.64
Javelin (womens)	0.001	0.60	26	0.025	0.4×10^5	0.85
Long jumper	0.600	75	10	1.8	12.0×10^5	0.03
Rugby football	0.028	0.42	30	0.19	3.8×10^5	0.75
Shot put (mens)	0.011	7.26	15	0.12	1.1×10^5	0.01
Shot put (womens)	0.008	4.00	12	0.10	0.8×10^5	0.008
Shuttlecock	0.001	0.005	35	0.04	0.2×10^5	27
Soccer football	0.038	0.42	30	0.22	4.4×10^5	1.02
Squash ball	0.001	0.02	50	0.04	1.3×10^5	3.52
Table tennis ball	0.001	0.002	25	0.04	0.7×10^5	8.80
Tennis ball	0.003	0.06	40	0.06	1.6×10^5	1.00
Water polo ball	0.038	0.42	15	0.22	2.2×10^5	0.23

A	=	projectile cross-sectional area
m	=	projectile mass
v_0	=	typical projectile launch velocity
l	=	the projectile representative length scale
R_e	=	projectile Reynold's number
v	=	kinetic viscosity of air taken as $1.5 \times 10^{-5}\,m^2\,s^{-1}$
ρ	=	air density, taken at sea level to be $1.23\,kg\,m^{-3}$
C_D	=	projectile drag coefficient
ε	=	drag-to-weight ratio
g	=	acceleration due to gravity taken as $9.81\,m\,s^{-2}$

The fourth-order Runge–Kutta method

You will have noted that, throughout this book, we have derived or otherwise stated the equations of motion for the various projectiles in the form of first-order differential equations. This happens when we take the dragless equations of motion, such as those derived at Equations 3.5 or 3.6, and start to incorporate elements of drag and lift to modify the parabolic trajectory. What we are saying by expressing the equations in this form, is that we can only calculate the position of the projectile at any point in its flight by knowing both its rate of change and its precise location of the projectile at the point immediate before it.

There is no one exact solution to an equation of this sort; nor would we expect one. We cannot state, for example, that the answer to a javelin being thrown at $20\,\mathrm{m\,s^{-1}}$ at an angle of $25°$ to the horizontal is 7.25. However, what we would like to say is: from those input values, I'd like to see the predicted plot of its trajectory. I can then obtain important results such as range, flight time, etc. There are, however, some barriers to this aspiration.

First, differential equations of the form we encounter in projectile analysis cannot be solved by simple algebraic manipulation (or analytically as it is called) in the way that we derived, say, Equations 3.5 or 3.6. We can only solve them by throwing some possible solutions at them and then seeing what it does with our suggestions. From the equation's response, we can then modify our attempts and have another go, repeating the process and hopefully refining the solutions. We stop when we have reached the stage where we cannot find a more accurate answer, or we have an answer that is realistically close to our needs. This is known as solving by numeric methods. It is not perfect, but it's all we've got.

There is a further problem. In the process of solving a differential equation we are, in fact, doing a form of integration, even though the action may be well hidden in the process. However, as in the case of integration, we have to crack the problem of the integration constant which is created out of the process. The solution of the differential equation may be a whole group of curves, rather than a single curve. Furthermore, these curves may not even have identical shapes to each other, but rather, a series of gradually evolving plots, as you might imagine a field diagram of the forms shown in Figures 9.4 or 13.1. Luckily, for our trajectory flight scenarios, an obvious boundary point can usually be located which locks the curve so that it passes through that single point in space and time from which we can say 'That is the one true trajectory plot I need, and I can ignore all the others'. That boundary point will normally be the x, y, z location of the projectiles initial launch, be it the point of impact of racket or boot, the bowler's trailing finger or the end of a gun barrel. That particular point in space is the location that the projectile was at, at time $t = 0$. This single location at a fixed point in time allows us to select the one curve that represents our particular trajectory path.

There are numerous algorithms which may be used to solve our family of equations of motion; each with individual advantages and disadvantages concerning matters such as speed of iteration, accuracy and ease of application. Most, by their nature, lend themselves to algorithmic computer-aided solutions. The most commonly used algorithm for solving these form of equations is the fourth-order Runge–Kutta method developed by the German mathematicians, Carl David Runge and Martin Wilhelm Kutta around 1900.

In fact the algorithm may be utilized to solve our equations using any number of professional mathematical programs such as MathCad, Mathematica or Maple (for which trial downloads may be available). There are even add-on tools to MS Excel which will allow you to input your equations and create plots and coordinate values from the plots. It may be the case that the free mathematical web browser WolframAlpha (created by the same Company who produce Mathematica) can be leveraged into use to solve these equations.

Notwithstanding, for completeness, what follows is an explanation of the fourth-order Runge–Kutta method of solving differential equations.

If our first-order differential equation is of the form:

$$\frac{dy}{dx} = f(x,y), \quad y(0) = y_0$$

Then the algorithm is based on the following:

$$y_{i+1} = y_i + \left(a_1 k_1 + a_2 k_2 + a_3 k_3 + a_4 k_4\right) h \tag{A1}$$

where, once we know the value of $y = y_i$ at x_i, we can obtain the value of $y = y_{i+1}$ at x_{i+1}, and $h = x_{i+1} - x_i$ the step size.

Equation A1 is now equated to the first five terms of the Taylor series:

$$y_{i+1} = y_i + \frac{dy}{dx}\Big|_{x_i, y_i} \left(x_{i+1} - x_i\right) + \frac{1}{2!}\frac{d^2 y}{dx^2}\Big|_{x_i, y_i} \left(x_{i+1} - x_i\right)^2 + \frac{1}{3!}\frac{d^3 y}{dx^3}\Big|_{x_i, y_i} \left(x_{i+1} - x_i\right)^3$$
$$+ \frac{1}{4!}\frac{d^4 y}{dx^4}\Big|_{x_i, y_i} \left(x_{i+1} - x_i\right)^4 \tag{A2}$$

Now, we know that $dy/dx = f(x, y)$ and $x_{i+1} - x_i = h$. So:

$$y_{i+1} = y_i + f\left(x_i, y_i\right) h + \frac{1}{2!} f'\left(x_i, y_i\right) h^2 + \frac{1}{3!} f''\left(x_i, y_i\right) h^3 + \frac{1}{4!} f'''\left(x_i, y_i\right) h^4 \tag{A3}$$

One solution of Equations A2 and A3 is:

$$y_{i+1} = y_i + \frac{1}{6}\left(k_1 + 2k_2 + 2k_3 + k_4\right) h \tag{A4}$$

where:

$$\left.\begin{aligned}
k_1 &= f\left(x_i, y_i\right) \\
k_2 &= f\left(x_i + \tfrac{1}{2}h, y_i + \tfrac{1}{2}k_1 h\right) \\
k_3 &= f\left(x_i + \tfrac{1}{2}h, y_i + \tfrac{1}{2}k_2 h\right) \\
k_4 &= f\left(x_i + h, y_i + k_3 h\right)
\end{aligned}\right\} \tag{A5}$$

This obviously solves only a single differential equation. However, the method can be extended to deal with a system of n first-order differential equations, by utilizing standard vector notation in an identical derivation to the one given.

Bibliography

Alam, F., Chee We, P., Subic, A. and Watkins, S. (2006) 'A comparison of aerodynamic drag of a rugby ball using EFD and CFD', *Sports Engineering: 6th International Conference on the Engineering of Sport (Munich)*, 9(3): 182. London: Blackwell Science.

Alam, F., La Brooy, R., Watkins, S. and Subic, A. (2007) 'An experimental study of cricket ball aerodynamics', *Proceedings of the International Conference on Mechanical engineering*. January 17. ICME 2007 Dhaka, Bangladesh.

Alaways, L.W. and Hubbard, M. (2001) 'Experimental determination of baseball spin and lift', *Journal of Sports Sciences*, 19: 349–358. London: E. & F.N. Spon.

Andrew, D.P.S., Chow, J.W., Knudson, D.V. and Tillman, M.D. (2003) 'Effect of ball size on player reaction and racket acceleration during the tennis volley', *Journal of Science and Medicine in Sport*, 6(1): 102–112. Oxford: Elsevier Science Ltd.

Aoyama, A. and Nakashima, M. (2006) 'Simulation analysis of maneuver in skydiving', Haake, S. (Ed.) *Engineering of Sport 6*. 2: 165–172. New York: Springer-Verlag.

Armenti, A. (1992) *The Physics of Sports*, New York: Springer-Verlag.

Asai, T., Carré, T., Akatsuka, T. and Haake, S.J. (2002) 'The curve kick of a football impact with the foot', *Sports Engineering*, 5: 183–192. London: Blackwell Science.

Asai, T., Seo, K., Kobayashi, O. and sakashita, R. (2006) 'Flow visualization on a real flight non-spinning and spinning soccer ball', Haake, S (Ed.) *Engineering of Sport*, 6: 327–332. New York: Springer-Verlag.

Baker, S.W. Hammond, L.K.F., Owen, A.G. and Adams, W.A. (2003) 'Soil physical properties of first class cricket pitches in England and Wales', *Journal of Sports Turf Research Institute*, 79: 13–21.

Bartlett, R.M. (1992) 'The biomechanics of the discus throw: a review', *Journal of Sports Sciences*, 10: 467–510. London: E. & F.N. Spon.

Bartlett, R.M. (2007) *Introduction to Sports Biomechanics*, London: Routledge.

Bartlett, R.M. and Best, R.J. (1988) 'The biomechanics of javelin throwing: a review', *Journal of Sports Sciences*, 6: 1–38. London: E. & F.N. Spon.

Barton, N.G. (1982) 'On the swing of a cricket ball in flight', *Proceedings of the Royal Society*, 379: 109–131.

Bearman, P.W. and Harvey, J.K. (1976) 'Golf ball aerodynamics', *Aeronaut Q*, 27: 112–122.

Bentley, K., Varty, P., Proudlove, M. and Mehta, R.D. (1982) 'An experimental study of cricket ball swing', *Aero Tech. Note*, pp. 82–106. London: Imperial College.

Best, R.J., Bartlett, R.M. and Morriss, C.J. (1993) 'A three-dimensional analysis of javelin throwing technique', *Journal of Sports Sciences*, 11: 315–328. London: E. & F.N. Spon.

Best, R.J., Bartlett, R.M. and Sawyer, R.A. (1995) 'Optimal javelin release', *Journal of Applied Biomechanics*, 11: 371–394. Illinois: Human Kinetics.

Binnie, A.M. (1976) 'The effect of humidity on the swing of cricket balls', *International Journal of Mechanical Science*, 18: 497–499. London: Pergamon Press.

Bown, W. and Mehta, R.D. (1993) 'The seamy side of swing bowling', *New Scientist*, 139(1887): 21–24.

Brancazio, P.J. (1981) 'Physics of basketball', *American Journal of Physics*, 49: 356–366. American Association of Physics Teachers.

Brancazio, P.J. (1985a) 'The physics of kicking a football', *The Physics Teacher*, 23: 403–407. American Association of Physics Teachers.

Brancazio, P.J. (1985b) 'Why does a football keep its axis pointing along its trajectory', *The Physics Teacher*, 23: 571–573. American Association of Physics Teachers.

Brancazio, P.J. (1987) 'Rigid-body dynamics of a football', *American Journal of Physics*, 55: 415–420. American Association of Physics Teachers.

Bray, K. and Kerwin, D.G. (2003a) 'Simplified flight equations for a spinning soccer ball', *Science and Football 5*, 5: 40–45. World Congress of Science and Football.

Bray, K. and Kerwin, D.G. (2003b) 'Modelling the flight of a soccer ball in a direct free kick', *Journal of Sports Sciences*, 21: 75–85. London: Taylor & Francis Ltd.

Briggs, L.J. (1959) 'Effect of spin and speed on the lateral deflection of a baseball; and the Magnus effect for smooth spheres', *American Journal of Physics*, 27: 589–596. American Association of Physics Teachers.

Brody, H. (1984) 'That's how the ball bounces', *The Physics Teacher*. 494–497. NASA/Smithsonian.

Brody, H. (1996) 'The modern tennis racket', *The Engineering of Sport* (Ed. Haake, S.), pp. 79–82. Rotterdam: Balkerma.

Brody, H. (2002) 'Improving your serve', *The Physics and Technology of Tennis*, pp. 193–200. California: RacquetTech Publishing.

Brody, H., Cross, R. and Lindsey, C. (1982) *The Physics and Technology of Tennis*, pp. 311–316. California: RacketTech Publications.

Buckingham, M-P, Marmo, B.A. and Blackford, J.R. (2006) 'Design and use of an instrumented curling brush', *Proc. IMechE Part L. Journal Materials: Design and Applications*, 220(L4): 199–205. IMechE.

Caputo, M. (1967) *The Gravity Field of the Earth, Classical and Modern Methods*. New York/London: Academic Press.

Carré, T., Asai, T., Akatsuka, T. and Haake, S.J. (2002) 'The curve kick of a football in flight through the air', *Sports Engineering*, 5: 193–200. London: Blackwell Science.

Collings, P. (2007) 'The influence of court surface upon elite male tennis strategy and metabolic demand: implications for enhanced training and performance'. Final year undergraduate report. Portsmouth: University of Portsmouth.

Cooke, A.J. (1996) 'Shuttlecock design and development', *The Engineering of Sport* (Ed. Haake, S.), pp. 91–95. Rotterdam: Balkerma.

Cooke, A.J. (1999) 'Shuttlecock aerodynamics', *Sports Engineering*, 2: 85–96. London: Blackwell Science.

Cooke, A.J. (2000) 'An overview of tennis ball aerodynamics', *Sports Engineering*, 3: 123–129. London: Blackwell Science.

Cooke, A.J. (2002) 'Computer simulation of shuttlecock trajectories', *Sports Engineering*, 5: 93–105. London: Blackwell Science.

Cooke, A.J. and Mullins, J. (1994) 'The flight of the shuttlecock', *New Scientist*. 1916, March: 40–42.

Cooke, J.C. (1955) 'The boundary layer and seam bowling', *The Mathematical Gazette*, 39: 196–199.

Cooper, L., Dalzell, D. and Silverman, E. (1959) 'Flight of the discus', Unpublished. Available in 1980 from Purdue University Library.

Coutis, P. (1998) 'Modelling the projectile motion of a cricket ball', *International Journal of Mathematical Education in Science and Technology*, 29(6): 789–798. London: Taylor & Francis Ltd.

Cross, R. (1999) 'The bounce of a ball', *American Journal of Physics*, 67(3): 222–227. Washington D.C.: American Association of Physics Teachers.

Cross, R. (2000) 'Effects of friction between the ball and the strings in tennis', *Sports Engineering 3*, 85–97. London: Blackwell Science.

Cubitt, A.C. and Bramley, A. (2006) 'Degradation of tennis balls and their recovery', *The Engineering of Sport 6* (Ed. Moritz, E.F. and Haake, S.), 63–68. New York: Springer-Verlag.

Cunningham, J. and Dowell, L. (1976) 'The effects of air resistance on three types of football trajectories', *Research Quarterly*, 47(4): 852–854.

Daish, C.B. (1972) *Ball Games, the Sportsmens' Guide to Impact, Swing, Stroke, Rolling and Bounce*. London: The English Universities Press Ltd.

Danby, J.M.A. (1998) *Computer Modeling From Sports to Spaceflight ... From Order to Chaos.* Willman-Bell, Inc.

De Mestre, N. (1991) *The Mathematics of Projectiles in Sport.* Cambridge: Cambridge University Press.

Edelmann-Nusser, J., Gruber, M. and Gollhofer, A. (2002) 'Measurement of on-target trajectories in Olympic archery', *Engineering of Sport 4* (Ed. Ujihashi, S. and Haak, S.), pp. 487–493. New York: Springer.

Elliott, B.C. (1989) 'Tennis strokes and equipment', *Biomechanics of Sport* (Ed. Vaughn, L.), 263–288. London: Informa Health Care.

Fernandes, E. and Matos, J.F. (1998) 'Goal!!! [The mathematics of a penalty shoot-out]', *Mathematical Modelling: Teaching and Assessment in Technology*, pp. 159–167. London: Ellis Horwood Ltd.

Frohlich, C. (1981) 'Aerodynamic effects on discus flight', *American Journal of Physics*, 49(12): 1125–1132. Washington D.C.: American Association of Physics Teachers.

Frohlich, C. (1985) 'Effect of wind and altitude on record performance in foot races, pole vault and long jump', *American Journal of Physics*, 53: 726–730. Washington D.C.: American Association of Physics Teachers.

Ganslen, R.V. (1964) 'Aerodynamic and mechanical forces in discus flight', *The Athletic Journal.* (Chicago), 44: 50.

Golden, M. (1998) *Sport and Society in Ancient Greece.* Cambridge: Cambridge University Press.

Goodwill, S.R. and Haake, S.J. (2001) 'Why were 'spaghetti string' rackets banned in the game of tennis?' *The Engineering of Sport 4* (Ed. Ujihashi, S. and Haake, S.), pp. 231–237. New York: Springer-Verlag.

Goodwill, S.R. and Haake, S.J. (2004) 'Ball spin generation of oblique impacts with a tennis racket', *Society for Experimental Mechanics*, 44(2): 195–206. Illinois: University of Illinois.

Grant, C., Anderson, A. and Anderson, J.M. (1998) 'Cricket ball swing: the Cooke–Lyttleton theory revisited', *The Engineering of Sport: Design and Development.* (Ed. Haake, S.J.), pp. 371–378. Oxford: Blackwell Publishing.

Groppel, J.L., Dillman, C.J. and Lardner, T.J. (1983) 'Derivation and validation of equations of motion to predict ball spin upon impact in tennis', *Journal of Sports Sciences*, 1: 111–120. London: E. & F.N. Spon.

Grund, T. and Senner, V. (2006) 'Traction testing of soccer boots under game relevant loading conditions', *The Engineering of Sport 6* (Ed. Moritz, E.F. and Haake, S.), pp. 339–344. New York: Springer-Verlag.

Haake, S.J., Carre, M.J. and Goodwill, S.R. (2003) 'The dynamic impact characteristics of tennis balls with tennis rackets', *Journal of Sports Sciences*, pp. 21: 839–850. London: E. & F.N. Spon.

Haake, S.J., Chadwick, S.G., Dignall, R.J., Goodwill, S. and Rose, P. (2000) 'Engineering tennis: slowing the game down', *Sports Engineering*, pp. 3: 131–143. London: Blackwell Science.

Hardisty, J., Taylor, D.M. and Metcalfe, S.E. (1993) *Computerised Environmental Modelling: A Practical Introduction Using Excel.* Chichester: J. Wiley & Sons.

Hart, D. and Croft, T. (1988) *Modelling with Projectiles.* Chichester: Ellis Horwood Ltd.

Hatton, L. and Parkes, B. *Javelin throwing: the Appliance of Science.* Online. Available HTTP: http://www.leshatton.org/Documents/AW_JavelinArticle_1105.pdf (accessed Feb, 2009).

Hay, J.G. (1993) *The Biomechanics of Sports Techniques* (4th Ed.). New Jersey: Prentice Hall.

Holmes, C., Jones, R., Harland, A. and Petzing, J. (2006) 'Ball launch characteristics for elite rugby union players', *The Engineering of Sport 6* (Ed. Moritz, E.F. and Haake, S.), pp. 211–216. New York: Springer-Verlag.

Hopkins, D.C. and Patterson, J.D. (1977) 'Bowling frames: paths of a bowling ball', *American Journal of Physics*, 45: 263–266. Washington D.C.: American Association of Physics Teachers.

Hubbard, M. (1984a) 'Optimal javelin trajectories', *Journal of Biomechanics*, 17(10): 777–787. London: Pergamon Press.

Hubbard, M. (1984b) 'Javelin trajectory simulation and its use in coaching', *Proc. 2nd Int. Symposium on Biomechanics in Sport.* Colorado: Colorado Springs.

Hubbard, M. and Alaways, L.W. (1987) 'Optimum release conditions for the new rules javelin', *International Journal of Sport Biomechanics*, 3: 207–221. Illinois: Human Kinetics.

Hubbard, M. and Alaways, L.W. (1989) 'Rapid and accurate estimation of release conditions in the javelin throw', *Journal of Biomechanics*, 22(6–7): 583–595. London: Pergamon Press.

Hubbard, M. and Cheng, K.B. (2007) 'Optimal discus trajectories', *Journal of Biomechanics*, 40: 3650–3659. Oxford: Elsevier Science Ltd.

Hubbard, M. and Rust, H.J. (1984) 'Simulation of javelin flight using experimental aerodynamic data', *Journal of Biomechanics*, 17(10): 769–776. London: Pergamon Press.

Hughes, M. and Franks, I.M. (2004) *Notational Analysis of Sport*, 2nd Ed. London: Routledge.

Huntley, I.D. and James, D.J.G. (1990) *Mathematical Modelling A Source book of Case Studies.* London: Oxford Science Publications.

James, D., Carré, M. and Haake, S. (2006) 'Studies on the oblique impact of a cricket ball on a cricket pitch', *The Engineering of Sport 6* (Ed. Moritz, E.F. and Haake, S.), pp. 235–240. New York: Springer-Verlag.

Jenkins, M. (Ed.) (2003) *Materials in Sports Equipment.* Cambridge: Woodhead Publishing.

Jennings-Temple, M., Leeds-Harrison, P. and James, I. (2006) 'An investigation into the link between soil physical conditions and the playing quality of winter sports pitch rootzones', *The Engineering of Sport 6* (Ed. Moritz, E.F. and Haake, S.), pp. 315–320. New York: Springer-Verlag.

Joseph, S.H. and Stewart, S. (1996) 'Mechanics of the modern target archery bow and arrow', *The Engineering of Sport* (Ed. Haake), pp. 205–210. Rotterdam: Balkerma.

Kao, S.S., Sellends, R.W. and Stevenson, J.M. (1994) 'A mathematical model for the trajectory of a spiked volleyball and its coaching applications', *Journal of Applied Biomechanics*, 10: 95–108. Illinois: Human Kinetics.

Kerwin, D.G. and Bray, K. (2006) 'Measuring and modelling the goalkeeper's diving envelope in a penalty kick', *The Engineering of Sport 6* (Ed. Moritz, E.F. and Haake, S.), pp. 321–326. New York: Springer-Verlag.

Klopsteg, P.E. (1943) 'Physics of Bows and Arrows', *American Journal of Physics*, 11: 175–192. Washington D.C.: American Association of Physics Teachers.

Landau, L.D. and Lifschitz, E.M. (1976) *Theoretical Physics.* International Mechanics Volume 1. London: Pergamon Press.

Lichtenberg, D.B. and Wills, J.G. (1978) 'Maximizing the range of the shot put', *American Journal of Physics*, 46: 546–549. Washington D.C.: American Association of Physics Teachers.

Marmo, B.A. and Blackford, J.R. (2004) 'Friction in the sport of curling', *The 5th International Sports Engineering Conference.* Davis, California, 1: 379–385. International Sports Engineering Association, Sheffield. UK.

Marmo, B.A., Farrow, I.S., Buckingham, M-P and Blackford, J.B. (2006) 'Frictional heat generated by sweeping in curling and its effect on ice friction', *Proc. IMechE Part L. Journal Materials: Design and Applications*, 220(L4): 189–197. Institute of Mechanical Engineers.

Mehta, R.D. (1985) 'Aerodynamics of sports balls', *Annual Review of Fluid Dynamics*, 17: 151–189. Annual Reviews Inc.

Mehta, R.D. (2000) 'Cricket ball aerodynamics: myth versus science', Subic, A.J., and Haake, S.J. (ed). *The Engineering of Sport: Research Development and Innovation*, pp. 153–167. London: Blackwell Science.

Mehta, R.D. (2005) 'An overview of cricket ball swing', *Sports Engineering*, 8: 181–192. London: Blackwell Science

Mehta, R.D. and Pallis, J.M. (2001) 'The aerodynamics of a tennis ball', *Sports Engineering*, 4: 177–189. London: Blackwell Science.

Mehta, R.D. and Wood, D.H. (1983) 'Aerodynamics of the cricket ball', *New Scientist*, 87(1213): 442–447.

Mehta, R.D., Bentley, K., Proudlove, M. and Varty, P. (1983) 'Factors affecting cricket ball swing', in *Nature*, 303: 787–788. London: Macmillan Journals Ltd.

Mizera, F. and Horváth, G. (2002) 'Influence of environmental factors on shot put and hammer throw range', *Journal of Biomechanics*, 35: 785–796. Oxford: Elsevier Science Ltd.

Neuff, A. (2008) 'Specification for shot and hammer'. Personal email.

Okubo, H. and Hubbard, M. (2004) 'Dynamics of basketball-rim interactions', *Sports Engineering.* 7(1): 15–29. London: Blackwell Science.

Ohgi, Y., Seo, K., Hirai, N. and Murakami, M. (2006) 'Measurement of jumper's body motion in ski jumping', *The Engineering of Sport 6* (Ed. Moritz, E.F. and Haake, S.), pp. 1: 275–280. New York: Springer-Verlag.

Panoutsakopoulos, V. (2006) 'Biomechanical analysis of the men's hammer throw in the Athens 2006 I.A.A.F. World Cup in athletics', *US Track Coaching Association Journal.*

Penrose, J.M.T., Hose, D.R. and Trowbridge, E.A. (1996) 'Cricket ball swing: a preliminary analysis using computational fluid dynamics', *The Engineering of Sport* (Ed. S.J. Haake), pp. 11–19. Rotterdam: Balkerma.

Podgayets, A., Rudakov, R. and Tuktamishev, V. (2002) 'Aerodynamic optimization of a ski jump', *The Engineering of Sport 4* (Ed. Ujihashi, S and Haake, S.), 4: 415–422. London: Blackwell Science.

Rae, W.J. (2003) 'Flight dynamics of an American football in a forward pass', *Sports Engineering*, 6: 149–163. London: Blackwell Science.

Rae, W.J. and Streit, R.J. (2002) 'Wind-tunnel measurements of the aerodynamic loads on an American football', *Sports Engineering*, 5: 165–172. London: Blackwell Science.

Rayleigh Lord (1877) 'On the irregular flight of a tennis ball', *Messenger of Mathematics*, 7: 14–16. Oxford, Cambridge & Dublin Universities.

Red, W.E. and Zogaib, A.J. (1977) 'Javelin dynamics including body interactions', *Journal of Applied Mechanics*, (Trans of the ASME), 44: 496–498.

Reilly, T. and Williams, A.M. (2003) *Science and Soccer*, 2nd Ed. Routledge. London

Ronkainen, J. and Harland, A. (2006) 'Soccer ball modal analysis using a scanning laser Doppler vibrometer (SLDV), *The Engineering of Sport 6* (Ed. Moritz, E.F. and Haake, S.), pp. 357–362. New York: Springer-Verlag.

Sayers, A.T. (2001) 'On the reverse swing of a cricket ball: modelling and measurements', Proceedings of the Institution of Mechanical Engineers, 215(C): 45–55. Institute of Mechanical Engineers.

Sayers, A.T. and Hill, A. (1999) 'Aerodynamics of a cricket ball', *Journal of Wind Engineering and Industrial Aerodynamics*, 79: 169–182. Oxford: Elsevier Science Ltd.

Seo, K., Kobayashi, O. and Murakami, M. (2006a) 'Flight dynamics of the screw kick in rugby', *Sports Engineering*, 9: 49–58. London: Blackwell Science.

Seo, K., Kobayashi, O. and Murakami, M. (2006b) 'Multi-optimization of three kicks in rugby', *The Engineering of Sport 6* (Ed. Moritz, E.F. and Haake, S.), pp. 223–228. New York: Springer-Verlag.

Seo, K., Ohta, K., Watanabe, I. and Murakami, M. (2002) 'The optimization of flight distance in ski jumping', *The Engineering of Sport 4* (Ed. Ujihashi, S and Haake, S.J.), pp. 408–414. London: Blackwell Science.

Shipton, P., James, I. and Vickers, A. (2006) 'The mechanical behaviour of cricket soils during preparation for rolling', *The Engineering of Sport 6* (Ed. Moritz, E.F. and Haake, S.), 6: 229–234. New York: Springer-Verlag.

Silvester, J. (2003) *Complete Book of Throws*. Illinois: Human Kinetic.

Soong, T-C. (1975) 'The dynamics of javelin throw', *Journal of Applied Mechanics*, 42: 257–262. California: American Society of Mechanical Engineers.

Soong, T-C. (1976) 'The dynamics of discus throw', *Journal of Applied Mechanics*, 98: 531–536. California: American Society of Mechanical Engineers.

Štěpánek, A. (1987) 'The aerodynamics of a tennis ball: the topspin lob', *American Journal of Physics*, 56(2): 138–142. American Association of Physics Teachers.

Stiffler, A.K. (1984) 'Friction and wear with a fully melting surface', *Journal of Tribology*, 106: 416–419. California: American Society of Mechanical Engineers.

Stronge, B. and Ashcroft, A. (2006) 'Large deflections during bounce of inflated balls', *The Engineering of Sport 6* (Ed. Moritz, E.F. and Haake, S.), 2: 109–114. New York: Springer-Verlag.

Suzuki, S., Togari, H., Isokawa, M., Ohashi, J. and Ohgushi, T. (1988) 'Analysis of the goalkeeper's diving motion', *Science and Football II* (Ed. Reilly, T., Clarys, J. and Stibbe, A.), pp. 468–475. London: E & FN Spon

Tan, A. and Miller, G. (1981) 'Kinematics of the free throw in basketball', *American Journal of Physics*, 49: 542–544. American Association of Physics Teachers.

Thompson, H. (2007) *Penguins Stopped Play*. Hodder Headlines.

Vaughn, L. (Ed.) (1989) *Biomechanics of Sport*. London: Informa Health Care.

Wakai, M. and Linthorne, N.P. (2002) 'Optimum takeoff angle in the standing long jump', *The Engineering of Sport 4* (Ed. Ujihashi, S. and Haake, S.J.), pp. 817–823. Oxford: Blackwell Science.

Watts, R.G. and Sawyer, E. (1975) 'Aerodynamics of a knuckleball', *American Journal of Physics*. 43: 960–963. American Association of Physics Teachers.

Woolmer, B. (2008) *The Art and Science of Cricket*. South Africa: New Holland Publishing.

Wu, T. and Gervais, P. (2008) 'An examination of slo-pitch pitching trajectories', *Sports Biomechanics*. 7(1): 88–99. London: Routledge.

Zanevskyy, I. (2001) 'Lateral deflection of archery arrows', *Sports Engineering*. 4: 23–42. London: Blackwell Science.

Zanevskyy, I. (2007) 'Mechanical and mathematical modeling of sports archery arrow ballistics', *International Journal of Computer Sciences in Sport*. 7(1): 40–49. International Association on Computer Science in Sport.

Zernicke, R.F. and Roberts, E.M. (1978) 'Lower extremity forces and torques during systematic variation of non-weight bearing motion', *Medicine and Science in Sport*. 10: 21–26.

Zernicke, R.F. and Roberts, E.M. (2008) *IAAF Competition Rules*. International Association of Athletics Federations.

Zernicke, R.F. and Roberts, E.M. (2009) *ITF Approved Tennis Balls and Classified Court Surfaces: a guide to products and test methods*. International Tennis Federation.

Selected answers

Chapter 2

Question 2   39.24 N on the slope, 78.48 N on the horizontal take-off section

Question 3 4.743×10^{-12} N

Question 4 $81 \, \mathrm{m \, s^{-1}}$

Question 5 $0.4 \, \mathrm{m \, s^{-1}}$

Chapter 3

Question 1   At $5.6 \, \mathrm{m \, s^{-1}}$; yes.

Question 2   (a) Below the serve line by 0.17 m. (b) Midway between serve line and out line at a height of 2.656 m.

Question 3   No. 45.57° is the optimum angle, in which case his throw would have been 94.4 m.

Question 4   At precisely 47° the ball will go over the net with 0.016 m to spare.

Question 5   No. The ball overshoots the crossbar by 10 cm.

Chapter 4

Question 1   The apparent CoR may appear greater than 1. It is a complex process which depends on the three-way energy exchanges between floor and lower ball, and lower ball and upper ball, including the timing of these exchanges. As an example, the author has witnessed a table tennis ball on top of a Superball placed inside a Perspex tube and dropped. The table tennis ball does indeed rebound much higher than its drop height.

Of course, this does not contravene energy conservation laws, as the excess energy imparted to the upper ball comes from the lower ball, which, as a consequence, does not bounce as high as it would normally.

Question 2 $e'.\tan \psi = e.\cotan \theta$. So, when $e' = e$, $\psi = 90° - \theta$

Question 4   Nice idea, but sadly; no. Remembering that the spin will reverse on each collision, when the ball hits the ground the first time, it will acquire a considerable amount of topspin. However, this will behave like backspin with reference to the underside of the table. It will therefore, return approximately along its original path, again with a spin reversal. The thrower may catch it, but on its final path the ball will possess backspin. The opponent will not ever see the ball and will quickly get bored!

Question 5 Although it may appear on first inspection that the mechanics of these two processes are similar, they are, in fact, very different. One can look at the processes from a momentum exchange perspective. In the more usual case of the club hitting the ball, the club losses a little momentum in the exchange; the ball gains it. In the case of the static club, the momentum exchange is simply a reversal of the sign of momentum vector following impact. The club gains no momentum from the ball. Totally different.

Question 6 A loft angle of approximately $47°$ should do the trick. This corresponds to, either a 9-iron or a pitching wedge. However, the stated horizontal velocity is a little high for the wedge so the 9-iron would seem to be the best bet.

Chapter 5

Question 1 The transition from laminar to chaotic flow is very much favoured in projectiles that have a convex trailing surface. A ball has a particular predisposition to the eddies working their way around from the back of the ball at relatively low velocities. The arrow, or javelin, on the other hand is long, sleek and streamlined. This prevents the initial eddies spawning at the back of the projectile, which are required before the eddies can creep towards the front of the ball creating the chaotic drag.

Question 2 For a torpedo kick (kicked through its long axis), $R_e = 3 \times 10^5$, $\varepsilon = 0.75$. For a kick side-on to the ball, $R_e = 5.72 \times 10^5$, $\varepsilon = 1.2$.

Question 3 $x = 18.2\,\text{m}$ $y = 14.4\,\text{m}$ (Excel Model 5.2 will calculate these equations).

Question 4 (i) $k = 1.46$ (ii) $k = 0.245$ (iii) $k = 1.09$.

Question 5 (i) $k = 0.22$ (ii) $k = 0.006$ (iii) $k = 0.12$.

Question 6 $24\,\text{m}\,\text{s}^{-1}$.

Chapter 6

Question 1 For the British golf ball,
at 2 seconds $s_x = 93.62\,\text{m}$, $s_y = 18.57\,\text{m}$
at 4 seconds $s_x = 150.83\,\text{m}$, $s_y = 18.69\,\text{m}$
For the American golf ball
at 2 seconds $s_x = 91.77\,\text{m}$, $s_y = 19.02\,\text{m}$
at 4 seconds $s_x = 146.57\,\text{m}$, $s_y = 19.97\,\text{m}$
The American ball will fly higher but not travel so far as the British ball.

Question 2 It has a positive second-order change in height rather than a linear rise. If the Tait variable κ_2 is too high, it will stall, although the model does not show this!

Question 3 For the top-spun tennis ball,
at 1 second $s_x = 43.15\,\text{m}$, $s_y = 2.5\,\text{m}$
at 2 seconds $s_x = 76.62\,\text{m}$, $s_y = 0.73\,\text{m}$
For the back-spun tennis ball
at 1 second $s_x = 43.15\,\text{m}$, $s_y = 3.73\,\text{m}$
at 2 seconds $s_x = 76.62\,\text{m}$, $s_y = 4.93\,\text{m}$

Broadly, the heights are similar for both topspin and backspin at the 1 and 2 second intervals. However, for the top-spun case, the ball has nearly reached the ground after 2 seconds. By contrast, the back-spun ball still has a way to go, as it is still nearly 5 m above the ground at this point in the flight.

Chapter 7

Question 1   Range = 21.8 m (remember h is negative!)
Time of flight = 2 s. (Use Equation 3.12 and ignore the negative time solution.)

Maximum height = 5.86 m. (Calculate for a ground level throw and then add the shot putter's height to your answer.)

Question 2   19.54 m.

Question 3   21.8 (answer to question 1) – 0.155 (drag reduction from Equation 7.15) = 21.64 m.

Question 4   21.8 (answer to question 1) – 0.164 (drag reduction from Equation 7.15) = 21.63 m (Not a practical value as the stated release velocity for men and women is the same – the greater drag is caused by the smaller value of m which increases the drag-to-weight ratio.)

Question 5   48° but the angle is less sensitive to range than the release velocity and therefore it is better to lower the release angle (to about 38–42°) so that greater release velocity can be achieved.

Question 6   As the weight of the shot putter's arm increases so that r_m ranges from, say, 1.1 to 2.0 (a very 'light' arm to a very heavy one!), the range drops in a linear fashion from 19.0 m down to 17.45 m. So even in the shot-putting event there's a pay-off for possessing too much muscle mass!

Chapter 8

Question 1   Bit of a trick question this one. The flying disc holds the record for the longest throw, currently standing at 406 m. However, the question asks 'the greatest distance through the air' and it is known that boomerangs regularly cover greater distances than this albeit not in a straight line. Whodya mean, not fair!

Question 2   29.46 m s^{-1} at an angle of 15.2° to the horizontal.

Question 3   When a projectile flies through the air it invariably hits the ground before it runs out of forward velocity (otherwise it would hit the ground vertically). Aerodynamic projectiles can create lift from the forward flight, keeping it in the air longer. It therefore travels further and hits the ground at a correspondingly greater angle. The optimum range will occur with a vertical impact with the ground. If the projectile stayed in the air longer after that, it will start its return to the thrower, reducing range.

Question 4   Stalling occurs when the air flow over the top of the aerofoil separates from the air flow under the aerofoil at the back of the projectile. This happens at the critical angle, where the lift force abruptly reduces to zero and drag greatly increases. In the case of the discus, the projectile will become totally unstable and somersault to the ground.

Question 5   16°.

Question 6   31.5 m. Assumes minimal drag and perfect angle of launch.

Question 7   Calculate the effective mass of a women's discus with diameter of 180 mm and mass of 1 kg. Assume standard temperature and pressure conditions and the diameter of a men's discus is 220 mm.

Question 8   1.5 kg.

Chapter 9

Question 1   The flights would probably be virtually the same. The lift and drag are largely defined by the weight to drag ratio which would be identical. However, the axial moment of inertia of the solid javelin would be greater and so, on release, the javelin would have negligible spin. This would affect the stability and the pitching moment to some degree. But then again, many athletes throw with little spin anyway so the difference would not be significant.

Question 2   The drag coefficient varies exponentially with angle of attack. However, for small angles of attack it is approximately linear. The opposite is true for the lift coefficient. For small angles of attack the variation is exponential but from about 1° onwards it approximates to a linear variation.

Question 3   3.37 m.

Question 4   Aerodynamicity $= 0.13$. This falls between the values for the shot/hammer and the discus/javelin.

Question 5   0.06 m only.

Chapter 11

Question 1   25 m s^{-1}. For these parameters the incoming and outgoing ball are the same speed. The loss in the coefficient of restitution is matched by the velocity of the incoming racket.

Question 2   0.0885 kg m^2.

Question 3   The analysis assumes that the ball is hit in the center of racket.
Also the calculation is based only on the moment of inertia of the racket at the point of impact as though there is no force provided by the player on the follow through. This follow-through action will considerably affect the accuracy, creating an effective moment of inertia which is much greater than that used in the calculations. The Groppel $et\ al.$ (1983) analysis ignores this. Later published analyses do include this effect.

Question 4 $k = 22\,500$, $c = 22.8$.

Question 5   The CPR is medium-fast with a value of 40.7 ($e = 0.65$, $\mu = 0.84$).

Chapter 12

Question 3   Originally, the definition for a bowl, rather than a throw, was that the elbow joint must not straighten during the bowling action. In 2005, the rule was changed to allow extensions or hyper extensions of the elbow of up to 15° before deeming the ball illegally thrown.

Question 7   Incoming ball at 30 m s^{-1} produces a return ball of 34.6 m s^{-1}
Incoming ball at 40 m s^{-1} produces a return ball of 37.51 m s^{-1}
Incoming ball at 60 m s^{-1} produces a return ball of 43.35 m s^{-1}.

Chapter 13

Question 1   The coefficient of restitution ball–boot is not linear with impulse. It reduces as the ball compression increases.

Question 3   22.435 m s^{-1} (obtained, using the wonderful – and free $w\alpha$ from alpha).

Question 4   The wet sand pitch has a hardness factor of 5.36 while the dry sand pitch has a hardness factor of only 4.78.

Chapter 14

Question 1 $v_0 = 18$ m s^{-1}, $\theta_0 = 47.86°$.

Question 2   The nearer the ball is to the center line from the goal, the more precise the kicker has to be in placing the ball. The smaller the offset, the sharper the angle versus distance function.

Question 3 $C_D = 0.6$, $C_L = 0.485$, $C_y = -0.307$ and $C_m = 0.333$.

Question 4   (a) 4.3 s   (b) 60 m (= 65 yds).

Question 5   (a) 86.21 m   (b) 41.38 m   (c) 39.66 m   (d) 37.93 m.

Chapter 15

Question 1 $v_{0m} = 7.4$ m s^{-1}, $\theta_{0m} = 50.7°$.

Question 2   Entry angle into middle of basket is approximately 40°
Margin of error is 1.35°.

Question 3 $C_L = -0.326$ $k_L = 0.0533$ N.

Question 4   0.165 N.

Question 5   2.32 m.

Question 6   13.317 k N.

Question 7   At 10° $C_L = 0.111$; $C_D = 0.103$.
At 20° $C_L = 0.264$; $C_D = 0.206$.

Question 8   The terminal velocity is never reached; it is only asymptotically approached. So strictly speaking, we can only give a time or distance for when it reaches some defined fraction of v_T.

Index